Full-Duplex Communications for Future Wireless Networks

Hirley Alves • Taneli Riihonen • Himal A. Suraweera

Editors

Full-Duplex Communications for Future Wireless Networks

 Springer

Editors
Hirley Alves
Centre for Wireless Communications
University of Oulu
Oulu, Finland

Taneli Riihonen
Unit of Electrical Engineering
Tampere University
Tampere, Finland

Himal A. Suraweera
Department of Electrical and Electronic
Engineering
University of Peradeniya
Peradeniya, Sri Lanka

ISBN 978-981-15-2971-9   ISBN 978-981-15-2969-6   (eBook)
https://doi.org/10.1007/978-981-15-2969-6

This Springer imprint is published by the registered company Springer Nature Singapore Pte Ltd.
The registered company address is: 152 Beach Road, #21-01/04 Gateway East, Singapore 189721, Singapore

Preface

This book is the capstone to a biannual series of workshops titled *Full-Duplex Communications for Future Wireless Networks*. Since 2017, the editors were part of the organizing committee of a series of workshops at the flagship conferences in communications engineering, namely *IEEE International Conference on Communications* and *IEEE Global Communications Conference*, with the objective of showcasing the most recent advances in full-duplex technologies. Throughout these years, the workshops have gathered leading-edge contributions evincing Full-Duplex research's maturity, novelty in overcoming key challenges, and a wide range of applications, besides practical demonstrations and field trials.

This book captures the essence of the state-of-the-art research, where we aim to provide a critical overview of the research challenges in full-duplex communications and solutions thereof—that enabled its feasibility—as well as novel applications that demonstrate the flexibility and potential of full-duplex for future wireless systems and networks. Therefore, we have split the book into two parts; notably

Part I: Self-interference Cancellation delves deep into self-interference characterization and cancellation in the complete transceiver chain from antennas and electronics up to digital-domain processing.

Part II: Applications and Future Trends scrutinizes established and trending full-duplex applications from radio resource management to energy harvesting systems and from security to test beds and trials.

Part I is composed of four chapters out of which Chaps. 1 and 2 review radio-frequency transceiver architectures for in-band full-duplex encompassing antenna design and radio-frequency cancellation of the self-interference produced by its own transmitter. Distinct isolation mechanisms and the use of multiple antennas for transmission and reception are discussed. Moreover, the authors overview single-antenna duplexing architectures, which are preferred due to their small form factors despite hardware additions that are needed for signal cancellation. Therefore, the authors then discuss passive—in which radio-frequency components mitigate the self-interference—and active cancellation techniques with respect to performance and complexity since channel estimation of residual self-interference is required and performed by adding transmit chains. Despite the advances in antenna design and analog domain, where self-interference is attenuated, it still prevails, which brings us to Chap. 3, where the authors delve into digital-domain solutions for efficient self-interference cancellation in low-cost full-duplex radios that are able to mitigate impairments caused by nonlinear distortion. The combined achievements in the antenna design and analog and digital domains suppress the self-interference, in some cases, by more than 100 dB, which is close to—and in some cases even below—the noise floor of some architectures. Further, Chap. 4 discusses the use of time-domain rather than frequency-domain digital filtering, thus allowing to relax constraints on waveforms and synchronization between transmitted and received signals, which once combined with spatial filters are able to suppress nonlinearities caused by circuitry impairments. Altogether, due to all the recent advancements in antenna design, analog and digital domain brought residual self-interference to tolerable levels, thus enabling several applications, which are the focus on *Part II: Future Trends and Applications*.

Part II encompasses seven chapters that describe the prospects, challenges, and solutions that are enabled by full-duplex technology, which has the potential of doubling the spectral efficiency of point-to-point transmissions. Nonetheless, since more nodes are potentially active due to simultaneous uplink and downlink transmissions, a full-duplex network perceives higher interference profile than its half-duplex counterparts, thus bringing challenges to radio resource management. In this context, Chap. 5 overviews the challenges of full-duplex operation under cellular networks and the solutions that unlock its potential. Then, Chap. 6 discusses novel solutions for enhanced mobile broadband and ultra-reliable low latency communication service classes introduced by the fifth generation of cellular networks. The authors discuss also the impact of network traffic asymmetry, which can hamper gains in throughput. The inherent characteristics of simultaneous transmission and reception renders full-duplex operation a suitable solution for latency-constrained applications. Likewise, in the recent years, non-orthogonal multiple access solutions have emerged as an alternative to increase spectral efficiency by allowing multiplexing signals on power or code domain. The envisaged spectral efficiency gains are magnified when non-orthogonal solutions are combined with full-duplex operation, especially through the use of techniques such as optimal relay and beamforming antenna selection in multiple antenna, relaying and cognitive systems as surveyed in Chap. 7. One more technology that has emerged in the recent years, whose potential gains are amplified in full-duplex setting, is wireless energy transfer in communication networks. Chapter 8 overviews energy harvesting conversion process and its relation to key full-duplex transceiver architectures and discusses the idea of self-energy recycling transceivers hinting the idea of a sustainable wireless network. The authors evaluate key scenarios that showcase the potential gains and performance of full-duplex wireless-powered systems. Next, we move from non-orthogonal and wireless-powered full-duplex systems to defense and security. In fact, full-duplex systems first historically emerged in continuous-wave radars already around mid-twentieth century. Conversely to cellular systems that focus predominantly on spectral efficiency enhancements, in the context of defense and security the full-duplex transceiver returns to its original function as a sensor as well as a key component in electronic warfare systems. Therefore, the authors in Chap. 9 overview the integration of tactical communications with full-duplex electronic warfare and explore concepts of securing critical applications in the form of a radio shield. Moving further, from a practical applied perspective to more theoretical security, in Chap. 10, the authors discuss the use of artificial noise jamming to secure uplink transmissions in cellular networks by proposing a multi-objective optimization framework securing transmissions while minimizing power consumption against potential eavesdroppers. Finally, we conclude this journey through the realm of full-duplex transceivers and applications with Chap. 11 that reviews recent research on integrated full-duplex radio systems, discusses design aspects related to the semiconductor technology, and revisits key ideas discussed throughout Parts I and II of this book. Therefore, Chap. 11 presents the reader a walk-through relevant concepts and a wide picture of integrated full-duplex radio through an experimental setup.

All in all, we hope that the readers enjoy the contributions and the compilation proposed by this book. We would like to point out that we have not attempted to provide a comprehensive survey on all full-duplex concepts and ideas, but rather a discerning compilation of carefully selected contributions that evince challenges, solutions, and the potential of full-duplex technologies in a wide range of applications. In addition, we would like to thank audience, authors, technical committee members, and reviewers that have been participating in the workshop series, and especially to the authors and the reviewers that have contributed to this book. Finally, we sincerely hope that this book will inspire students, practitioners, and researchers to contribute, perhaps even more, to future developments of full-duplex communications.

Oulu, Finland   Hirley Alves
Tampere, Finland   Taneli Riihonen
Peradeniya, Sri Lanka   Himal A. Suraweera
December 2019

Contents

Part I
Self-Interference Cancellation

Chapter 1
Antennas and Radio Frequency Self-Interference Cancellation

Leo Laughlin and Mark A. Beach

Abstract This chapter presents and reviews radio frequency (RF) transceiver architectures for in-band full-duplex, spanning both antenna techniques and RF cancellation loops. In the antenna domain, isolation can be achieved using separate transmit and receive antennas, exploiting propagation loss for transmit-to-receive isolation, or through more complicated multi-antenna arrangements which exploit propagation domain cancellation. Single antenna duplexing architectures are also discussed, covering circulators and electrical balance duplexers. Passive and active feedforward cancellation techniques are reviewed, discussing the advantages and disadvantages of various cancellation architectures in terms of their complexity, and their performance in cancelling noise and multipath self-interference. The chapter concludes with a discussion on combining antenna and radio frequency cancellation techniques in full-duplex transceiver front-end architectures.

1.1 Introduction

Radio systems typically require transmit signal powers which are many orders of magnitude higher than the receive signal powers (often by over 100 dB), due to the high path loss between a transmitter and a distant receiver. Due to this basic property, it has long been held that a radio system cannot transmit and receive on the same frequency at the same time, as the higher powered transmit signal would be unavoidably coupled to the receiver circuitry resulting in comparatively strong *self-interference* (SI), thereby obscuring the receive signal and preventing its reception.

L. Laughlin (✉) · M. A. Beach
University of Bristol, Bristol, UK
e-mail: leo.laughlin@bristol.ac.uk; m.a.beach@bristol.ac.uk

© Springer Nature Singapore Pte Ltd. 2020   3
H. Alves et al. (eds.), *Full-Duplex Communications for Future Wireless Networks*,
https://doi.org/10.1007/978-981-15-2969-6_1

Until now, radio systems have achieved duplex operation by simply circumventing self-interference; time division duplexing (TDD) and frequency division duplexing (FDD) are widely used techniques for doing this, separating transmit and receive signals in the time domain, avoiding SI altogether (TDD), or separating them in the frequency domain, allowing the SI to be removed using filters (FDD). The concept of in-band full-duplex does away with this division altogether, reusing the same frequency spectrum for simultaneously transmitting and receiving, and employing various methods to reduce and cancel SI in order to suppress it to below the receiver noise floor, such that it does not significantly impact on the receiver signal-to-noise ratio. In essence, cancellation is simple—the transmit signal is known, and therefore it can be subtracted at the receiver. However, in reality, it is far from easy—the "known" transmit signal is corrupted by noise and non-linearities in the transmitter, and circuit imperfections in the cancellation hardware will limit its effectiveness.

Every radio system requires at least one antenna, this being the interface between electrical signals in the radio circuits, and radio waves propagating in space. Therefore, when designing in-band full-duplex transceivers to isolate the receiver from the transmitter, the antenna domain is an obvious place to start. Indeed, a substantial amount of transmit-to-receive (Tx-Rx) isolation is required in the radio frequency domain, i.e., prior to the receiver input, in order to prevent the high powered transmit signal from overloading (or even destroying) the receiver front-end, and to provide adequate suppression of non-linearities and noise components in the Tx signal. Depending on the design, antenna based techniques alone may not provide the necessary levels of isolation, and it is common for IBFD transceivers to deploy a further stage of RF cancellation. Thus, RF-domain isolation techniques can broadly be divided into antenna based isolation techniques and RF cancellation techniques, with transceiver architectures using combinations thereof.

This chapter addresses both of these areas, presenting and reviewing concepts and recent advancements in RF-domain Tx-Rx isolation techniques. Section 1.2 discusses requirements for radio frequency domain isolation, deriving equations for the minimum isolation in terms of transmit power, receiver sensitivity, and various imperfections in the transceiver, and providing a quantitative example based on some typical transceiver parameters. Antenna systems designed to provide propagation domain isolation are reviewed in Sect. 1.3, and Sect. 1.4 gives and introduction to passive feedforward cancellation circuit architectures. Section 1.5 addresses the electrical balance duplexer, which implements a form of feedforward cancellation at the antenna interface, and Sect. 1.6 provides an overview of active RF self-interference cancellation, which uses additional active RF circuitry to generate a cancellation signal. Section 1.7 discusses the combination of antenna and RF cancellation techniques, and Sect. 1.8 concludes this chapter.

1.2   Radio Frequency-Domain Isolation Requirements

To understand RF-domain isolation requirements, it is necessary to consider not only the desired transmit signal, but also the various sources of noise and distortion introduced by the Tx chain.

Various imperfections in the Tx chain cause noise and distortion; all of the components will contribute thermal noise, the digital-to-analog converter (DAC) will add quantization noise, and the local oscillator (LO) will introduce phase noise. Furthermore, many components, but in particular the power amplifier (PA), will introduce non-linear distortion. Thus, the Tx signal at the output of the power amplifier comprises these components:

- The wanted transmit signal transmitted at the intended power level.
- The non-linear distortion, which may typically be some 30 dB below the Tx power and is both in-band, and out-of-band (often referred to as *"spectral regrowth"*).
- The Tx noise floor, which may typically be 50–60 dB below the Tx power.

Both the desired Tx signal and the distortion must be mitigated, either in the analogue domain or the digital domain, or both; however, different types of distortion present different requirements. The requirements of the RF-domain isolation are threefold:

- To suppress the desired Tx signal sufficiently to avoid overloading. To avoid saturation, the SI signal at the Rx input must never be above the maximum Rx input power, and thus this requirement must be based on the **peak** power. Thus, for non-constant envelope Tx signals, this depends on the mean (Tx power) and the peak to average power ratio (PAPR). Mathematically, this criterion can be expressed as

$$ISOL_{Signal} > P_{tx} + PAPR - P_{Rx,Max} \tag{1.1}$$

  where $ISOL_{Signal}$ is the RF-domain isolation for the intended Tx signal, P_{tx} is the Tx power, $PAPR$ is the peak to average power ratio, and $P_{Rx,Max}$ is the maximum receiver input power.
- To suppress the non-linear components of the Tx signal sufficiently to avoid overloading. Assuming non-linear digital cancellation is used, the non-linear components need only to be suppressed within the receiver dynamic range. In this case the RF-domain isolation requirement for non-linear components, $ISOL_{NL}$, is calculated as

$$ISOL_{NL} > P_{NL} - P_{Rx,Max} \tag{1.2}$$

- To suppress the Tx noise below the Rx noise floor. Standard digital cancellation schemes generate the cancellation signal from the digital baseband transmit signal. This can potentially cancel linear SI, and non-linear distortion, but not noise (which is non-deterministic). This gives us the requirement to suppress the transmitter thermal noise below the receiver's noise floor in the RF domain. The RF-domain isolation requirements for noise components, $ISOL_{Noise}$, is calculated as

$$ISOL_{Noise} > P_{Tx,Noise} - P_{Rx,Noise} \tag{1.3}$$

where $P_{Tx,Noise}$ and $P_{Rx,Noise}$ are the Tx noise power and Rx noise power, respectively (in the band of interest).

Figure 1.1 shows an example Tx spectrum at the PA output, with the different SI components labelled, along with the SI spectra at the receiver, and the criteria given above indicated with reference to the labelled powers.

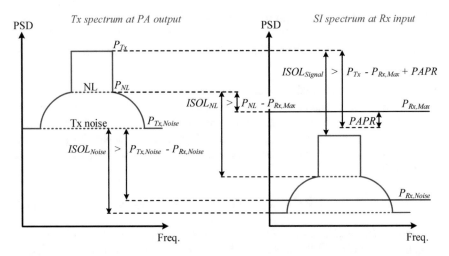

Fig. 1.1 Typical spectra at the PA output and Rx input. Vertical axis is power spectral density (PSD), however indicated power quantities are the integrated powers across the bandwidth of the intended Tx signal. Note: This example assumes that all SI components have been suppressed by the same amount, however not all SI suppression techniques provide this feature

Table 1.1 Example
transceiver system parameters

Parameter	Value
P_{Tx}	25 dBm
$PAPR$	10 dB
$P_{Rx,Max}$	−30 dBm
P_{NL}	−5 dBm
$P_{Tx,Noise}$	−35 dBm
$P_{Rx,Noise}$	−90 dBm

A quantitative example of RF-domain isolation requirements is given here, using some typical transceiver system parameters as given in Table 1.1. The Tx signal isolation criterion is

$$ISOL_{Signal} > P_{tx} + PAPR - P_{Rx,Max} \tag{1.4}$$

$$ISOL_{Signal} > 25\,\text{dBm} + 10\,\text{dB} - (-)30\,\text{dBm} \tag{1.5}$$

$$ISOL_{Signal} > 65\,\text{dB}. \tag{1.6}$$

The non-linear isolation criterion is

$$ISOL_{NL} > P_{NL} - P_{Rx,Max} \tag{1.7}$$

$$ISOL_{NL} > -5\,\text{dBm} - (-)30\,\text{dBm} \tag{1.8}$$

$$ISOL_{NL} > 25\,\text{dB}. \tag{1.9}$$

The noise isolation criterion is

$$ISOL_{Noise} > P_{Tx,Noise} - P_{Rx,Noise} \tag{1.10}$$

$$ISOL_{Noise} > -30\,\text{dBm} - (-)90\,\text{dBm} \tag{1.11}$$

$$ISOL_{Noise} > 60\,\text{dB}. \tag{1.12}$$

Therefore, an IBFD transceiver front-end design which achieves >65 dB isolation for all types of SI would fulfil all requirements. However, as will be seen in later sections, some canceller architectures *do not* cancel all types of SI, and thus it may be necessary to consider each of these requirements individually for some designs.

1.3   Antenna Based Isolation

Antenna based isolation can broadly be divided into three types: single antenna
systems, which share the same antenna for transmitting and receiving, and implement some form of duplexing based on the direction of travel of signals at the
antenna port; multi-antenna systems which use separate transmitting and receiving
antennas and obtain isolation due to the propagation loss between the antennas;
and antenna cancellation systems which use multiple antennas to obtain isolation
through propagation domain cancellation of transmit signals.

1.3.1   Separate Transmit and Receive Antennas

A simple and effective method of providing isolation between the transmitter and
receiver is to use separate antennas for transmitting and receiving, as shown in
Fig. 1.2. This achieves isolation by *avoiding* the interference, aiming to reduce
the self-interference power at the receiver by exploiting the limited electromagnetic coupling between the transmitting and receiving antennas. Achieving greater
isolation in the antenna domain reduces the level of self-interference suppression
which must be achieved in further stages of radio frequency cancellation and digital
baseband cancellation, thereby reducing the requirement for high dynamic range
signal processing hardware in the receiver [1].

Due to propagation loss, the isolation between the Tx and Rx antennas is dependent on the separation distance [2], and therefore this technique is less effective
where device form factor limits the antenna separation [3]. However, this technique
is well suited to infrastructure applications, where large physical separations are
permissible. The isolation can be further increased by exploiting directional and/or
cross-polar antenna configurations to further reduce electromagnetic coupling, and
by using absorptive shielding to block the direct electromagnetic coupling between
them [3–7].

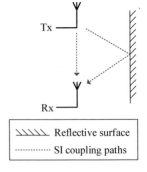

Fig. 1.2 Using separate
transmit and receive antennas
to provide passive
propagation based isolation in
the antenna domain.
Self-interference is coupled
via direct and reflected paths

Where space allows, this technique can be extremely effective, as demonstrated by the system in [7], which achieved 72 dB of isolation between two cross-polar directional antennas, separated by 50 cm and with shielding material between them. This method presents no fundamental limit on bandwidth or tunability, these being limited by the antennas, rather than by the antenna separation method itself. A potential drawback of designing the antenna polarisations and/or patterns to minimise self-interference coupling is that the resulting antenna design might also reduce the wanted receive signal, depending on the application. For example, isolation between directional antennas can be increased by pointing the antennas in opposite directions. In the case of a full-duplex link between two devices, with one of the antennas correctly aligned (i.e., pointing toward the distant transceiver), the other would be facing in the wrong direction, significantly increasing the loss of that link; however, in a relaying application, having the antennas facing opposite directions could be beneficial for both increasing isolation *and* mitigating path loss [8].

Antenna based isolation techniques can be affected by reflections in the local environment—the transmitted signal leaves the transmitting antenna, is reflected from nearby objects, and arrives at the receiving antenna as self-interference (see Fig. 1.2). This is clearly demonstrated in [7]: the antenna arrangement which achieved 72 dB of Tx-Rx isolation only did so when placed inside an anechoic chamber. When measured in a reflective indoor environment, the isolation was reduced to just 46 dB. This is because the absorptive shielding used in this arrangement can only block the direct coupling path between the two antennas, which is effective in the anechoic chamber as this is the only coupling path, however in the reflective environment, multipath propagation results in substantially increased Tx-Rx antenna coupling. Moreover, this results in a multipath *self-interference channel*; this complicates the design of subsequent stages of cancellation, which must be able to cancel these multipath SI components.

1.3.2 Circulators

In the traditional single-input-single-output (SISO) wireless communication paradigm, a device typically uses the same antenna for transmitting and receiving, which reduces the size and cost of the device compared to using dedicated antennas for transmitting and receiving. Circulators have long been used for duplexing in radar systems, and, more recently, various in-band full-duplex wireless communication designs have used circulators to couple a shared antenna to transmitter and receiver, whilst achieving some isolation [9, 10].

Fig. 1.3 A circulator used
for duplexing a single
antenna. Self-interference is
coupled via circulator
leakage, reflection due to
antenna mismatch, and
multipath reflections from the
environment

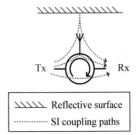

A signal entering any port of a circulator is transmitted to the next port in rotation
only, and isolated from the others. A 3-port circulator can be used for duplexing, as
shown in Fig. 1.3. Ferrite circulators exploit Faraday rotation of electromagnetic
waves propagating in a magnetic field to arrange constructive interference at the
coupled port, and destructive interference at the isolated port, and in that sense, the
isolation achieved is based on a form of SI cancellation.

Circulators can typically provide around 20–30 dB of isolation, limited by direct
signal leakage through the device. However, if a portion of the Tx signal is
reflected due to mismatch at the antenna port, this will arrive at the receiver as
self-interference. Whilst very good matching can be achieved in high cost radar
deployments, in consumer communication systems antenna matching is seldom
perfect, and often as bad as −6 dB in multi-band antennas. Thus, in communication
applications, the antenna match is often the limiting factor in determining the
isolation provided by a circulator system, and can result in relatively low isolation.
Like the multi-antenna systems described above, the Tx-Rx isolation provided by
a circulator can also be impacted by environmental reflections. Since the circulator
separates signals based on direction of travel at the antenna port, transmit energy
which is reflected back to the antenna cannot be distinguished from the wanted
receive signal, and therefore energy reflected from the environment is coupled to
the receiver port as self-interference, along with energy reflected due to impedance
mismatch with the antenna itself, and the direct leakage through the circulator. The
direct, antenna mismatch and environmental reflection self-interference coupling
mechanism are indicated in Fig. 1.3. Furthermore, circulators can be large and
expensive, have a limited bandwidth, and are not tunable. Thus, although circulators
have been included in various IBFD prototypes reported in the literature, there are
many commercial applications where they may not be suitable. However, recent
advances in electronic circulator technology may be able to address these drawbacks
in the future [11].

1.3.3 Propagation Domain Cancellation

Another antenna based isolation method is *antenna cancellation*, which involves
using multiple transmitting antennas, and arranging the receiving antenna(s) such

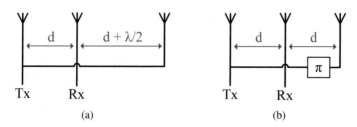

Fig. 1.4 Two different antenna cancellation arrangements: (**a**) half-wavelength propagation path difference [12], (**b**) anti-phase transmission from equidistant antennas [13]

that they are located in positions where the signals from the transmit antennas interfere destructively. Two simple antenna cancellation arrangements, as reported in [12] and [13], are depicted in Fig. 1.4. This method is effective at cancelling self-interference in the propagation domain; for example, the system in [13] achieves up to 45 dB of antenna based isolation. One drawback is that the technique is sensitive to the placement of the antennas and to slight differences in the characteristics of the individual antennas (e.g., pattern and efficiency), which means that manufacturing tolerances will limit performance [14, 15].

The arrangement in Fig. 1.4a relies on a half-wavelength path difference; thus even in theory the cancellation will only occur perfectly at particular frequency points, and this serves to reduce the overall isolation as signal bandwidths increase [12, section 3.1]. This is overcome using the architecture proposed in [13] and depicted in Fig. 1.4b, which uses equal spacing between the Rx antenna and each of the Tx antennas, but instead transmits one of the Tx signals in anti-phase in order to contrive destructive interference at the receiving antenna. This removes the dependence on antenna geometry, however achieving wideband cancellation requires a wideband phase shifter: a delay based phase shifter implementation will exhibit the same bandwidth limitations as the asymmetric antenna placement (Fig. 1.4a), however a wideband inverter based on a transformer or balun can facilitate much wider cancellation bandwidths [16].

An unwanted by-product of antenna cancellation systems such as these is that, since antenna cancellation techniques rely on destructive interference at the receive antenna, this will also cause the transmit signal to destructively interfere at other points in space, (i.e., creating multiple nulls in the aggregate antenna pattern) potentially reducing the power received at the other end of the radio link. This can be mitigated by using different transmit powers at each Tx antenna, whilst ensuring the Rx antenna is at a point of destructive interference through asymmetric antenna placement, as shown in [12].

1.3.4 Adaptive Propagation Domain Cancellation

Substantial performance improvements can be obtained when adjustable antenna weightings are applied. A basic architecture for adaptive antenna cancellation is depicted in Fig. 1.5. This architecture allows the antenna weightings to be adjusted to compensate for any variations in the performance and placement of the antennas, increasing isolation to as much as 60 dB [15]. Moreover, more complex arrays of antennas can enable antenna cancellation in multiple-input-multiple-output (MIMO) systems [13, 14, 17, 18].

An antenna cancelling MIMO array, as first proposed in [17], is depicted in Fig. 1.6. This system divides the antenna elements into transmit elements and receive elements, and uses transmitter pre-coding to minimise the SI coupling from the transmit elements to the receive elements, combining this with digital cancellation only on the receive side (requiring sufficient RF-domain isolation from the antenna cancellation). This technique necessarily sacrifices spatial multiplexing gain to achieve cancellation instead, as some of the antennas are effectively used for cancellation instead of MIMO transmission. The complexity of this architecture

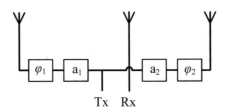

Fig. 1.5 A basic architecture for adaptive antenna based cancellation

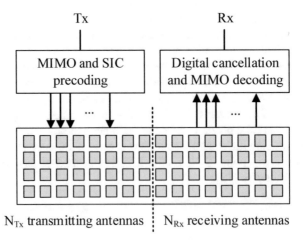

Fig. 1.6 Antenna based isolation based on array beamforming

is suitable for infrastructure applications only (i.e., not mobile devices) and the effectiveness depends on the propagation environment. That is, rich multipath SI coupling is more difficult to cancel—the opposite of what is required for spatial multiplexing, where rich multipath increases the link capacity. However, despite this contradiction in requirements, experimental analysis of this system as reported in [18] shows capacity gains above that of half-duplex MIMO. It is worth noting here that to provide a fair comparison, the number of antennas must be conserved; i.e., half-duplex MIMO will use **all** antennas for transmitting, and then **all** antennas for receiving, whereas the IBFD MIMO system depicted in Fig. 1.6 divides the array into transmit only and receive only elements ([18] does provide this fair comparison). Another potential drawback is that, since each antenna element uses a separate Tx chain, the Tx noise from each antenna element will be uncorrelated and therefore will not cancel. Thus, this system may not achieve the Tx noise reduction requirement [Eq. (1.3)], impacting on latter stages of digital cancellation. This is not addressed in [18].

The isolation provided by antenna cancellation systems can also be substantially degraded by environmental reflections. This is demonstrated in [13], the system achieved 45 dB of Tx-Rx isolation only when measured in an outdoor environment, however, when measured in a reflective indoor environment, the isolation reduced to just 15 dB [13]. Non-adaptive antenna cancellation systems are based on placing the Rx antenna at the **predicted** point of destructive interference based on the cancellation of the direct propagation paths, and thus, the multipath components are not cancelled, thereby reducing the Tx-Rx isolation. However, adaptive systems can mitigate multipath adjusting antenna weightings to compensate for the change in the coupling channel. But, since multipath channels are frequency selective, the resulting cancellation also varies with frequency, limiting the cancellation bandwidth (and therefore the overall cancellation). The cancellation achieved by the MIMO SIC system in [18] (shown in Fig. 1.6) was reduced by around 10 dB in an indoor environment compared to an outdoor environment.

Table 1.2 compares the characteristics and reported isolation for various types of IBFD antenna systems. It is notable the different designs of these antenna systems exhibit a wide range of disparate characteristics in terms of size, complexity, and performance. Thus the choice of antenna system design has a large impact on the features and performance of the overall full-duplex transceiver, and must be considered carefully at the design stage. The designer should note the requirements and constraints of the intended application, and the implications of the particular antenna system design on the overall system design.

Table 1.2 Comparison of antenna based isolation techniques

Ref.	Type	Isolation	Notes
[19]	Antenna separation	25 dB	Mobile phone size
[3]	Antenna separation	57 dB	Laptop size with shielding
[8]	Antenna separation	60 dB	Access point size, back-to-back patch antennas (relay application)
[7]	Antenna separation	72 dB	Base station size with shielding and x-polar antennas
[13]	Antenna cancellation	45 dB	Basestation size. Sensitive to multipath
[15]	Antenna cancellation	60 dB	Mobile phone size. Adaptive antenna weighting
[18]	Antenna cancellation	30–80 dB	IBFD MIMO system. Isolation depends on environment and number of antennas used

1.4 Passive Feedforward Cancellation

Passive self-interference cancellation techniques [10, 12, 14, 19–25] tap the Tx signal at the power amplifier (PA) output and apply analogue signal processing in order to generate a cancellation signal. The cancellation signal is then injected into the Rx path prior to the low-noise amplifier (LNA) using a coupler. Alternatively the cancellation signal can be injected using a specially adapted LNA design which facilitates cancellation and low-noise amplification in the same circuit [26], or can be down-converted to baseband using a separate mixer to cancel self-interference in analogue baseband, as shown in [24]. The technique is passive in the sense that all of the analogue signal processing is passive, applying variable delays, attenuations, and phase shifts to the transmit signal to generate the cancellation signal; however, the signal processing is adaptive, with control algorithms running in digital signal processing (DSP) to set the control inputs to the analogue signal processing components, iteratively measuring the self-interference power and adjusting the control inputs to maximise isolation.

Since some isolation will already have been provided by the antenna based isolation, the power of the cancellation signal will be significantly below the transmit power, and therefore this method does not require a significant portion of the transmit power to be tapped; however, the insertion loss of the couplers in the transmit and receive paths will reduce the transmitter efficiency and receiver sensitivity, respectively.

Active analog processing[1] can also be used, e.g., using a vector modulator to process the tapped signal; however, this introduces noise, negating one of the key benefits of analogue cancellation.

---

[1] This is distinct from the active cancellation techniques discussed below in Sect. 1.6, which tap the Tx signal in the digital baseband domain instead.

A drawback of passive feedforward cancellation is that the characteristics of the self-interference channel must be known or assumed in the cancellation circuit design process [10], such that the delay line(s) used in the adaptive analogue processing are selected with suitable lengths, and the variable attenuators operate over a suitable range. This would require the cancellation circuitry to be co-designed with the antenna system and antenna feed transmission lines, and thus reduces the flexibility of the resulting system (e.g., prohibiting the same modem hardware being used with different length antenna feed cables).

A significant advantage of passive self-interference cancellation is that, since the Tx signal is sampled at the output of the PA, the technique is able to cancel all noise and distortion introduced by the transmitter [1, 10], and therefore provides cancellation of linear SI, non-linear SI, and noise. However, since the required signal processing must be implemented in RF hardware, the cost and size of the circuitry are comparatively high.

1.4.1 Single Loop Cancellation

The simplest form of feedforward passive cancellation is to employ a single feedforward loop [19]. An IBFD transceiver architecture using *single loop cancellation* is depicted in Fig. 1.7, combining the cancellation loop with separate Tx and Rx antennas (see Sect. 1.3.1). Single loop cancellation applies a delay, amplitude, and phase shift to the tapped transmit signal in order to generate the cancellation signal. This may be described as a *narrowband cancellation* process, as the application of frequency invariant signal processing in the cancellation signal generation implicitly assumes a frequency invariant self-interference channel (i.e., a *narrowband assumption*).

An implementation of this architecture [19] achieved 27 dB of passive antenna separation isolation; however, the level of cancellation achieved, given in Table 1.3, depends heavily on the bandwidth. The figures quoted are from passive self-interference cancellation only, and do not include the additional isolation obtained from the antenna separation; including antenna separation, the system achieved a total of 70–80 dB isolation (depending on bandwidth). As can be seen in Table 1.3, single loop cancellation performs well at narrow bandwidths; however, the isolation deteriorates at wider bandwidths due to the frequency variant characteristics of the self-interference channel (due to multipath propagation and the resonant characteristics of antennas) *and* the cancellation circuitry (the couplers, antennas, phase shifter, etc. will not have perfectly flat frequency responses). For this reason, single

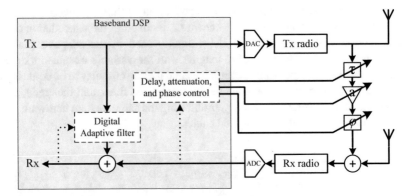

Fig. 1.7 An IBFD transceiver architecture using single loop passive RF cancellation, as proposed in [19]

Table 1.3 Measured single loop passive RF cancellation reported in [19]

Bandwidth	Passive cancellation
30 kHz	51 dB
200 kHz	49 dB
2000 kHz	42 dB

This is the amount of *cancellation*, not including the antenna isolation

loop cancellation is generally not suitable where substantial isolation is needed over wide bandwidths (e.g., in wideband IBFD radio systems).

1.4.2 Multi-Loop Cancellation

For the reasons given above, achieving wideband cancellation requires frequency selective signal processing in the cancellation loop. This allows the frequency variant nature of the self-interference channel to be replicated with greater accuracy, and mitigates frequency selectivity in the cancellation components themselves, but increases the complexity of the signal processing hardware. For passive RF cancellation, this is achieved by increasing the number of cancellation loops, effectively constructing an adaptive finite impulse response (FIR) filter in RF hardware, as shown in Fig. 1.8. IBFD transceiver front ends using multi-loop passive RF self-interference cancellation have been reported in [10, 25], demonstrating impressive levels of cancellation; however, in both of these systems the cost and size of the hardware implementation are high. A further drawback of multi-loop cancellation techniques is the high computational complexity of the optimisation processes which are required to tune the weightings of the filter taps to obtain cancellation [10, section 3.3].

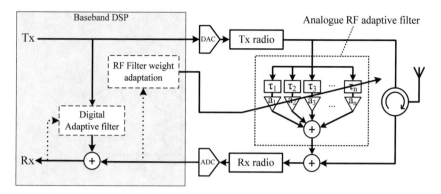

Fig. 1.8 A division free duplex transceiver architecture using multi-loop passive RF cancellation, as reported in [10]

The system reported in [10] combines a circulator with an adaptive analogue filter comprising 16 taps with fixed delay and variable attenuation (but without phase control). This system achieves a total isolation of 72 dB over a 20 MHz bandwidth, reducing to 62 dB over an 80 MHz bandwidth [10, Figure 8]. In this system, the circulator provides approximately 15 dB of isolation [10, section 3.1], and thus the total isolation equates to 57 dB of passive RF cancellation over 20 MHz and 47 dB of isolation over 80 MHz. The reduction in isolation at the wider bandwidth can be explained by considering that, in the 80 MHz case, the same number of filter taps are used to mimic the self-interference channel frequency response over a wider bandwidth, and therefore the cancellation accuracy is reduced.

The multi-loop self-interference cancellation system reported in [25, 27] comprises a two tap filter, but allows for adjustment of the amplitude *and phase* of each tap, thus significantly improving the utility of each tap compared to the amplitude only control of the filter taps in [10]. When combined with a circulator, this technique achieved a total isolation of 63 dB over an 80 MHz bandwidth [27], this being similar to [10].

1.5   Electrical Balance Duplexers

Interest in self-interference cancellation for duplexing has led to a renewed interest in an old duplexing technology based on *electrical balance in hybrid junctions* [28, 29]. The *electrical balance duplexer* (EBD) [30–37] facilitates simultaneous transmission and reception from a single antenna whilst providing high Tx-Rx isolation in both the transmit and receive bands, and being tunable over wide frequency ranges.

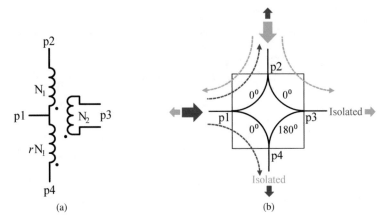

Fig. 1.9 (a) Hybrid transformer with ports and winding ratios labelled. (b) Circuit symbol for hybrid junction with corresponding port labels and annotated with signal coupling behaviour

The hybrid junction[2] is a four port lossless reciprocal network in which opposite pairs of ports are isolated from each other [28, p. 246–249]. Its operation is depicted in Fig. 1.9b: assuming all ports are terminated with the correctly matched characteristic impedance, a signal arriving at any one of the hybrid's ports is divided between the two adjacent ports, but not coupled to the opposite port. This property of "electrical balance" can be exploited to isolate transmitter and receiver circuitry, but allow use of a shared antenna.

1.5.1 EBD Operation

In a wireless electrical balance duplexer the transmit current, supplied by the PA, enters the hybrid at the centre tap of the primary winding (as shown in Fig. 1.10b) and is split between two paths flowing in opposite directions, with one component flowing to the antenna (and being transmitted), and the other to the balancing impedance. The relative magnitudes of these currents are determined by the transformer tapping ratio, r, as shown in Fig. 1.9a, and by the values of the antenna and balancing impedances. The balancing impedance is a tunable impedance which is adjusted such that these two currents create equal but opposite magnetic fluxes that cancel, and therefore zero current is induced in the secondary winding of the transformer—the receiver is isolated from the transmitter. A signal received at the antenna, however, causes current to flow through the primary winding in one direction only, thereby coupling it to the receiver winding.

---

[2]Also known as a "hybrid transformer," "hybrid coil," "bridge transformer," "magic tee," or simply a "hybrid."

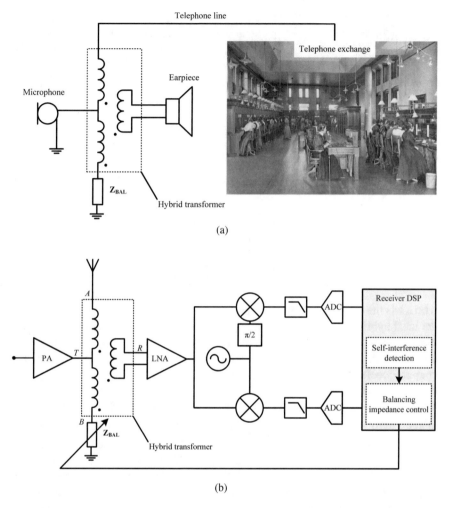

Fig. 1.10 Electrical balancing duplexing applied to (**a**) wired communication and (**b**) wireless communication. Inset photograph is of Montreal telephone exchange (c. 1895), image credited to [38]

Electrical balance duplexing is by no means new, but has been used since the early days of wired telephony [29]. In a telephone system, the microphone and earpiece must both be connected to the telephone line, but must be isolated from one another to prevent the users own speech deafening them to the much weaker incoming audio signal. This was achieved using a hybrid transformer connected as shown in Fig. 1.10a. The wireless EBD duplexing

(continued)

application [30] is entirely analogous to the telephone duplexer, but substitutes
the microphone, earpiece, and telephone line with the transmitter, receiver,
and antenna, respectively, as shown in Fig. 1.10b.

1.5.1.1   Tx-Rx Isolation

The isolating property of the EBD stems from the coupling behaviour of signals
within a hybrid junction. The S-matrix equation describing a 4-port lossless hybrid
is

$$
\begin{bmatrix} b_T \\ b_R \\ b_A \\ b_B \end{bmatrix} = \begin{bmatrix} 0 & 0 & k & l \\ 0 & 0 & l & -k \\ k & l & 0 & 0 \\ l & -k & 0 & 0 \end{bmatrix} \begin{bmatrix} a_T \\ a_R \\ a_A \\ a_B \end{bmatrix}
\tag{1.13}
$$

where k is the coupling coefficient, and $l = \sqrt{1 - k^2}$ (since energy is conserved and
the ideal hybrid is lossless), a_T, a_R, a_A, a_B are the incident signals at the transmit,
receive, antenna, and balance ports, respectively [ports T, R, A, and B as annotated
on Fig. 1.10b], and b_T, b_R, b_A, b_B are the corresponding scattered signals. The
coupling coefficient quantifies the proportions by which power at an input port is
divided between the two adjacent ports. For a hybrid transformer implementation
as depicted in Fig. 1.9a, the coupling coefficient is determined by the transformer
tapping ratio according to $k = \frac{\sqrt{r}}{\sqrt{1+r}}$. As shown in [28], by expanding (1.13), and
noting that the incident signals at the antenna and balance ports are the reflections of
the scattered signals at those ports, such that $a_A = b_A \Gamma_A$ and $a_B = b_B \Gamma_B$ where Γ_A
and Γ_B are the antenna reflection coefficients and balancing reflection coefficients,
respectively, and assuming matched impedances at the Tx and Rx ports, the Tx-Rx
gain, G, can be calculated as

$$
G = \frac{b_R}{a_T} = kl(\Gamma_A - \Gamma_B)
\tag{1.14}
$$

and therefore the Tx-Rx gain of the EBD is directly proportional to the difference
between the antenna and balancing reflection coefficients. For a symmetrical hybrid,
$k = l = 1/\sqrt{2}$, and thus $kl = 1/2$, and for an asymmetrical hybrid, $kl < 1/2$.
Mathematically, balancing the duplexer involves setting the value of the balancing
reflection coefficient, Γ_B, such that there is zero gain

$$
G\big|_{\Gamma_B = \Gamma_A} = kl(\Gamma_A - \Gamma_B)\big|_{\Gamma_B = \Gamma_A} = 0.
\tag{1.15}
$$

Physically, this is achieved by adjusting the balancing reflection coefficient using the tunable impedance circuit at the balancing port, to achieve this state of electrical balance, resulting in very high (theoretically infinite) isolation between the transmitter and receiver. It is pertinent to note here that the EBD can be considered a form of passive RF cancellation, and the distinction between antenna based isolation and RF cancellation, as described in the introduction, does not strictly apply. Although this technique is a type of passive RF cancellation, it can only be applied at the antenna port and cannot be combined with another form of antenna based isolation in the same way as feedforward cancellation techniques can be combined with antenna separation or antenna cancellation.

Physically, this cancellation occurs in the opposing magnetic fluxes generated by the hybrid transformer coils (or the coupled transmission lines in the case of a microstrip hybrid coupler implementation [39, section 5.4]). Mathematically, this cancellation is manifest in the subtraction of the two terms in (1.14). Conceptually, this cancellation can also be considered as the superposition of two copies of the transmit signal, which, when $\Gamma_B = \Gamma_A$, "arrive" at the receiver port in perfect anti-phase. In fact, by considering the self-interference channel in this way, it is possible to directly obtain an expression for the scattered signal at the Rx port in terms of the incident signal at the Tx port, and thereby derive (1.14) by inspection, as shown in Fig. 1.11.

1.5.2 Tx and Rx Insertion Loss

Since signals entering the hybrid junction are divided between two signal paths, there are, of course, some inherent losses associated with the EBD. Some of the transmit power is dissipated by the balancing impedance, instead of being transmitted from the antenna and not all of the power received at the antenna is coupled to the receiver. These losses depend on the symmetry of the hybrid junction, which for a hybrid transformer is determined by the tapping ratio, and can be traded

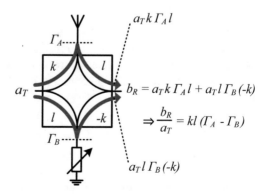

Fig. 1.11 Derivation of the Tx-Rx scattering by inspection of the Tx-Rx signal coupling paths in the EBD

Fig. 1.12 Relationship
between the tapping ratio, r,
and the Tx and Rx insertion
losses

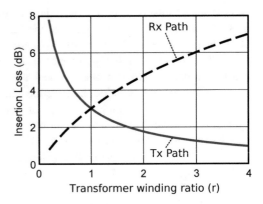

off against one another by using a hybrid which is skewed in favour of either the
transmit or receive path, as shown in Fig. 1.12.

This relationship between the losses in the transmit and receive paths results from
the reciprocity of the hybrid junction. For example, if the circuit were designed to
deliver all of the transmit power to the antenna then, since the circuit is reciprocal,
all power arriving at the antenna would be delivered back to the transmitter (and
therefore not to the receiver) [30]. Consequently, for the system to function in
both transmit and receive modes, there is some unavoidable loss, which for a
symmetrical hybrid (i.e., when $r = 1$), is 3 dB in both the transmit and receive paths,
reducing the transmitter efficiency and receiver sensitivity. However, these losses
can be mitigated through co-design of the EBD and LNA, using *noise matching* to
minimise the noise power being coupled to the receiver. For example, the prototype
reported in [30] skewed the hybrid junction to reduce Tx insertion loss, reducing
this to 2.2 dB. The high Rx insertion loss which results was mitigated through noise
matching to achieve an antenna referred Rx noise figure of just 5 dB.

1.5.3 Balancing Limitations

As shown above in Sect. 1.5.1.1, to balance the EBD the balancing impedance
is adjusted such that the balancing reflection coefficient is equal to the antenna
reflection coefficient. The balancing reflection coefficient at the balance port is
determined by the impedance presented to the balance port according to the well-
known equation [39, p. 74]

$$\Gamma_B = \frac{Z_B - Z_C}{Z_B + Z_C} \tag{1.16}$$

where Z_B is the balancing impedance, and Z_C is the characteristic impedance of
the system (e.g., 50 Ω). In the telephone duplexer (Fig. 1.10a), the telephone line

has a fixed characteristic impedance, and therefore, in this application, balancing the duplexer requires only that the balancing port is terminated with the same fixed impedance. However, in the wireless application, balancing the duplexer is considerably more complicated due to the fact that, in practice, the antenna is not an ideal 50 Ω resistor. Antennas are resonant devices—they are inherently reactive and therefore the impedance changes with frequency. Similarly, since the tunable balancing impedance must be able to balance a complex antenna impedance, it must itself also contain reactive components, and the resulting balancing reflection coefficient is therefore also a function of frequency. Taking this into account, the Tx-Rx transfer function (1.14) can be re-written as

$$G(\omega) = kl\big(\Gamma_A(\omega) - \Gamma_B(\omega)\big). \tag{1.17}$$

Due to the frequency domain variation in the antenna and balancing reflection coefficients, the resulting Tx-Rx isolation is also a function of frequency. To obtain perfect (theoretically infinite) balancing over a given band, the antenna and balancing reflections coefficients need to be equal at all frequencies within that band.

> In practice, wide isolation bandwidths are difficult to achieve using EBDs. The balancing circuit used can only accurately mimic the antenna reflection coefficient over a very narrow bandwidth, limiting the isolation bandwidth.

With a single pole balancing circuit, such as the parallel RC circuits used in [30] and [31] [see Fig. 1.13], perfect balancing may only be obtained at one frequency point. This is illustrated in Fig. 1.14a, which shows smith chart plots of a measured antenna reflection coefficient, and a theoretically calculated balancing reflection coefficient assuming a single pole balancing network and transmission line (see [40] for full details). The resulting Tx-Rx isolation can be calculated using (1.17), as shown in Fig. 1.14b, which exhibits a band-limited "notch-like" isolation. A single

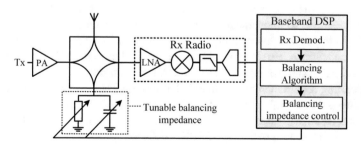

Fig. 1.13 A transceiver architecture using an electrical balance duplexer with resistor-capacitor (RC) balancing network, and adaptive balancing control

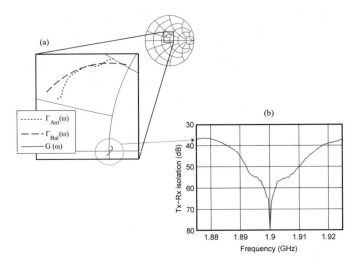

Fig. 1.14 (**a**) Measured antenna reflection coefficient and theoretical single pole balancing reflection coefficient, and the resulting Tx-Rx gain, on the smith chart. (**b**) The resulting Tx-Rx isolation

pole balanced EBD is akin to single loop cancellation as discussed in Sect. 1.4.1, and typically achieves 40–50 dB isolation over a 20 MHz bandwidth [40, 41], depending on how well the particular reflection coefficients of the antenna and balancing network match. More complicated balancing network designs, which include a larger number of tunable components, have also been demonstrated. For example, the system in [42] achieves an average of 50 dB of isolation over a wider bandwidth of 80 MHz bandwidth; however, this wider isolation bandwidth comes at the expense of higher RF complexity in the balancing network. This is akin to multi-loop passive cancellation as discussed in Sect. 1.4.2.

In addition to the dependence of the isolation on the frequency domain antenna characteristics as described above, the antenna is also time variant due to interaction with the local environment. Transmitted energy leaves the antenna, but can be reflected back to the antenna by nearby objects, being received as self-interference. This adds multipath to the SI channel, complicating SI cancellation, but is also time variant due to movement of the device and/or environmental reflectors. To obtain electrical balance, and maintain optimum balance in the presence of dynamic reflections, EBDs typically use adaptive balancing algorithms [43, 44] (see Fig. 1.13). The effects of environmental reflections on EBD isolation and corresponding requirements for dynamic balancing have also been addressed [45–47], showing that in highly dynamic hand-held and vehicular scenarios, EBDs must update the balancing impedance to its optimum value at intervals of the order of milliseconds in order to maintain ∼50 dB Tx-Rx isolation.

1.6 Active Cancellation

Active cancellation techniques [2, 3, 7, 48–53] apply digital signal processing to the baseband transmit signal, generating a digital baseband cancellation signal. The baseband cancellation signal is then upconverted to RF using a second transmitter chain, and coupled in to the receive path, cancelling the self-interference prior to the receiver input. IBFD transceivers combining antenna separation, active cancellation, and digital baseband cancellation have been reported in [2, 3, 7], and this architecture is depicted in Fig. 1.15. Like passive cancellation techniques, active techniques can be frequency flat or frequency selective. The active canceller in [2] applies a frequency invariant complex gain to the Tx signal to generate the cancellation signal, providing 31 dB of self-interference cancellation over a relatively narrow bandwidth of 625 kHz, however, as is the case with passive single loop cancellation, the amount of cancellation achieved will deteriorate rapidly as bandwidth increases, due to frequency selective SI coupling. However, a key advantage of active cancellation is that, due to the high availability of signal processing resources in the digital baseband domain, *high order filtering can readily be applied.*

Since the Tx signal is tapped in the digital baseband domain, prior to the Tx radio, the cancellation process has no knowledge of the imperfections introduced by the Tx hardware, and therefore cannot cancel them. Indeed the second transmitter chain itself will be a source of further noise and non-linear SI. As mentioned above (see Sect. 1.4), passive techniques have the benefit of being able to cancel the noise and non-linear components introduced by the Tx chain, however with active cancellation, these imperfections remain as residual self-interference. The limited the accuracy of a Tx output signal due to the noise and hardware imperfections in the transmit chains is quantified by the error vector magnitude (EVM) [54]. As a

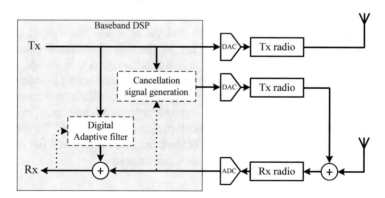

Fig. 1.15 A division free duplex transceiver architecture using active RF cancellation, as reported in [2, 3, 7]

rule-of-thumb, the level of cancellation which can be achieved is roughly the same as the EVM.

> Whereas the limitation of passive cancellation is the restricted self-interference modelling accuracy due to the limited signal processing resources, active cancellation allows the self-interference to be modelled with a much greater (practically unlimited) accuracy. Instead, the cancellation is limited by Tx radio hardware imperfections, which limit the accuracy of the translation of signals from baseband, where the cancelling signals are generated, to RF, where the cancellation actually occurs.

1.6.1   Hardware Cost

Active cancellation has the drawback of requiring an additional transmit chain, which increases the cost and power consumption of the system. But, since some significant isolation will have been obtained from some form of antenna based isolation, the signal to be cancelled will have significantly lower power compared to the Tx signal. The cancellation Tx chain will therefore function at a lower power compared to the main Tx, reducing the cost and power consumption of this subsystem as compared to the main transmitter (e.g., a power amplifier stage may not be required). Furthermore, in comparison to the analogue signal processing components required for passive cancellation, the cost and size of the additional hardware may be relatively small, and the additional Tx chain can potentially be implemented within the same radio frequency integrated circuit (RFIC) as the primary transmitter and the receiver: a distinct advantage for low cost, small form factor devices.

The active architecture also scales to multiple-input-multiple-output (MIMO) with a linear increase in complexity [52], as the digital baseband cancellation signal generation easily allows for a single cancellation signal to be composed from multiple Tx baseband signals in order to concurrently cancel multiple sources of SI from multiple antennas. Whereas passive cancellation in a MIMO transceiver would require a cancellation loop from every transmit antenna to every receive antenna (i.e., a dedicated cancellation loop for each permutation of Tx and Rx antennas), with active cancellation only one cancellation transmitter is required per receiving antenna, regardless of the number of transmitting antennas.

1.6.2 Wideband Cancellation

Like passive cancellation systems, achieving wideband active cancellation requires higher order digital signal processing; however, unlike passive systems, the required order of processing can easily be achieved without substantial increases in hardware complexity.

Wideband IBFD radio transceivers implementing active cancellation have been reported in [3, 7], combining this with antenna separation. These systems extend the narrowband cancellation technique presented in [2] to cancel wideband self-interference by exploiting an orthogonal frequency division multiplexing (OFDM) physical layer. OFDM modems use the fast Fourier transform (FFT) and the inverse fast Fourier transform (IFFT) to move between the time and frequency domain, allowing the signals to be composed of a large number of modulated subcarriers, and facilitating frequency domain digital baseband signal processing. However this modulation technique can also be exploited for wideband cancellation by estimating and cancelling self-interference on a per-subcarrier basis.

This architecture is depicted in Fig. 1.16, showing the RF hardware subsystems and digital baseband signal processing operations. In this case, the cancellation signal generation function can be implemented as a frequency domain equalizer (FDE), processing the Tx signal in the frequency domain by applying a complex multiplication to each subcarrier. This does however increase the computational complexity, requiring additional FDE, IFFT, and cyclic prefix generation operations. The technique has been shown to be effective, achieving around 25 dB of RF self-interference cancellation over a 20 MHz bandwidth [7], this being limited by the

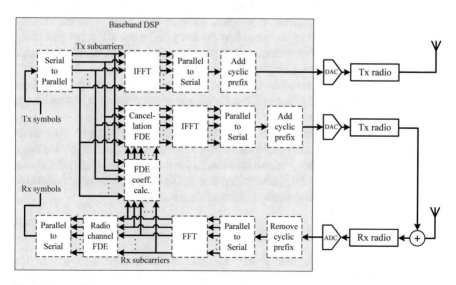

Fig. 1.16 An IBFD transceiver architecture using wideband active RF cancellation based on an OFDM physical layer, as reported in [3, 7]

EVM of the Tx radios. The relationship between subcarrier spacing and achievable self-interference cancellation is also investigated in [7]. Results demonstrate that increased frequency domain variation in the self-interference channel frequency response, for example, due to multipath propagation between the Tx and Rx antennas, requires higher order filtering to maintain wideband cancellation. However the prototype reported in [7], which used 64 subcarriers across a 20 MHz bandwidth (i.e., a subcarrier spacing of 312.5 kHz) achieved levels of active cancellation similar to the Tx EVM, and was therefore not limited by modelling accuracy. This frequency resolution is significantly lower than the subcarrier spacing typically used in OFDM systems (3GPP Long Term Evolution uses a 15 kHz subcarrier spacing), and therefore the required self-interference channel modelling accuracy can be achieved with the subcarrier spacing of typical wireless systems.

1.6.3   Equaliser Function Calculation

In order for cancellation to be effective, the required equaliser function must be accurately determined in order that an accurate cancellation signal can be generated. Furthermore, the transfer functions of the RF front-end components must also be taken into account, as for the cancellation to occur, the two signals must be arranged such that they arrive at the coupling point with identical amplitude and in perfect anti-phase. Therefore, in addition to characterising the self-interference channel itself (for example, between the Tx and Rx antennas of an antenna separation architecture), any amplitude and phase offsets introduced by the hardware (e.g., transmitters, receiver, transmission lines, coupler, etc.…) must also be accounted for when generating the cancellation signal, using inverse filtering to compensate for the transfer functions of the non-ideal circuitry. With the RF front end being such a complex system with numerous components and interactions, on the surface it would appear that counteracting the effect of these components would be difficult. However, this can in fact be achieved very simply by considering the problem from the perspective of the equivalent baseband only, and characterising the behaviour of the RF system as a "black box" [48].

Disregarding the desired receive signal (which at this stage of the receiver chain remains several orders of magnitude below the self-interference), and ignoring noise and non-linear system behaviour, we may model the total received signal as the sum of the signals from the primary transmitter and auxiliary (cancellation) transmitter, multiplied by corresponding transfer functions, such that

$$S_{Rx}(\omega) = S_{Tx}(\omega)\Lambda(\omega) + E_{Cx}(\omega)S_{Tx}(\omega)\Theta(\omega) \qquad (1.18)$$

where $S_{Tx}(\omega)$ is the transmit signal, $E_{Cx}(\omega)$ is the transfer function of the cancellation equalizer (which generates the cancellation signal from the transmit signal), $\Lambda(\omega)$ is the self-interference channel, this being the channel between the primary transmitter and the receiver, $\Theta(\omega)$ is the "cancellation channel," this being the channel between the cancellation transmitter and receiver (see Fig. 1.17), $S_{Rx}(\omega)$ is the received signal, and all quantities are frequency domain equivalent complex baseband signals or transfer functions. Mathematically, cancellation occurs when the two terms of (1.18) sum to zero. Therefore, setting (1.18) to zero and solving for $E_{Cx}(\omega)$ give the cancellation equaliser function which will result in cancellation as

$$E_{Cx}(\omega) = \frac{-\Lambda(\omega)}{\Theta(\omega)}. \tag{1.19}$$

Evidently, this requires $\Lambda(\omega)$ and $\Theta(\omega)$ to be known. In the case of $\Lambda(\omega)$, the self-interference channel, this can be measured by transmitting a signal from the main transmitter and receiving the self-interference. This is much like a pilot based channel measurement, however it is not necessarily a pilot, as this can be performed when the device is transmitting a regular transmit signal with payload data. The caveat in the measurement of $\Lambda(\omega)$ is that the cancellation transmitter must be inactive, such that the cancellation signal does not interfere with SI channel measurement. Therefore there will be no active cancellation, meaning that any receive signals will most likely be corrupted due to the SI, and to avoid this the self-interference channel measurement must be performed when the device is in transmit only mode. To determine the cancellation channel, a pilot can be sent from the cancellation transmitter when the main transmitter is inactive, measuring the channel in the same manner and likewise causing uncancelled SI. Therefore there can be no simultaneous transmission *or* reception during the cancellation channel measurement. These conditions will limit the opportunity to run these training processes, however for many wireless systems, which operate with discontinuous transmission and reception anyway, as determined by the medium access control (MAC) protocols, it may be feasible that these equaliser training measurements may be performed opportunistically within the transmit/receive scheduling of the device.

It is pertinent to note that, since all of the quantities are measured and processed in equivalent baseband DSP, the cancellation signal which is generated will inherently compensate for the transfer functions of the non-ideal circuit components discussed above. This method makes no distinction between the Tx-Rx transfer function between the Tx and Rx antennas, and the transfer functions of the other components in the RF front end such as the transmitters and receiver. As depicted in Fig. 1.17, the measured self-interference channel is the aggregate channel response between the baseband primary transmitter output and the baseband receiver input, and similarly, the measured cancellation channel is the aggregate channel response between the baseband cancellation transmitter output and the baseband receiver input. The equaliser function calculation method given above simply characterises these channels and determines the cancellation signal required to force the receive baseband signal to be zero, thereby forcing cancellation to occur in the RF domain.

The example given above is for a general case with continuous frequency. Where an OFDM physical layer is used, as shown in Fig. 1.16, this method can be applied at discrete frequencies. In this special case, the equaliser function becomes a set of

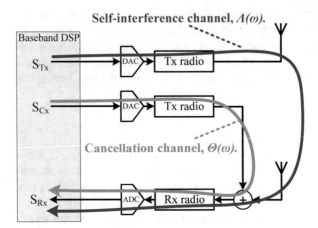

Fig. 1.17 The self-interference channels and cancellation channel in an active self-interference cancellation transceiver front-end

equaliser coefficients (one complex quantity for each transmit subcarrier), which are calculated as

$$\Xi[k] = \frac{-\Lambda[k]}{\Theta[k]},\tag{1.20}$$

where $k \in [0, K-1]$ is the index of the K system subcarriers.

1.7   Combining Antenna and RF Cancellation Techniques

In-band full-duplex systems typically combine multiple stages of cancellation to obtain high isolation, and many IBFD transceiver architectures combine some form of antenna based isolation with a further stage of RF cancellation. Moreover, depending on the isolation provided by the antenna system, additional RF-domain cancellation may be necessary to fulfil the RF-domain isolation requirements (see Sect. 1.2).

However, not all antenna and RF cancellation techniques can be directly combined. The electrical balance duplexer (see Sect. 1.5), for example, is inseparable from the single antenna shared for Tx and Rx, and thus combining EBDs with multi-antenna isolation techniques (see Sects. 1.3.1 and 1.3.3) is not a straightforward design problem.

The effectiveness of a particular cancellation technique depends on the characteristics of the SI channel, including the effect of any preceding stages of cancellation. For example, where the SI channel is dominated by a single strong coupling path, as would be the case for separate omnidirectional Tx and Rx antennas without shielding [2], single loop cancellation can provide substantial cancellation by cancelling that dominant path. However, if the SI channel is a multipath channel without a dominant component, for example, in an antenna system where the direct path is blocked by shielding, then single loop cancellation may provide very limited cancellation, as it can only cancel one of the many SI components; multi-loop cancellation would be required in this case.

It is notable that cancelling the dominant components of multipath (or indeed avoiding them in the case of using shielded antennas) increases the overall isolation, but can also increase the frequency selectivity of the SI channel as seen by the subsequent stage of cancellation. The lower powered multipath components which were previously insignificant compared to dominant components become significant once the dominant components are removed. Thus in general, as isolation increases, so does the requirement for frequency selective cancellation.

Figure 1.18 plots simulation results which show the Tx-Rx isolation of an electrical balance duplexer, along with results for the EBD combined with a further stage of cancellation of two different types: electrical balance and single loop cancellation (EBSLC), and electrical balance and active cancellation (EBAC). The simulation embeds a measured antenna reflection coefficient, thereby capturing the multipath self-interference returning to the antenna in the indoor measurement environment at 1.9 GHz (see [33]). As shown, a further stage of single loop cancellation has very little impact on the isolation (only 0.5 dB of additional cancellation)—this is because the EBD has already cancelled the dominant SI component, leaving behind multipath SI for which the single loop canceller is ineffective. In contrast, combining the electrical balance duplexer with a frequency selective active canceller increased the isolation by 28 dB, as the active canceller *is* able to cancel the multipath SI.

Table 1.4 summarises various combinations of antenna and RF cancellation techniques as reported in the literature. When designing an IBFD transceiver, the

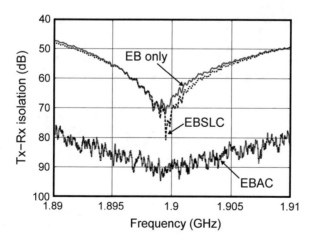

Fig. 1.18 Simulated Tx-Rx isolation for an electrical balance duplexer (EBD), the electrical balance duplexer with a second stage of single loop cancellation (EBSLC), and the electrical balance duplexer with a second stage of active cancellation (EBAC)

Table 1.4 Measured performance of various prototypes combining antenna and RF cancellation reported in the literature

Group	Ref.	BW	Ant. isol.		RF Clc.			Comments
			Type	Sup.	Type	Sup.	Total	
Bristol	[19]	200 kHz	AS	27 dB	SLC	49 dB	76 dB	Narrowband only
	[43]	80 MHz	EBD	45 dB	Active	36 dB	81 dB	Wide frequency tunability
Rice	[2]	625 kHz	AS	45 dB	Active	31 dB	76 dB	Narrowband only
	[3]	10 MHz	AS	56 dB	Active	24 dB	81 dB	OFDM. Tablet form factor
	[7]	20 MHz	AS	72 dB	Active	23 dB	95 dB	OFDM. Basestation sized. System
								Cross-polar directional antenna with shielding
Stanford	[12]	2 MHz	AC	30 dB	SLC	20 dB	60 dB	Relies on half-wavelength path difference
	[10]	20 MHz	Circ.	≈15 dB	MLC	57 dB	72 dB	16 taps (variable amplitudes only)
		80 MHz	Circ.	≈15 dB	MLC	47 dB	62 dB	Hardware size and cost are high. Requires co-design with antenna and feed. RF tuning is computationally expensive
NEC	[13]	625 kHz	AC	45 dB	none		45 dB	Very susceptible to multipath
EPFL	[52]	20 MHz	Circ.	≈18 dB	Active	48 dB	66 dB	Implements 2 × 2 MIMO
Duplo	[25]	20 MHz	Circ.	26 dB	MLC	33 dB	59 dB	Two taps, variable amplitude and phase. Hardware size and cost is high (but lower than [10])
		100 MHz	Circ.	23 dB	MLC	18 dB	41 dB	

Abbreviations: *Ant. isol.* antenna based isolation, *RF Clc.* RF cancellation, *AS* antenna separation, *AC* antenna cancellation, *EBD* electrical balance duplexer, *Circ.* circulator, *Active* active cancellation, *SLC* single loop cancellation, *MLC* multi-loop cancellation, *Sup* self-interference suppression

choice of which RF cancellation technology to pair with the selected antenna system design depends on many factors. As discussed above, the choice of antenna based isolation and the resulting characteristics of the SI coupling must be considered, along with the cost, size, complexity, power consumption, linearity, and power handling capabilities of any hardware implementation. The designer must also consider the RF-domain isolation requirements as given earlier in Sect. 1.2, as in some instances this will limit the choice of the type of RF cancellation which can be applied. For example, since active RF cancellation cannot cancel Tx noise, it can only be used if adequate isolation of Tx noise has already been obtained in the antenna domain, or else this will limit the effectiveness of the subsequent stage of digital cancellation.

1.8 Conclusions

Radio frequency domain antenna and circuit techniques are a fundamental part of any IBFD radio system. This chapter has provided an overview of antennas and RF cancellation, deriving requirements for RF-domain isolation, and discussing a range of antenna systems and circuits for avoiding and cancelling self-interference.

Many antenna systems use separate Tx and Rx antennas to provide transmit-to-receive isolation, through either propagation loss between the antennas, or through more advanced designs which cancel SI in the propagation domain. The amount of isolation obtained in the antenna system may also not be sufficient to fulfil RF cancellation requirements, and a second stage of RF cancellation is typically required. The design of the second stage cancellation must be carefully selected based on the characteristics of the residual SI after the first stage of antenna based isolation, with reference to RF-domain isolation requirements, and considering the implications on the overall system design. Various cancellation techniques have been developed and applied in recent years, differing in their characteristics and capabilities in terms of cancellation of multipath components and Tx noise.

Whilst there have been many advances in antenna and cancellation technologies in recent years, as discussed in this chapter, further work remains to bring these designs from the lab bench in to products. In-band full-duplex technologies have already been commercialised for point-to-point microwave links [55], where separate Tx/Rx antennas with high directivity can be used, and the operating environment is relatively static. However, substantial challenges remain, especially where space, complexity, and power consumption limit design choices, and dynamic environments will place challenging requirements on the dynamic adaptation of coefficients in the cancellation loops.

References

1. A. Sabharwal, P. Schniter, D. Guo, D. W. Bliss, S. Rangarajan, and R. Wichman, "In-Band Full-Duplex Wireless: Challenges and Opportunities," *IEEE Journal on Selected Areas in Communications*, vol. 32, no. 9, pp. 1637–1652, Sep 2014.
2. M. Duarte and A. Sabharwal, "Full-duplex wireless communications using off-the-shelf radios: Feasibility and first results," in *Signals, Systems and Computers (ASILOMAR), 2010 Conference Record of the Forty Fourth Asilomar Conference on*, 2010, pp. 1558–1562.
3. A. Sahai, G. Patel, and A. Sabharwal, "Pushing the limits of full duplex wireless: design and real-time implementation," Rice Univ. Houston, TX, Tech. Rep. TREE1104, 2011.
4. S. Chen, "Division-free Duplex for Wireless Applications," Ph.D. dissertation, University of Bristol, UK, 1997.
5. C. Anderson, S. Krishnamoorthy, C. Ranson, T. Lemon, W. Newhall, T. Kummetz, and J. Reed, "Antenna Isolation, Wideband Multipath Propagation Measurements, and Interference Mitigation for On-frequency Repeaters," in *IEEE SoutheastCon, 2004. Proceedings*. IEEE, 2004, pp. 110–114.
6. E. Everett, M. Duarte, C. Dick, and A. Sabharwal, "Empowering full-duplex wireless communication by exploiting directional diversity," in *2011 Conference Record of the Forty Fifth Asilomar Conference on Signals, Systems and Computers (ASILOMAR)*, 2011, pp. 2002–2006.
7. E. Everett, A. Sahai, and A. Sabharwal, "Passive Self-Interference Suppression for Full-Duplex Infrastructure Nodes," *IEEE Transactions on Wireless Communications*, vol. 13, no. 2, pp. 680–694, 2014.
8. M. Heino, D. Korpi, T. Huusari, E. Antonio-Rodriguez, S. Venkatasubramanian, T. Riihonen, L. Anttila, C. Icheln, K. Haneda, R. Wichman, and M. Valkama, "Recent advances in antenna design and interference cancellation algorithms for in-band full duplex relays," *IEEE Communications Magazine*, vol. 53, no. 5, pp. 91–101, May 2015.
9. N. Phungamngern, P. Uthansakul, and M. Uthansakul, "Digital and RF interference cancellation for single-channel full-duplex transceiver using a single antenna," in *2013 10th International Conference on Electrical Engineering/Electronics, Computer, Telecommunications and Information Technology (ECTI-CON)*, 2013, pp. 1–5.
10. D. Bharadia, E. McMilin, and S. Katti, "Full Duplex Radios," in *Proc. 2013 ACM SIGCOMM*, Hong Kong, 2013.
11. N. A. Estep, D. L. Sounas, J. Soric, and A. Alù, "Magnetic-free non-reciprocity and isolation based on parametrically modulated coupled-resonator loops," *Nature Physics*, vol. 10, no. 12, pp. 923–927, Nov 2014.
12. J. I. Choi, M. Jain, K. Shrinivasan, P. Levis, and S. Katti, "Achieving Single Channel, Full Duplex Wireless Communication," in *Proc. ACM Int. conf. Mobile Comput. & Netw.*, Chicago, IL, 2010, pp. 1–12.
13. E. Aryafar, M. A. Khojastepour, and S. Rangarajan, "MIDU: Enabling MIMO Full Duplex," in *Proc. ACM Int. conf. Mobile Comput. & Netw.*, 2012, pp. 257–268.
14. C. Jung Il, S. Hong, M. Jain, S. Katti, P. Levis, and J. Mehlman, "Beyond full duplex wireless," in *2012 Conference Record of the Forty Sixth Asilomar Conference on Signals, Systems and Computers (ASILOMAR)*, 2012, pp. 40–44.
15. G. Pedersen, E. de Carvalho, O. Franek, E. Foroozanfard, E. Tsakalaki, and A. Tatomirescu, "Full-duplex MIMO system based on antenna cancellation technique," *Electronics Letters*, vol. 50, no. 16, pp. 1116–1117, Jul 2014.
16. M. Jain, J. I. Choi, T. M. Kim, D. Bharadia, S. Seth, K. Shrinivasan, P. Levis, S. Katti, and P. Sinha, "Practical Real-Time Full Duplex Wireless," in *Proc. ACM Int. conf. Mobile Comput. & Netw.*, Las Vegas, NV., 2011, pp. 301–312.
17. T. Riihonen, S. Werner, and R. Wichman, "Mitigation of Loopback Self-Interference in Full-Duplex MIMO Relays," *IEEE Transactions on Signal Processing*, vol. 59, no. 12, pp. 5983–5993, Dec 2011.

18. E. Everett, C. Shepard, L. Zhong, and A. Sabharwal, "Softnull: Many-antenna full-duplex wireless via digital beamforming," *IEEE Transactions on Wireless Communications*, vol. 15, no. 12, pp. 8077–8092, Dec 2016.

19. S. Chen, M. A. Beach, and J. P. McGeehan, "Division-free duplex for wireless applications," *Electronics Letters*, vol. 34, no. 2, pp. 147–148, 1998.

20. S. Kannangara and M. Faulkner, "Adaptive duplexer for multiband transreceiver," in *2003. RAWCON '03. Proceedings Radio and Wireless Conference*, 2003, pp. 381–384.

21. ——, "Analysis of an Adaptive Wideband Duplexer With Double-Loop Cancellation," *IEEE Transactions on Vehicular Technology*, vol. 56, no. 4, pp. 1971–1982, 2007.

22. D. Korpi, S. Venkatasubramanian, T. Riihonen, L. Anttila, S. Otewa, C. Icheln, K. Haneda, S. Tretyakov, M. Valkama, and R. Wichman, "Advanced self-interference cancellation and multiantenna techniques for full-duplex radios," in *2013 Asilomar Conference on Signals, Systems and Computers*, Nov 2013, pp. 3–8.

23. D. Korpi, L. Anttila, V. Syrjala, and M. Valkama, "Widely Linear Digital Self-Interference Cancellation in Direct-Conversion Full-Duplex Transceiver," *IEEE Journal on Selected Areas in Communications*, vol. 32, no. 9, pp. 1674–1687, sep 2014.

24. D.-J. van den Broek, E. A. M. Klumperink, and B. Nauta, "An In-Band Full-Duplex Radio Receiver With a Passive Vector Modulator Downmixer for Self-Interference Cancellation," *IEEE Journal of Solid-State Circuits*, vol. 50, no. 12, pp. 3003–3014, Dec 2015.

25. T. Huusari, Y.-S. Choi, P. Liikkanen, D. Korpi, S. Talwar, and M. Valkama, "Wideband Self-Adaptive RF Cancellation Circuit for Full-Duplex Radio: Operating Principle and Measurements," in *2015 IEEE 81st Vehicular Technology Conference (VTC Spring)*. IEEE, may 2015, pp. 1–7.

26. J. Zhou, A. Chakrabarti, P. R. Kinget, and H. Krishnaswamy, "Low-Noise Active Cancellation of Transmitter Leakage and Transmitter Noise in Broadband Wireless Receivers for FDD/Co-Existence," *IEEE Journal of Solid-State Circuits*, vol. 49, no. 12, pp. 3046–3062, Dec 2014.

27. D. Korpi, J. Tamminen, M. Turunen, T. Huusari, Y.-S. Choi, L. Anttila, S. Talwar, and M. Valkama, "Full-duplex mobile device: pushing the limits," *IEEE Communications Magazine*, vol. 54, no. 9, pp. 80–87, sep 2016.

28. H. J. Carlin and A. B. Giordano, *Network theory: an introduction to reciprocal and non-reciprocal circuits*. Prentice-Hall, 1964.

29. G. A. Campbell and R. M. Foster, "Maximum Output Networks for Telephone Substation and Repeater Circuits," *American Institute of Electrical Engineers, Transactions of the*, vol. XXXIX, no. 1, pp. 231–290, 1920.

30. M. Mikhemar, H. Darabi, and A. A. Abidi, "A Multiband RF Antenna Duplexer on CMOS: Design and Performance," *IEEE Journal of Solid-State Circuits*, vol. 48, no. 9, pp. 2067–2077, 2013.

31. S. H. Abdelhalem, P. S. Gudem, and L. E. Larson, "Hybrid Transformer-Based Tunable Differential Duplexer in a 90-nm CMOS Process," *IEEE Transactions on Microwave Theory and Techniques*, vol. 61, no. 3, pp. 1316–1326, 2013.

32. ——, "Tunable CMOS Integrated Duplexer With Antenna Impedance Tracking and High Isolation in the Transmit and Receive Bands," *IEEE Transactions on Microwave Theory and Techniques*, vol. 62, no. 9, pp. 2092–2104, sep 2014.

33. L. Laughlin, M. A. Beach, K. A. Morris, and J. L. Haine, "Optimum Single Antenna Full Duplex Using Hybrid Junctions," *IEEE Journal on Selected Areas in Communications*, vol. 32, no. 9, pp. 1653–1661, sep 2014.

34. B. van Liempd, J. Craninckx, R. Singh, P. Reynaert, S. Malotaux, and J. R. Long, "A Dual-Notch +27dBm Tx-Power Electrical-Balance Duplexer," in *ESSCIRC 2014 - 40th European Solid State Circuits Conference (ESSCIRC)*. IEEE, sep 2014, pp. 463–466.

35. B. van Liempd, B. Hershberg, S. Ariumi, K. Raczkowski, K.-F. Bink, U. Karthaus, E. Martens, P. Wambacq, and J. Craninckx, "A +70-dBm IIP3 Electrical-Balance Duplexer for Highly Integrated Tunable Front-Ends," *IEEE Transactions on Microwave Theory and Techniques*, pp. 1–13, 2016.

36. L. Laughlin, C. Zhang, M. A. Beach, K. A. Morris, J. L. Haine, and M. K. Khan, "A 700–950 MHz Tunable Frequency Division Duplex Transceiver Combining Passive and Active Self-interference Cancellation," in *2018 IEEE MTT-S International Microwave Symposium (IMS)*, Philadelphia, PA. USA., 2018, pp. 1–4.

37. L. Laughlin, C. Zhang, M. A. Beach, K. A. Morris, J. L. Haine, M. K. Khan, and M. McCullagh, "Tunable Frequency-Division Duplex RF Front End Using Electrical Balance and Active Cancellation," *IEEE Transactions on Microwave Theory and Techniques*, vol. 66, no. 12, pp. 5812–5824, Dec 2018.

38. Various photographers for Cassel & co., "image from The Queen's Empire. Volume 3. Cassell & Co. London. Public domain."

39. R. Ludwig and G. Bogdanov, *RF Circuit Design*, 2nd ed. New Jersey: Pearson Education Inc., 2009.

40. L. Laughlin, C. Zhang, M. A. Beach, K. A. Morris, and J. Haine, "A Widely Tunable Full Duplex Transceiver Combining Electrical Balance Isolation and Active Analog Cancellation," in *2015 IEEE 81st Vehicular Technology Conference (VTC Spring)*. IEEE, may 2015, pp. 1–5.

41. C. Zhang, L. Laughlin, M. Beach, K. Morris, and J. Haine, "Micro-Electromechanical Impedance Control for Electrical Balance Duplexing," in *European Wireless Conference*, 2016.

42. E. Manuzzato, J. Tamminen, M. Turunen, D. Korpi, F. Granelli, and M. Valkama, "Digitally-Controlled Electrical Balance Duplexer for Transmitter-Receiver Isolation in Full-Duplex Radio," in *European Wireless Conference*, 2016.

43. L. Laughlin, C. Zhang, M. A. Beach, K. A. Morris, and J. L. Haine, "Passive and Active Electrical Balance Duplexers," *IEEE Transactions on Circuits and Systems II: Express Briefs*, vol. 63, no. 1, pp. 94–98, Jan 2016.

44. G. Castellano, D. Montanari, D. De Caro, D. Manstretta, and A. G. M. Strollo, "An Efficient Digital Background Control for Hybrid Transformer-Based Receivers," *IEEE Transactions on Circuits and Systems I: Regular Papers*, vol. 64, no. 12, pp. 3068–3080, Dec 2017.

45. L. Laughlin, C. Zhang, M. A. Beach, K. A. Morris, and J. L. Haine, "Dynamic Performance of Electrical Balance Duplexing in a Vehicular Scenario," *IEEE Antennas and Wireless Propagation Letters*, vol. 16, pp. 844–847, 2017.

46. L. Laughlin, C. Zhang, M. Beach, K. Morris, and J. Haine, "Electrical Balance Duplexer Field Trials in High-Speed Rail Scenarios," *IEEE Transactions on Antennas and Propagation*, vol. 65, no. 11, 2017.

47. L. Laughlin, M. A. Beach, K. A. Morris, and J. L. Haine, "Electrical balance duplexing for small form factor realization of in-band full duplex," *IEEE Communications Magazine*, vol. 53, no. 5, pp. 102–110, may 2015.

48. W. Schacherbauer, T. Ostertag, C. C. W. Ruppel, A. Springer, and R. Weigel, "An Interference Cancellation Technique for the Use in Multiband Software Radio Frontend Design," in *30th European Microwave Conference*, 2000, pp. 1–4.

49. A. Sahai, G. Patel, C. Dick, and A. Sabharwal, "Understanding the impact of phase noise on active cancellation in wireless full-duplex," in *2012 Conference Record of the Forty Sixth Asilomar Conference on Signals, Systems and Computers (ASILOMAR)*, 2012, pp. 29–33.

50. ———, "On the Impact of Phase Noise on Active Cancelation in Wireless Full-Duplex," *IEEE Transactions on Vehicular Technology*, vol. 62, no. 9, pp. 4494–4510, Nov 2013.

51. M. Duarte, C. Dick, and A. Sabharwal, "Experiment-Driven Characterization of Full-Duplex Wireless Systems," *IEEE Transactions on Wireless Communications*, vol. 11, no. 12, pp. 4296–4307, Dec 2012.

52. A. Balatsoukas-Stimming, P. Belanovic, K. Alexandris, and A. Burg, "On self-interference suppression methods for low-complexity full-duplex MIMO," in *2013 Asilomar Conference on Signals, Systems and Computers*. IEEE, Nov 2013, pp. 992–997.

53. Z. Zhan, G. Villemaud, and J.-M. Gorce, "Analysis and reduction of the impact of thermal noise on the Full-Duplex OFDM radio," in *2014 IEEE Radio and Wireless Symposium (RWS)*. IEEE, Jan 2014, pp. 220–222.

54. M. McKinley, K. Remley, M. Myslinsky, J. Kenney, D. Schreurs, and B. Nauwelaers, "EVM Calculation for Broadband Modulated Signals," in *Proc. 64th ARFTG Conf. Dig.*, Orlando, FL. USA, 2004, pp. 45–52.
55. Geoff Carey (Mimotech), "Air Division Duplexing doubles Transmission Capacity for Microwave Backhaul," in *Cambridge Wireless Radio Technology Special Interest Group*, Bristol, 2015.

Chapter 2
Antenna/RF Design and Analog Self-Interference Cancellation

Jong Woo Kwak, Min Soo Sim, In-Woong Kang, Jaedon Park, and Chan-Byoung Chae

Abstract The main obstacle to full-duplex radios is self-interference (SI). To overcome SI, researchers have proposed several analog and digital domain self-interference cancellation (SIC) techniques. Digital cancellation has the following limitations: (1) It is only possible if the SI is sufficiently removed in the analog domain to fall within the dynamic range of an analog-to-digital converter (ADC). (2) It cannot mitigate the transmitter noise. Thus, analog cancellation plays an important role in a SIC scenario. This chapter provides an overview of current research activities on the analog cancellation scheme. Analog cancellation can be categorized into two classes—passive and active. In the passive analog cancellation, an RF component suppresses the SI. This can be implemented using a circulator or antenna separation. Leakages are cancelled by the active analog cancellation, which is based on a channel estimation of residual SI channel. The leakage from the passive cancellation can be matched by a signal generated from a tunable circuit or an auxiliary transmit chain. A key issue then in active analog cancellation is designing a circuit and optimization algorithm.

2.1 Introduction

In the demand of increasing spectral efficiency and data rates, full-duplex has emerged as a highly promising technique for 5G wireless communications. Full-duplex radio has the potential to double the spectral efficiency by transmitting and receiving simultaneously on the same frequency. The main challenge in the full-duplex radio is self-interference (SI)—a phenomenon where a transmit signal is received by its own receiver. Without the self-interference cancellation (SIC),

J. W. Kwak · M. S. Sim · C.-B. Chae (✉)
Yonsei University, Seoul, Korea
e-mail: kjw8216@yonsei.ac.kr; simms@yonsei.ac.kr; cbchae@yonsei.ac.kr

I.-W. Kang · J. Park
Agency for Defense Development, Daejeon, Korea
e-mail: iwkang@add.re.kr; jaedon2@add.re.kr

© Springer Nature Singapore Pte Ltd. 2020
H. Alves et al. (eds.), *Full-Duplex Communications for Future Wireless Networks*,
https://doi.org/10.1007/978-981-15-2969-6_2

the SI significantly degrades the signal-of-interest. To alleviate this problem, the traditional wireless communication systems operate in half-duplex which separates the uplink and downlink transmission in either time domain (TDD) or frequency domain (FDD).

Recently, many SIC methods have been proposed. SIC can be categorized, roughly, into two classes: (1) intrinsic cancellation and (2) SI channel estimation-based cancellation. In the intrinsic cancellation, the SI is weakened in a passive manner using antenna separation, antenna cancellation, or an isolator. The intrinsic cancellation is often called passive analog cancellation since it is done in the analog domain. The remaining SI from passive analog cancellation is mitigated by the SI channel estimation-based cancellation. Since the TX signal is known perfectly, the SI can be reconstructed through SI channel estimation. The reconstructed SI is then subtracted from the received signal leaving the signal-of-interest. Conventional channel estimation methods can be easily applied for the SIC in the baseband. The greatest hurdle in the SIC is the fact that ADC has to convert the SI and signal-of-interest simultaneously. To avoid ADC saturation, an additional cancellation is adopted in the analog domain; this is called active analog cancellation. The active analog cancellation regenerates the destructive SI in the analog domain. It is also based on the SI channel estimation which is done in the digital domain.

Because the SI channel estimation in the digital cancellation is dependent on the result of the analog cancellation, the analog cancellation plays a crucial role in the SIC. The realm of analog cancellation research covers the designing of the antenna/RF components and digital signal processing. The size of analog canceller, power/computational costs must be designed so as to be implementable. Another hurdle is the limited bandwidth of the RF component. Recently, researchers have proposed a photonics-based analog cancellation system to provide a broadband cancellation.

This chapter provides an overview of the state-of-the-art analog cancellation methods. The rest of this chapter is organized as follows: In Sect. 2.2, the main obstacles in full-duplex system are introduced. Section 2.3 presents the passive cancellation methods and Sect. 2.4 presents the analog cancellation methods. In Sect. 2.5, a numerical analysis of the SIC is provided through simulation. The digital cancellation method is briefly introduced in Sect. 2.5.

2.2 Requirements for a Full-Duplex System

To carry out full-duplex communication, SI must be suppressed at the receiver thermal noise floor. In a multiple-antenna system, the passive analog cancellation can be achieved by using a directive antenna or an antenna cancellation technique. If the transmitter and the receiver share the single antenna, an isolator is required to separate the SI.

The remaining SI from passive analog cancellation consists of linear components and nonlinear components. Linear components are caused by multi-path propagation between transmitter and receiver. In the single-antenna system with a circulator, leakages from the circulator are modeled as linear. Nonlinear components mainly come from power amplifier nonlinearity, ADC quantization noise, transmitter noise, and other RF imperfections.

The channel estimation-based cancellation can be made in both the analog and digital domains. This cancellation involves three steps. (1) Construct a model of the SI channel. (2) Estimate the SI channel using perfect knowledge of transmit signal. (3) Reconstruct and subtract the SI from the received signal. Linear components of SI can be easily mitigated by the existing channel estimation methods. The power amplifier nonlinearity and I/Q imbalance are modeled in [8, 9]. The main obstacle of SIC is a required ADC dynamic range to acquire the SI and signal-of-interest simultaneously. A dynamic range of a q-bit ADC is calculated as

$$6.02 \times q + 1.76 \text{ [dB]}. \tag{2.1}$$

In the case of the 14-bit ADC, the dynamic range is 86 dB. This means that the SI power after analog cancellation should not be higher than the receiver noise floor by 86 dB. Figure 2.1 depicts an example of the power levels in the successful SIC scenario. Essentially, the ADC dynamic range determines a required analog cancellation amount. Typically, it is not possible to achieve this with the intrinsic cancellation alone. Hence, most of the full-duplex systems adopt an additional active cancellation in the analog domain, which is also based on the SI channel estimation. Figure 2.2 shows a full-duplex system adopting a combination of analog and digital domain SIC.

Fig. 2.1 An example of the power levels in the successful SIC scenario

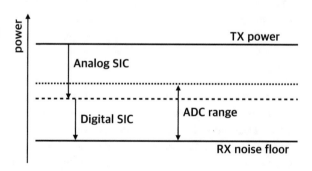

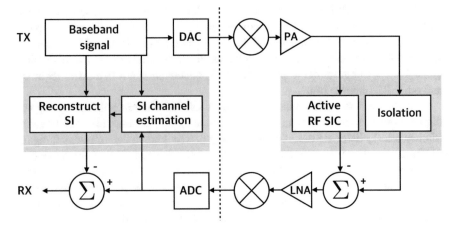

Fig. 2.2 A block diagram of a full-duplex system with SIC

2.3 Passive Analog Cancellation

In the propagation stage, an isolator separates the desired signal and the SI. An active RF component is not needed here. Therefore, it is often called passive analog cancellation. The most intuitive way is to separate the antenna for the transmit chain and the receive chain. The path loss between two antennas can be described as

$$L = 10n \log 10d + C, \tag{2.2}$$

where L, n, d, and C denote the path loss, the path loss exponent, the distance between two antennas, and a system-dependent constant, respectively. As (2.2) shows, the cancellation amount determined by a size of the transceiver.

Another solution is antenna cancellation, which generates a π-phase rotated SI signal at the receiver. Figure 2.3 describes an asymmetric antenna cancellation setup. The wavelength of the transmit signal is denoted as λ. The distance between TX1 and RX (d_1) is $\lambda/2$ larger than the distance between Tx2 and Rx (d_2). At RX, the two transmit signals will have a phase difference of π. To compensate for the different path loss, we have to allocate more power to the TX2. Ideally, the SI will be perfectly mitigated whereas the additional transmit signal acts as interference for the other receiver. The authors in [3] combined the interference cancellation methods for the antenna cancellation. The asymmetric antenna cancellation method is inherently available in the narrow bandwidth (due to λ).

To alleviate the bandwidth dependency, a symmetric antenna cancellation method is adopted in [4]. Figure 2.4 depicts a symmetric antenna cancellation. A π-phase shifter is employed instead of having a difference in the distance. The passive cancellation methods using polarized antenna are presented in [5, 6]. In [6], a dual-polarized antenna is implemented in a real-time full-duplex LTE prototype.

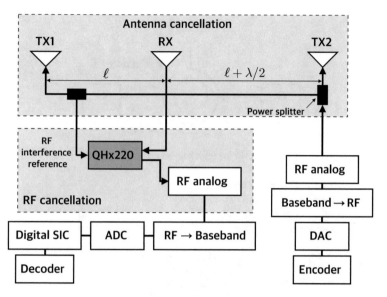

Fig. 2.3 System architecture proposed in [3]

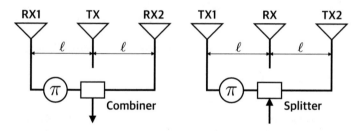

Fig. 2.4 Symmetric antenna cancellation using a π-phase shifter [4]

In single-antenna full-duplex radios, to achieve a passive cancellation, one needs a circulator. A circulator is a 3-port device that steers the signal entering any port is transmitted to the next port only. Ferrite circulators are often used in the communication systems. When a signal enters to port 1, the ferrite changes a magnetic resonance pattern to create a null at port 3. The circulator can provide isolation in a narrow bandwidth. Figure 2.5 shows the SI components in the circulator setup.

The strongest SI component is direct leakage. Typically, a circulator provides 20–30 dB isolation for the direct leakage. The remaining SI components are mitigated by an active analog cancellation. The second component is the signal reflected by an antenna. The reflected power (return loss) can be computed as

$$L_{\text{ret}} = -20 \log \left| \frac{Z_L - Z_S}{Z_L + Z_S} \right|, \tag{2.3}$$

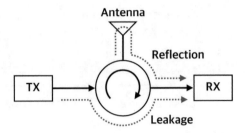

Fig. 2.5 Remaining SI components in the circulator setup

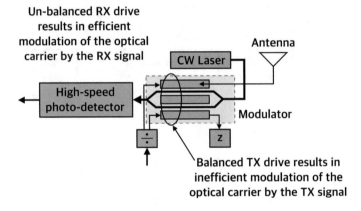

Fig. 2.6 A block diagram of the optical circulator proposed in [10]

where Z_L is a load impedance and Z_S is a source impedance. The power of the third component is an environmental-reflected signal that depends on the objects near the transceiver.

An optical circulator is proposed by a company called Photonic Systems Incorporated [10]. This optical circulator achieves better isolation and bandwidth dependency compared to the conventional RF circulator. Figure 2.6 depicts a block diagram of the optical circulator.

The received signal from the antenna is fed to a balanced optical modulator, whereas the transmit signal is just conveyed as an electrical signal. Since the transmit signal propagates in the opposite direction to the optical carrier, the modulation—the response of the transmit signal—is extremely low. The effect of the transmit signal reflected by the antenna—which propagates in the same direction as the optical carrier—can be compensated by a balance port. A principle of the optical modulator is in [2]. An experimental result of the optical circulator prototype is provided in [11]. Roughly, the optical circulator achieves 30–40 dB isolation from 2.5–20 GHz.

A fundamental limitation of passive analog cancellation is that it cannot suppress the SI reflected from the environment. The authors in [20] compare the remaining SI after passive cancellation in an anechoic chamber and a highly reflective room with

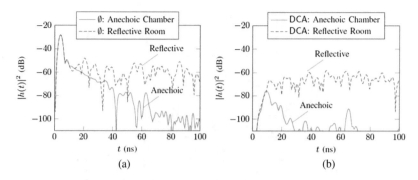

Fig. 2.7 Comparison between the time responses measured in an anechoic chamber and a reflective room. © [2014] IEEE. Reprinted, with permission, from ref. [20]. (**a**) No passive analog cancellation. (**b**) Directionals, Crosspol, Absorber (DCA)

metal walls (Fig. 2.7). Combinations of directional isolation, absorptive shielding (i.e., place an RF-absorptive material between transceivers to increase the pathloss), and cross polarization are applied for the comparison. For the absorptive shielding, Eccosorb AN-79 is used as an absorber which consists of discrete layers of lossy material. An impedance of the incident layer of Eccosorb AN-79 is designed to be close to air. The impedance gradually increases from the incident layer to the rear layer. This graded multilayer structure offers broadband electromagnetic wave absorption capability compared to the single layer structure [20]. Figure 2.7a shows the time responses of the SI channel without SI cancellation. The initial part of the time response corresponds to the direct path and the tail corresponds to the reflective paths. As Figure 2.7b shows that passive analog cancellation only suppresses the direct path, and the remaining SI from the reflective paths can be severe in the reflective environment.

2.4 Active Analog Cancellation

2.4.1 Adaptive RF Circuits

The basic concept of the active analog cancellation is to generate a signal that matches to the leakage from passive analog cancellation. The signal can be generated using a tunable circuit and auxiliary transmit chain. In the tunable circuit-based active analog cancellation, a small copy of transmitted signal is fed to the circuit. Figure 2.8 shows a general multi-tap canceller consisting of M delay(τ_i) lines.

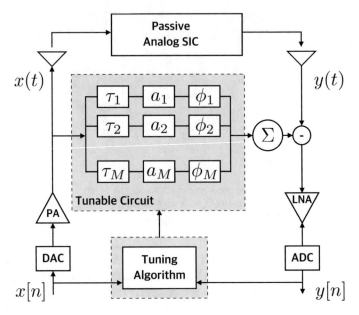

Fig. 2.8 A general design of an adaptive circuit for the active analog cancellation

Each delay line comprises an attenuator (a_i) and a phase shifter (ϕ_i). The parameters of the circuit are tuned via solving an optimization problem,

$$\min_{\tau,a,\phi} \left(y(t) - \sum_{i=1}^{M} x_{\tau_i,a_i,\phi_i}(t) \right)^2, \qquad (2.4)$$

where $x_{\tau_i,a_i,\phi_i}(t)$ is an output signal of the each delay line, $y(t)$ is the leakage, $x(t)$ is the transmit signal. The optimization is performed in the digital domain (i.e., the optimization algorithm receives as input variables the baseband signal ($x[n]$, $y[n]$). Note that the aim of analog cancellation is to avoid the ADC saturation. Therefore, at the initial stage, the SI signal is transmitted at weak power to carry out optimization while avoiding ADC saturation. After the parameters are tuned, full-duplex transmission is performed.

The authors in [13] proposed a novel optimization algorithm and architecture of adaptive circuit. The authors implemented the full-duplex WiFi radio in SISO scenario. Figure 2.9 shows the proposed full-duplex communication system. A circulator is used as a passive isolator. As discussed in Section (circulator), the leakage of the circulator consists of two primary components (i.e., direct leakage and reflected signal). Accordingly, an adaptive circuit should be designed to be suitable to cancel those two primary leakages.

The key idea is to fix the delays of the circuit using the characteristics of the leakage. Each delay line contains a variable attenuator. Theoretically, we can

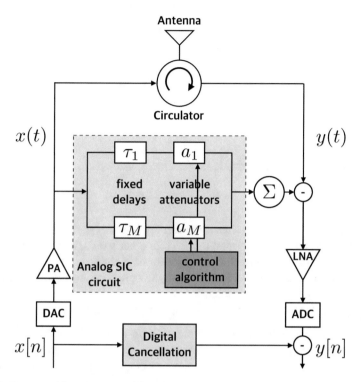

Fig. 2.9 System architecture proposed in [13]

regenerate the actual SI with appropriately chosen fixed delays even though the actual SI has an arbitrary delay within a certain range. Since the proposed method uses fixed delays, it does not require hard-to-implement high-resolution delays.

Let the actual SI ($y(t)$) have a delay d, an attenuating value a, and a Nyquist rate f_s. Assume that we set the delays (d_i) and attenuating values (a_i) of the circuit as (2.5),

$$d_i = d_{i-1} + \frac{1}{f_s},$$
$$a_i = \text{sinc}\,(f_s(d - d_i)),$$

(2.5)

where $\text{sinc}(\cdot)$ is a sinc function (Fig. 2.10). This virtual circuit consists of infinitely many delay lines. Then the summation of the each delay line's output at time instant t_1 is

$$\sum_{i=-\infty}^{\infty} a_i x(t_1 - d_i) = \sum_{i=-\infty}^{\infty} \text{sinc}(f_s(d - d_i))x(t_1 - d_i),$$

(2.6)

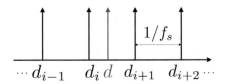

Fig. 2.10 An arrangement of the fixed delay

where actual SI at time instant t_1 is

$$y(t_1) = x(t_1 - d). \tag{2.7}$$

Since $x(t)$ is a band-limited signal, we can apply the Nyquist-sampling theorem and easily show that (2.6) is equal to (2.7). In other words, if we know the actual delay d with an infinite number of delay lines, the circuit can perfectly reconstruct the SI. Since the actual delay d is not known, an optimization algorithm adjusts the attenuator values to cancel the leakage.

To implement this algorithm in practice, there are some hurdles.

1. We have to configure the adaptive circuit with finite delay lines. It is obvious that the cancellation amount will increase as the number of delay lines increases. To make the full-duplex system implementable on mobile devices, however, it is preferable that the size of an adaptive circuit prefer to be small. The authors in [13] proposed a 16-tap (delay lines) 10×10 cm circuit.

   Figure 2.11 depicts the alignment of the fixed delays, where d_{dir} and d_{ref} denote the actual delays of the two primary leakages. To cancel the primary leakage $y_{\text{dir}}(t)$, the 8 equidistant delays are chosen over the actual delay d_{dir}. The authors in [13] investigated the variation range of the actual delay. They then picked the four delays below that range and the four delays above that range. Other delays are picked in the same way for the $y_{\text{ref}}(t)$ (i.e., reflection from the antenna).

2. A periodic optimization for the adaptive circuit causes a time overhead. The overhead depends heavily on the design of an optimization algorithm. A main

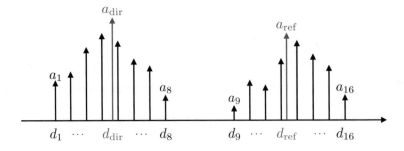

Fig. 2.11 An alignment of the fixed delays in [13]

challenge in the adaptive circuit-based analog cancellation is to reduce the time consumption for the optimization.

3. The variable components of the circuit have a finite resolution. The attenuator characteristics determine a feasible set of the optimization problem.

Given these limitations, the authors in [13] proposed a frequency response-based optimization algorithm. The basic concept of this algorithm is to represent the distortion ($H(f)$) introduced by the passive analog cancellation in the frequency domain as (2.11).

$$Y(f) = H(f)X(f), \qquad (2.8)$$

where $Y(f)$ and $X(f)$ are the frequency domain representation of the leakage ($y(t)$) and tapped signal ($x(t)$), respectively. The frequency response $H(f)$ (i.e., an FFT the SI channel) can be measured using pilot symbols. After measuring $H(f)$, the circuit is tuned by solving

$$\min_{a_1,\ldots,a_N} \left(H(f) - \sum_{i=1}^{N} H_i^{a_i}(f) \right)^2, \qquad (2.9)$$

where $H^{a_i}(f)$ is the frequency response for delay line i for attenuation setting of a_i. The problem is two-fold. First, the $H_i^{a_i}(f)$ should be measured for every possible attenuation value a_i. Since the circuit is well connected, it is impossible to emulate the delay line individually to measure $H_i^{a_i}(f)$. To isolate the target delay line as much as possible, the authors set the highest attenuator values for all of the delay lines except the target one. The frequency response with another attenuation values can be computed using this initial measurement and a S-parameter data of the attenuator, which provide the relative change of the frequency response with the changing attenuation value. Second, we have to find an optimal attenuation setting to match the actual frequency response $H^{a_i}(f)$. If the attenuator can take 128 different values, there are 128^{16} possible cases in total. In this situation, an exhaustive search is not feasible. Therefore, the authors relaxed it to a linear program and then random rounding to get a feasible solution. The authors achieved 45–50 dB analog cancellation using their testbeds. The design above is extended to full-duplex MIMO system [14]. To achieve full-duplex MIMO, interference from a neighboring transmitter (i.e., cross talk) has to be mitigated. We can simply replicate the SISO design, as shown in Fig. 2.12.

The SISO replication-based design requires M^2 times more taps than the SISO design, where M is the number of transmit antenna. Hence, the computational complexity with respect to M increases exponentially. The authors in [14] proposed a cascaded cancellation design based on the following insight: since the antennas are close to one another, the crosstalk and self-talk will experience similar channels. The relationship between two channels can be modeled as

$$H_{ct}(f) = H_c(f)H_s(f), \qquad (2.10)$$

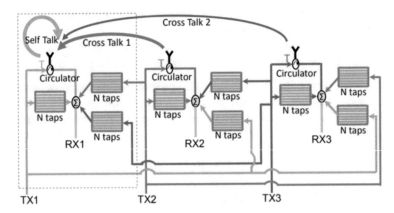

Fig. 2.12 SISO replication design proposed in [14]. Self-talk corresponds to the SI and cross talk corresponds to the interference from the neighbor transmitter

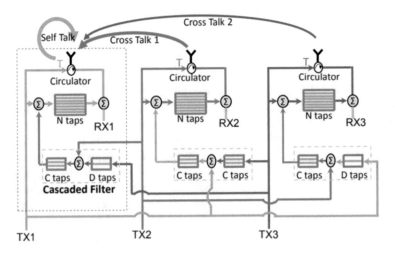

Fig. 2.13 Cascaded design proposed in [14]

where $H_{ct}(f)$ is a frequency response of the crosstalk channel, $H_s(f)$ is a frequency response of the self-talk channel, and $H_c(f)$ is a simple cascade transfer function.

Figure 2.13 illustrates a cascade cancellation design ($M = 3$). For the cascade transfer function of crosstalk channel 1 and 2, the adaptive circuit consists of C and D taps, respectively. Since the cascade transfer function is simple, we allocate the number of taps relatively small compared to N (i.e., $N >> C > D$). The number of taps for cascade transfer functions (C, D) is empirically chosen to provide sufficient cancellation. The authors in [14] compared the SISO replication-based design and cascade cancellation design for a 3×3 MIMO full-duplex radio operating a WiFi PHY in a 20 MHz band at 0 dBm TX power. Figure 2.14 depicts a comparison of the two designs with the same number of taps. The results show that the cascade design is more efficient than the SISO replication-based design.

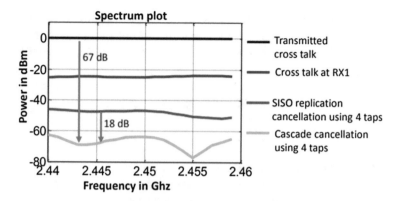

Fig. 2.14 Comparison of the SISO replication design and the cascaded design [14]

One of the other state-of-the-art analog cancellation technique is proposed in [7]. An adaptive circuit comprises fixed delay lines with variable phase shifters and attenuators in [7]. The main idea is to formulate the optimization problem as a convex problem. Consider an OFDM system with K subcarriers. Denote the sampling rate as $1/T_S$. The k-th frequency component of the adaptive circuit frequency response is modeled as

$$H_{\text{cir}}[k] = \sum_{n=1}^{N} a_n e^{\phi_n} e^{jk\delta_w \tau_n} = \boldsymbol{\Phi}_k^T \mathbf{W} \tag{2.11}$$

where a_n, ϕ_n, τ_n, and δ_w are attenuation, phase shift, fixed delay of n-th path, and sampling interval over the bandwidth of interest, respectively. Note that the previous active cancellation method uses measured frequency response of delay line $(H^{a_i}(f))$. On the other hand, the authors in [7] theoretically modeled the frequency response of the adaptive circuit, enabling the analysis of residual SI power. The sampling interval $\delta_w = 2\pi B/K$, where B is the bandwidth of the $x(t)$. For convenience, we define the following two vectors:

$$\begin{aligned}
\boldsymbol{\Phi}_k &= \left[e^{-jw\delta_w k\tau_1}, e^{-jw\delta_w k\tau_2}, \ldots, e^{-jw\delta_w k\tau_N} \right]^T, \\
\mathbf{W} &= \left[a_1 e^{-j\phi_1}, a_2 e^{-j\phi_2}, \ldots, a_N e^{-j\phi_N} \right].
\end{aligned} \tag{2.12}$$

With a given fixed delay alignment $(\tau_1, \tau_2..\tau_N)$, the optimization algorithm minimizes the difference between two sampled frequency response that can be formulated as

$$\min_{w} \mathbb{E}\left[\left(H_{\text{chan}}[k] - \boldsymbol{\Phi}_k^T \mathbf{W} \right) \left(H_{\text{chan}}[k] - \boldsymbol{\Phi}_k^T \mathbf{W} \right)^H \right], \tag{2.13}$$

where H_{chan} and $(\cdot)^H$ denote the measured frequency response of the SI channel and Hermitian operation, respectively. Conceding that $H_{\text{chan}}[k]$ and Φ_k are jointly stationary, the object function is convex. Therefore, the global optimum $(\hat{W})$ can be analytically obtained as

$$
\begin{aligned}
\hat{W} &= \mathbb{E}\left[\overline{\Phi_k}\Phi_k^T\right]^{-1} \mathbb{E}\left[H_{\text{chan}}[k]\overline{\Phi_k}\right], \\
&= R^{-1}P
\end{aligned}
\tag{2.14}
$$

where R is a complex correlation matrix operating on a set of sampled complex exponentials and P is a complex cross-correlation matrix operating on the measured channel response and set of complex exponentials. Note that R^{-1} can be precomputed. The main advantage of this algorithm is that we can relieve the matrix inversion operation. The time complexity of this algorithm depends on the calculation of P and the multiplication of R^{-1} and P.

In [18], the author analyzed the residual SI power after cancellation for the case of a time-invariant SI channel and a time-variant SI channel. For each case the authors considered the CSI imperfection and an imperfect delay alignment. The baseband equivalent time domain SI channel can be described as a uniformly spaced TDL model,

$$
h_{\text{SI}}^b(t) = \sum_{i=0}^{\infty} h_i \left(t - \frac{i}{B}\right),
\tag{2.15}
$$

where h_i is the i-th tap coefficient. The authors in [18] set the fixed delays of an adaptive circuit as $(0, \frac{1}{B}, ..\frac{N-1}{B})$, where N is the number of delay lines. The estimated channel $(\hat{H}(k))$ is modeled as

$$
\hat{H}[k] = H[k] + \tilde{N}[k],
\tag{2.16}
$$

where $\tilde{N}[k]$ is the circularly symmetric complex Gaussian (CSCG) noise with zero mean and variance $\tilde{\sigma}^2$. The reconstructed frequency response of the circuit is $\hat{H}_{cir} = [\Phi_1^T \hat{W}, \ldots, \Phi_K^T \hat{W}]^T$. Then the average power of the residual (ρ_{rd}) is represented as

$$
\begin{aligned}
\rho_{rd} &= \frac{1}{K}\mathbb{E}\left\{\left|X\left(H - \hat{H}_{\text{cir}}\right)\right|^2\right\} \\
&= \frac{1}{K}\mathbb{E}\left\{tr\left[\left(H - \hat{H}_{\text{cir}}\right)\left(H - \hat{H}_{\text{cir}}\right)^H X^H X\right]\right\} \\
&= \frac{1}{K}tr\left\{\mathbb{E}\left[\left(H - \hat{H}_{\text{cir}}\right)\left(H - \hat{H}_{\text{cir}}\right)^H\right]\right\},
\end{aligned}
\tag{2.17}
$$

Table 2.1 Residual SI power on the time-varying channel for the 3 cases

Cases	Description	Power of Residual SI
Case 1	• Perfect SI channel CSI and perfect delay alignment • Time-varying SI channel with processing delay t_d	$2 - \frac{2}{K_f+1}\{K_f + J_0[2\pi \Delta T f_d K T_S]\}$
Case 2	• Imperfect SI channel CSI and perfect delay alignment • Time-varying SI channel with processing delay t_d	$\frac{N}{K}\tilde{\sigma}^2 + 2 - \frac{2}{K_f+1}\{K_f + J_0[2\pi \Delta T f_d K T_S]\}$
Case 3	• Imperfect SI channel CSI and imperfect delay alignment with $N = K$ and $\Delta t = 1/B$ • Time-varying SI channel with processing delay t_d • Time-varying SI channel with processing delay t_d	$\tilde{\sigma}^2 + 2 - \frac{2}{K_f+1}\{K_f + J_0[2\pi \Delta T f_d K T_S]\}$

where $\boldsymbol{X} = \text{diag}\{\boldsymbol{X}(0), \boldsymbol{X}(1), \ldots, \boldsymbol{X}(K)\}$ with $\mathbb{E}\left[|\boldsymbol{X}(k)|^2\right] = 1$ and $\mathbb{E}\left[|\boldsymbol{X}(k_1)\boldsymbol{X}(k_2)^*|^2\right] = 0, k_1 \neq k_2$, $\text{tr}(\cdot)$ denotes the trace of a matrix. With the notation $\Omega = [\boldsymbol{\Phi}_1^T, \ldots, \boldsymbol{\Phi}_K^T]^T$ the authors in [18] derived the following,

$$\rho_{rd} = \frac{N}{K}\tilde{\sigma}^2 + \frac{N}{K}\text{tr}\left(\boldsymbol{H}\boldsymbol{H}^H - K\Omega\boldsymbol{R}\Omega^H\boldsymbol{H}\boldsymbol{H}^H\right). \quad (2.18)$$

The first term results from the imperfect SI CSI, while the second term results from the imperfect delay alignment.

Over the time-varying SI channel, the processing delay (i.e., time consumed in calculating the variable and tuning) comes up as an issue. In [18], the authors assume the processing delay t_d to be a multiple of the KT_S (i.e., $t_d = \Delta T K T_S$, ΔT is an integer). In this case, the power of residual SI is analyzed in Table 2.1 [18].

2.4.2 Micro Photonic Canceller

To achieve broadband cancellation, researchers have proposed several micro-photonic cancellers (MPC). Similar to the RF adaptive circuit, the SI channel is reconstructed using tunable delays and attenuators. Essentially, the delay comes from changing the propagating group velocity using carrier dynamics. A set of optical time delay lines (OTDL) and optical variable attenuators is used in [12] to reconstruct the SI channel. This system achieves 40 dB cancellation over 50 MHz bandwidth. However, tuning the OTDL takes a great deal of time. To obtain a low-latency tunable delay, the authors in [16] used a semiconductor optical

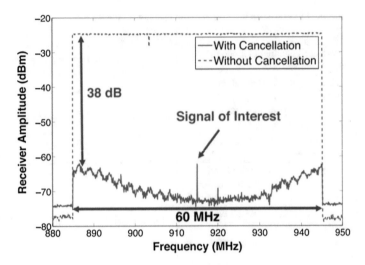

Fig. 2.15 Broadband cancellation result of the optical system. © [2016] IEEE. Reprinted, with permission, from ref. [16]

amplifier (SOA), which has 200 ns latency. The proposed system has a single SOA. Therefore, to tune the SOA involves a simple 2-dimensional optimization problem. The Nelder–Mead simplex algorithm [1] is used for the optimization. Figure 2.15 depicts a broadband cancellation result. The system achieves 38 dB cancellation over 60 MHz.

In [17], the delays and attenuations are obtained using tunable lasers with a dispersive element instead of the set of OTDLs and OVAs. This enhances the compactness of the system. Also, a dispersion-induced RF power fading can be easily compensated in this system.

The MPCs above use discrete fiber-optics. The authors in [19] first demonstrated an integrated microwave photonic circuit (IMPC), which requires no fiber. Optical components for the cancellation are monolithically integrated onto a substrate. As the IMPC has no fiber, the cost is significantly reduced. Figure 2.16 illustrates a block diagram and a microscopic image of the IMPC. The IMPC has RF inputs and outputs; it is thus appropriate to be implemented in the RF circuit board. The IMPC achieves 30 dB cancellation over within 400 MHz–6 GHz, which covers all existing FDD LTE and WiFi bands.

2.4.3 Auxiliary Transmit Chain

Suppose that an auxiliary transmit chain generates a copy of the SI signal. This copied SI signal goes through a different channel (h_{AUX}) from the SI channel (h_{SI}). The frequency response of the k-th subcarrier of these two different channels are

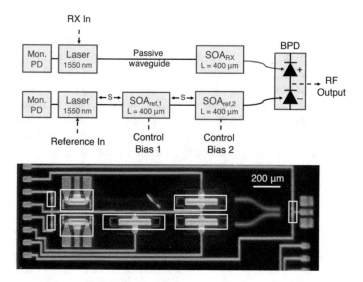

Fig. 2.16 Block diagram and microscope image of the IMPC. © [2017] IEEE. Reprinted, with permission, from ref. [19]

denoted as ($H_{AUX}[k]$) and ($H_{SI}[k]$), respectively. The basic concept is similar to the adaptive circuit-based cancellation, which generates the copy of the SI signal and subtracts it. First, estimate the two different channels (i.e., $\hat{H}_{AUX}$ and $\hat{H}_{SI}$). Using the estimated CSI, adjust the input of the auxiliary transmit chain X_{AUX} as (2.19)

$$X_{AUX}[k] = \frac{\hat{H}_{SI}[k]}{\hat{H}_{AUX}[k]} X_{SI}[k],$$ (2.19)

where X_{SI} is the transmitted signal. Note that this method assumes a linear model. Therefore, this method cannot suppress the nonlinear SI components such as transmitter noise.

2.5 Numerical Analysis and Discussions

In this section, we provide a numerical analysis of the SIC methods in an OFDM system. We build a simulator which can analyze the analog–digital integrated SIC performance. For the simulation of analog cancellation, we have to generate the leakages from passive cancellation. Generally, a time-invariant passband channel $h_p(t)$ is represented as

$$h_p(t) = \sum_{i=0}^{L} h_p^i \delta(t - \tau_i),$$ (2.20)

where L is the number of taps, h_p^i is the i-th tap gain, τ_i is the i-th tap delay. Note that the leakage from a circulator consists of three components—direct leakage, reflection from antenna, and reflections from the objects near transceiver.

We assume that the direct leakage has 15 dB attenuation and 300 ps delay, and the reflection from antenna has 17.5 dB attenuation and 3 ns delay. The simulation period is set to 10 ps. A delay of the reflection from the object near transceiver is assumed to be a multiple of 10 ps. Table 2.2 depicts the simulated tap delays and attenuations. A delay line of the adaptive circuit can be modeled as a single tap in $h_p(t)$. In the simulation, we transform $h_p(t)$ to a baseband equivalent form.

Figure 2.17 shows a full-duplex system in the simulation. A simple adaptive circuit is adopted for the active cancellation. The adaptive circuit consist of variable

Table 2.2 Attenuations and delays of the remaining SI after passive cancellation

Source	Attenuation	Delay
Direct leakage	−15 dB	300 ps
Reflection from antenna	−17.5 dB	3 ns
Reflection from objects	−60 dB	20 ns
Reflection from objects	−90 dB	60 ns
Reflection from objects	−100 dB	90 ns
Reflection from objects	−100 dB	120 ns

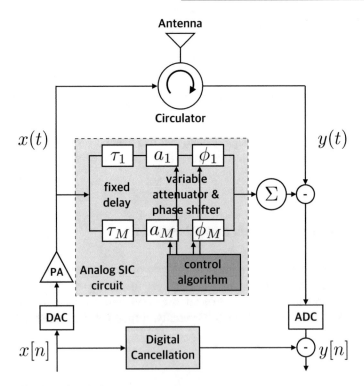

Fig. 2.17 Simulated full-duplex system

Table 2.3 Simulation parameters

System parameter	Notation	Values
Simulation sampling period	T_{sim}	10 ps
Center frequency	f_{c}	1.5 GHz
Bandwidth	B	20 MHz
FFT size		64
Used subcarrier		52
CP length		16
ADC resolution		14 bit
TX power		−5 dBm
PA gain		15 dB
$P_{1\,\text{dB}}$		23.09 dBm
Receiver noise floor		−85 dBm
Adaptive circuit attenuator resolution		∞

attenuators, variable phase shifters, and fixed delays. The fixed delay values are depicted in Table 2.3. The attenuations and phases are set to the solution of the following optimization problem ($\hat{a}$). Resolution of the variable component is not considered here:

$$\min_{a} |H(f) - H_{\text{cir}}(f)a|^2$$
$$\text{where } i\text{-th column of } H_{\text{cir}}(f) = H_i^{a_i=1}(f),$$

(2.21)

where $H_i^{a_i=1}(f)$ is the frequency response of the i-th fixed delay response with attenuation $a_i = 1$. As discussed in Sect. 2.2, during the initial stage, the SI signal is transmitted with little power so as to avoid the ADC saturation without the active cancellation.

After the analog cancellation, the residual SI is mitigated by digital cancellation. The greatest obstacle to the digital cancellation is to cancel out the nonlinear SI components, which come mainly from the power amplifier nonlinearity. Several nonlinear digital cancellation methods have been proposed to handle the nonlinearity [8, 9]. A parallel Hammerstein model is widely used to describe the nonlinearity as,

$$x_{\text{PA}}[n] = \sum_{k=0}^{K-1} \sum_{p=0}^{P-1} \psi_{k,p} |x[n-p]|^{2k} x[n-p],$$

(2.22)

where $x[n]$ and $x_{\text{PA}}[n]$ are the transmitted and power amplifier output signals on time n, $2K-1$ is the highest order of the model, P is the number of memory taps of the power amplifier, and $\psi_{k,p}$ are the nonlinear coefficients.

Figure 2.18 depicts the power amplifier characteristics which is modeled with the 5-th order parallel Hammerstein (PH) model. In the digital cancellation, we use a simple linear digital cancellation method [15]. The simulation parameters are

Fig. 2.18 Power amplifier characteristics in the simulator (blue line)

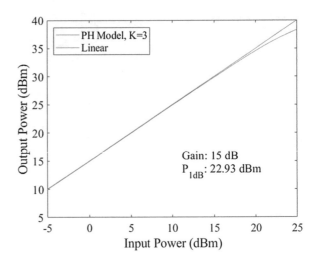

Table 2.4 Analog and digital SIC performance

	10 bit ADC	14 bit ADC
1-tap	Analog cancellation: 44.91 dB	Analog cancellation: 45.36 dB
(1.5 ns)	Digital cancellation: 48.11 dB	Digital cancellation: 48.17 dB
2-tap	Analog cancellation: 48.24 dB	Analog cancellation: 52.76 dB
(0.5 ns, 2.5 ns)	Digital cancellation: 45.13 dB	Digital cancellation: 40.85 dB

presented in Table 2.3. Transmitter noise is not considered here. Analog–digital integrated SIC performance results are depicted in Table 2.4.

The 2-tap adaptive circuit shows a better cancellation performance compared to the 1-tap adaptive circuit. The integrated cancellation amounts are almost the same for the two circuits since the transmitter noise is ignored.

2.6   Conclusion

In this chapter, we introduce an overview of analog cancellation methods in full-duplex radios. First, the necessity of the analog cancellation is explained in the two different aspects: (1) to meet up the ADC dynamic range and (2) to mitigate the nonlinear SI components which cannot be removed in the digital domain (e.g., transmitter noise).

A fact that the power of the transmitter noise in the received SI signal must be lower than the receiver noise floor gives us the insight to design the analog cancellation technique. The transmitter noise is random, therefore, we have only two possible ways to mitigate it: (1) weak the SI signal in a passive manner (passive analog cancellation) and (2) copy the transmitted SI signal and using it as a reference signal to recreate the received SI signal (active analog cancellation, except for the auxiliary transmit chain method). Various analog cancellation techniques are developed to provide enough cancellation with low complexity (i.e., size of the SI canceller, computational, and power cost) and low bandwidth dependency.

Still, there are some challenges in those issues. Size and power consumption of the adaptive circuit is one of the main bottlenecks to implement the full-duplex in mobile devices. In full-duplex MIMO systems, these practical issues become more severe since the adaptive circuit has to be extended to mitigate both self-talk (i.e., SI) and crosstalk (i.e., interference from the neighbor transmitters).

References

1. J. A. Nelder and R. Mead, "A simplex method for function minimization," *The Computer Journal*, vol. 7, no. 4, pp. 308–313, 1965.
2. C. Cox, *Analog optical links*. Cambridge, U.K: Cambridge Univ.Press, 2004.
3. J. I. Choi, M. Jain, K. Srinivasan, P. Levis, and S. Katti, "Achieving single channel, full duplex wireless communication," in *ACM Mobicom*, Chicago, Illinois, USA, September 2010, pp. 20–24.
4. E. Aryafar, M. A. Khojastepour, K. Sundaresan, S. Rangarajan, and M. Chiang, "MIDU: enabling mimo full duplex," in *ACM Mobicom*, Istanbul, Turkey, August 2012, pp. 22–26.
5. T. Oh, Y. Lim, C. Chae, and Y. Lee, "Dual-polarization slot antenna with high cross-polarization discrimination for indoor small-cell mimo systems," *IEEE Antennas and Wireless Propagation Letters*, vol. 14, pp. 374–377, 2015.
6. M. Chung, M. S. Sim, J. Kim, D. K. Kim, and C. Chae, "Prototyping real-time full duplex radios," *IEEE Communications Magazine*, vol. 53, no. 9, pp. 56–63, 2015.
7. J. G. McMichael and K. E. Kolodziej, "Optimal tuning of analog self-interference cancellers for full-duplex wireless communication," in *Allerton conference on communication, control and computing*, Monticello, IL, USA, October 2012, 2012, pp. 1–5.
8. L. Anttila, D. Korpi, V. Syrjälä, and M. Valkama, "Cancellation of power amplifier induced nonlinear self-interference in full-duplex transceivers," in *Asilomar conference on signals, systems and computers*, Pacific Grove, CA, November 2013, 2013, pp. 3–6.
9. A. Sahai, G. Patel, C. Dick, and A. Sabharwal, "On the impact of phase noise on active cancelation in wireless full-duplex," *IEEE Transactions on Vehicular Technology*, vol. 62, no. 9, pp. 4494–4510, 2013.
10. C. Cox and E. Ackerman, "Demonstration of a single-aperture, full-duplex communication system," in *Radio and wireless symposium (RWS)*, vol. 2013, Austin, TX, USA, January 2013, pp. 20–23.
11. C. Cox and E. Ackerman, "Tiprx: a transmit-isolating photonic receiver," *Journal of Lightwave Technology*, vol. 32, no. 20, pp. 3630–3636, 2014.
12. J. Chang and P. R. Prucnal, "A novel analog photonic method for broadband multipath interference cancellation," *IEEE Microwave and Wireless Components Letters*, vol. 23, no. 7, pp. 377–379, 2013.
13. D. Bharadia, E. McMilin, and S. Katti, "Full duplex radios," pp. 12–16, August 2013.
14. D. Bharadia and S. Katti, "Full duplex mimo radios," in *USENIX symposium on networked systems design and implementation*, Seattle WA, April 2014, 2014, pp. 2–4.
15. J. Kim, M. S. Sim, M. K. Chung, D. K. Kim, and C. Chae, *Signal processing for 5G: algorithms and implementations*. New York, NY, USA, p 539-560: Wiley, 2016.
16. M. P. Chang, C. Lee, B. Wu, and P. R. Prucnal, "Adaptive optical self-interference cancellation using a semiconductor optical amplifier," *IEEE Photonics Technology Letters*, vol. 27, no. 9, pp. 1018–1021, 2016.

17. W. Zhou, P. Xiang, Z. Niu, M. Wang, and S. Pan, "Wideband optical multipath interference cancellation based on a dispersive element," *IEEE Photonics Technology Letters*, vol. 28, no. 8, pp. 849–851, 2016.
18. D. Liu, Y. Shen, S. Shao, Y. Tang, and Y. Gong, "On the analog self-interference cancellation for full-duplex communications with imperfect channel state information," *IEEE Access*, vol. 5, pp. 9277–9290, 2017.
19. M. P. Chang, E. C. Blow, J. J. Sun, M. Z. Lu, and P. R. Prucnal, "Integrated microwave photonic circuit for self-interference cancellation," *IEEE Transactions on Microwave Theory and Techniques*, vol. 65, no. 11, pp. 4493–4501, 2017.
20. E. Everett, A. Sahai, and A. Sabharwal, "Passive self-interference suppression for full-duplex infrastructure nodes," *IEEE Transactions on Wireless Communications*, vol. 13, no. 2, pp. 680–694, 2014.

Chapter 3
Digital Self-Interference Cancellation for Low-Cost Full-Duplex Radio Devices

Dani Korpi, Lauri Anttila, Taneli Riihonen, and Mikko Valkama

Abstract Wireless inband full-duplex communications, where individual radio devices transmit and receive simultaneously on the same frequency band, has recently been proposed as another step towards the full utilization of the available spectral resources. This chapter concentrates on solving the greatest challenge in wireless inband full-duplex communications, i.e., the self-interference, which refers to the interference produced by the own transmitter. To this end, this chapter provides digital-domain solutions for efficient self-interference cancellation in low-cost full-duplex radios. The proposed digital cancellers are capable of modeling the most prominent radio circuit impairments, in particular the nonlinear distortion produced by the transmitter power amplifier. The digital cancellers are evaluated using an actual inband full-duplex prototype, which contains also other self-interference suppression mechanisms operating in the analog domain. The obtained measurement results show that, with the help of these digital cancellers, the self-interference can be cancelled almost perfectly, proving that true full-duplex operation is indeed possible. Altogether, the own transmit signal is shown to be suppressed in some cases by more than 100 dB, which is one of the highest reported self-interference cancellation performances to date.

3.1 Introduction

In order to achieve the immense requirements for the spectral efficiency of future wireless systems, several techniques and solutions have been proposed by the research community, one of which is wireless inband full-duplex (IBFD) communications [1–5]. What IBFD refers to is simultaneously transmitting and receiving

D. Korpi (✉)
Nokia Bell Labs, Espoo, Finland
e-mail: dani.korpi@nokia-bell-labs.com

L. Anttila · T. Riihonen · M. Valkama
Unit of Electrical Engineering, Tampere University, Tampere, Finland
e-mail: lauri.anttila@tuni.fi; taneli.riihonen@tuni.fi; mikko.valkama@tuni.fi

© Springer Nature Singapore Pte Ltd. 2020
H. Alves et al. (eds.), *Full-Duplex Communications for Future Wireless Networks*,
https://doi.org/10.1007/978-981-15-2969-6_3

radio signals on the *same center frequency within the same device*. Considering the fact that practically all the current systems operate in a half-duplex (HD) manner, dividing transmission and reception within the device either in time with time-division duplex (TDD) or in frequency with frequency-division duplex (FDD), IBFD-capable radios can as much as double the spectral efficiency. The reason for this is simply that neither the temporal nor spectral resources need to be shared between transmission and reception, meaning that the whole available time-frequency resource can be used for both. As a result, the effective resources are doubled.

However, the inherent challenge of wireless IBFD communications is the problem of self-interference (SI). Namely, any device transmitting and receiving simultaneously on the same frequency band will produce extremely powerful interference to its own RX chain. Moreover, unlike in FDD systems where transmission and reception occur on different frequency bands with a wide separation, in IBFD radios the own transmission can obviously not be filtered out with a duplexer. If not properly managed, this SI greatly reduces the signal-to-interference-plus-noise ratio (SINR) of the received signal in an IBFD transceiver when compared to a HD transceiver. Therefore, the central research challenge for IBFD systems is to develop methods and techniques for canceling the SI by some means. Moreover, the accuracy of the SI cancellation solutions must be extremely high since the signal emitted from the own TX chain can easily be over 100 dB stronger than the desired signal of interest [5, 6].

In principle, the SI can be cancelled rather easily: since the transceiver obviously knows its own transmit signal, it can simply subtract it from the received signal. Assuming that the possible channel effects up to the point of subtraction are known, the SI could in fact be perfectly cancelled with this simple principle. What makes SI cancellation challenging in reality is obtaining sufficiently accurate knowledge about the overall coupling channel, i.e., knowing exactly how the SI signal is distorted while propagating from the transmitter to the receiver. In particular, while the effects of the wireless coupling channel between the transmitter and the receiver can be compensated for in a relatively straightforward matter, in many cases the SI signal is distorted also by the transmitter (TX) and receiver (RX) circuitry. Such distortion, resulting from various analog impairments within the transceiver, cannot usually be captured by the same models that apply to wireless propagation, thereby making accurate SI cancellation rather cumbersome.

In this chapter, the emphasis is on the SI cancellation performed in the *digital domain*, that is, after the analog-to-digital converter (ADC) of the RX chain. As will be discussed in more detail below, the purpose of digital SI cancellation is to fully suppress the residual SI that still remains after the different analog or RF cancellation schemes. The benefit of digital-domain cancellation is the increased flexibility in terms of modeling and parameter estimation, which facilitates the use of advanced SI signal models. This means that the significant analog impairments can be explicitly included in the modeling within the digital canceller, and consequently they do not pose a limit for the cancellation performance. Considering that in many cases the error vector magnitude (EVM) of the transmitter is dominated by the nonlinear behavior of its power amplifier (PA) [6, 7], incorporating a model of this nonlinearity source into the digital canceller can provide a significant improvement in the digital cancellation performance [6, 8, 9]. This chapter is mainly based on the solutions and findings reported in [10] and the references therein.

3.1.1 Basic Full-Duplex Device Architecture

To facilitate wireless IBFD operation and the forthcoming discussion about digital SI cancellers, Fig. 3.1 illustrates a possible overall SI cancellation architecture for a direct-conversion transceiver. Firstly, it includes two alternatives for physically isolating the transmitter and the receiver: separate TX and RX antennas, or a shared TX/RX antenna. In the former option, the SI isolation simply stems from the path loss between the antennas. In the latter solution, on the other hand, physical isolation is obtained with a so-called circulator, which is a three-port device where each port is connected directly to the next port, while being isolated from the previous port, or vice versa, depending on the direction of rotation (i.e., clockwise or counterclockwise) [11, p. 487]. This means that the signal propagating in the opposite direction is heavily attenuated, resulting in a certain amount of isolation between the TX and RX chains when using such a circulator in an IBFD transceiver as shown in Fig. 3.1. The isolation provided by a circulator is typically in the order of 20–40 dB, depending on its size and cost as well as on the used bandwidth, while the attenuation in the desired direction is usually less than half a decibel [9]. It should be noted that the essential modeling of the SI signal is not affected by the adopted antenna architecture, and hence the signal models and algorithms reported in this chapter can be readily applied to both types of systems.

After the antenna interface, RF cancellation is performed on the received signal to reduce the SI power entering the actual RX chain. The IBFD transceiver architecture depicted in Fig. 3.1 utilizes an RF cancellation solution where the transmitter PA

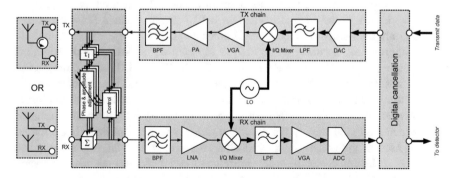

Fig. 3.1 The considered direct-conversion IBFD transceiver architecture where RF cancellation is performed using the PA output signal

output signal is used to form the cancellation signal, which is then subtracted from the received signal after proper manipulation. The benefit of this type of a solution is that all the TX-induced impairments are implicitly included in the cancellation signal, and are consequently suppressed by the RF canceller. In principle, this type of an RF canceller aims at modeling and subtracting the strongest SI components, while the multipath reflections are more easily cancelled in the digital domain [8, 12, 13].

Altogether, with sufficiently high RF cancellation performance, the SI is attenuated such that neither the low-noise amplifier (LNA) nor the ADC are saturated by the SI power, them being typically the critical components with regard to the highest tolerated power entering the RX chain [9]. Having then amplified, downconverted, and digitized the total signal consisting of the residual SI and the signal of interest, digital cancellation is performed. The digital cancellation signal is typically constructed from the original transmit data, using a predefined signal model and an estimate of the overall coupling channel. Ideally, the residual SI can be perfectly cancelled in the digital domain, after which only the received signal of interest, alongside with the noise, remains. The different digital cancellation solutions are described in detail in Sect. 3.3.

3.1.2 Related Work

Modeling of the residual SI signal in the digital domain is a crucial aspect for an IBFD transceiver as the objective of the digital canceller is typically to suppress the SI signal below the receiver noise floor. Furthermore, as shown in [10], in many cases this requires the modeling of some of the RF impairments since otherwise the accuracy of the cancellation signal is not sufficiently high. Nevertheless, in many of the related works, a linear signal model has been assumed in the digital cancellation stage [2, 3, 14–18], and consequently none of the RF impairments have been modeled. This represents a baseline for the signal model used within a digital canceller, and it is also described in detail in Sect. 3.3.1. However, as opposed to some works where the SI is cancelled in the frequency domain [15–17], in this chapter only time-domain cancellation is considered.

To improve the accuracy, many of the reported digital cancellation solutions incorporate also a model for the nonlinear TX PA [7–9, 13, 19–28]. Considering that in most systems the PA-induced nonlinearities are indeed the dominant source of distortion, such a nonlinear digital canceller is typically capable of highly efficient SI cancellation [8, 9, 13]. Furthermore, the works in [19, 20] utilize the nonlinear signal model for predistorting the PA output signal, which means that a linear model can be used at the actual digital cancellation stage. Addressing the PA-induced nonlinear distortion by intentionally introducing a polarization mismatch between the TX output and the RX digital domain has also been considered [29]. Moreover, in [22], also the nonlinear distortion produced by the RX chain is incorporated into the overall digital cancellation signal model, in addition to the TX nonlinearities,

Table 3.1 The key specifications and performance figures of the most notable IBFD prototype implementations

Prototype	Year	Frequency	Bandwidth	Structure	Total isolation
Rice I [2, 33]	2010	2.4 GHz	0.625 MHz	Two antennas	80 dB
Stanford I [3]	2010	2.48 GHz	5 MHz	Three antennas	100 dB
Rice II [34, 35]	2012	2.4 GHz	20 MHz	2×1 MISO	85 dB
Stanford III [13]	2013	2.45 GHz	80 MHz	Shared TX/RX antenna	110 dB
Rice III [36]	2014	2.4 GHz	20 MHz	Directional antennas	95 dB
Stanford IV [37]	2014	2.45 GHz	20 MHz	3×3 MIMO	104 dB
Yonsei [17]	2015	2.52 GHz	20 MHz	Dual-polar. antenna	103 dB
TUT+Intel [9]	2016	2.46 GHz	80 MHz	Shared TX/RX antenna	88 dB
TUT+Aalto [8]	2017	2.56 GHz	80 MHz	Compact relay antenna	100 dB

Note that all the prototypes are SISO transceivers, unless otherwise mentioned in the structure-column

albeit only 3rd-order distortion is considered for simplicity. Recently it has also been shown that machine learning can be used to aid the modeling of the SI signal in the digital domain [30–32].

In the recent years, various IBFD transceiver prototypes or demonstrator implementations have also been reported in the literature. While the key specifications and performance figures of the most notable prototype implementations are collected and discussed in [10], Table 3.1 provides an overview about the most prominent prototypes, including also the ones presented in [8, 9] that utilize the digital cancellers presented in this chapter. Note that, unless otherwise mentioned, the total amount of SI isolation listed in Table 3.1 for each prototype includes also the passive suppression, i.e., it is calculated as a difference between the transmit power and the residual SI power after all the cancellation stages.

The first IBFD prototype implementations utilizing both RF and digital cancellation were developed independently at Rice University [2] (reported in more detail later in [33]) and at Stanford University [3]. Even though the general architectures of these prototypes are rather similar, the RF cancellation technique is fundamentally different in these two implementations. Namely, in [2], the RF canceller utilizes an additional auxiliary TX chain to upconvert the cancellation signal from the baseband, while in [3] the TX output signal is used to generate the RF cancellation signal. Including also the suppression provided by linear digital cancellation, the total amounts of SI cancellation are 80 dB and 100 dB in [2] and [3], respectively, while [33] reports 74 dB of overall SI cancellation for the same Rice prototype.

Improved versions of Rice University's initial prototype are then reported in [34–36]. The prototype demonstrated in [34, 35] is a MISO IBFD transceiver with two TX antennas and one RX antenna, which are positioned around a device to provide a higher amount of passive isolation, measured to be between 60 and 70 dB. When complemented with active RF and digital cancellation, the overall amount of SI suppression for this prototype is 85 dB with a total transmit power of 5 dBm. The prototype reported in [36], on the other hand, utilizes directional antennas with cross-polarization and absorptive shielding to provide 70 dB of physical isolation between the TX and RX chains. After applying also RF and digital cancellation, the overall amount of SI suppression over 20 MHz is in the order of 95 dB with this solution, using a transmit power of 7 dBm.

The third prototype implementation of the Stanford group, reported in [13] already in 2013, represents still in many respects the state of the art. There, a shared-antenna architecture is adopted, the TX and RX chains being isolated by a circulator. The RF cancellation is done using the TX output signal, while the final cancellation stage in the digital domain utilizes a nonlinear signal model to reconstruct also the nonlinear distortion within the residual SI. The overall amount of SI cancellation with a 20-dBm transmit power and an instantaneous bandwidth of 80 MHz is reported to be 110 dB, consisting of 62 dB of analog SI suppression and 48 dB of digital cancellation.

The authors of [13] have also extended their SI cancellation architecture to support a 3 × 3 MIMO transceiver [37]. The proposed IBFD MIMO prototype consists of three TX–RX pairs where each pair shares a common antenna and the passive isolation is achieved using a circulator. The RX chains are preceded by a 56-tap 3 × 3 RF canceller, followed by a nonlinear digital canceller after the analog-to-digital conversion in each receiver. Using a sum transmit power of 20 dBm (divided equally between the three transmitters), the SI is cancelled in total by roughly 104 dB. This is sufficient to cancel the residual SI practically to the level of the receiver noise floor in each RX chain.

Considering then finally the prototype in [17], there the passive TX–RX isolation is improved by utilizing a dual-polarized antenna while also performing active RF and digital cancellation to further suppress the SI. In particular, a dual-polarized antenna transmits and receives signals of different polarization, resulting in a greatly attenuated leakage from the TX port to the RX port [38]. Overall, the proposed architecture in [17] is capable of canceling the SI by 103 dB over 20 MHz.

The above prototypes represent the current state of the art in the literature, and thereby constitute the proper context for the SI cancellation performance achieved using the digital cancellation solutions presented in this chapter and reported in [8, 9]. Table 3.1 presents also the key performance figures of these prototype implementations, while further details are provided in Sects. 3.5.1 and 3.5.2.

3.2 Challenges in Digital Cancellation

Due to the stringent cancellation requirements and the high power of the SI signal, many of the imperfections produced within the transceiver must be considered when modeling and regenerating the SI waveform for digital cancellation. This stems from the high-power difference between the received SI signal and the signal of interest: even a mild distortion component in the former can be extremely powerful compared to the latter, which is weakened due to the much longer propagation distance. Hence, if such a distortion component is ignored in the cancellation processing, it will remain unaffected and therefore results in a heavily decreased SINR upon detection. Below, the most relevant analog imperfections are described and discussed.

3.2.1 I/Q Imbalance

IQ imbalance is a prevalent issue in direct-conversion transceivers, stemming from the inherent phase and amplitude mismatches between the I- and Q-branches [39, 40]. As a result of these mismatches, a so-called image component is generated on top of the original signal, Fig. 3.2a illustrating this phenomenon in the frequency domain. The image component produced within a direct-conversion transceiver is in fact the original signal, whose spectrum has been inverted with respect to the frequency axis. In time domain, the image component is correspondingly the complex conjugate of the original signal. The magnitude of the image component in relation to the original signal is dictated by the severity of the IQ imbalance, although typically it is clearly weaker than the original signal [41].

As opposed to the direct-conversion architecture, in the older *superheterodyne* transceivers, where an intermediate frequency is used during up- and downconversion, the mirror images are in fact in the adjacent frequency bands, and they can simply be filtered out either in the intermediate frequency (TX) or in the RF domain (RX). Consequently, the IQ imbalance of a direct-conversion transceiver can be considered equivalent to the problem of image frequencies in the superheterodyne architecture. While the direct-conversion architecture avoids the need for the possibly bulky analog filters, it must carefully match the phases and amplitudes of the I- and Q-branches to sufficiently suppress the image component.

Although IQ imbalance affects most low-cost radio devices, it is negligibly weak in the laboratory equipment upon which the prototypes used in this chapter are based. Therefore, it is omitted also from the described signal models. For further information about the modeling of IQ imbalance in IBFD devices, please refer to [10, 42].

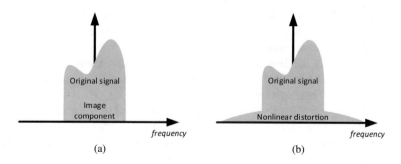

Fig. 3.2 Frequency domain illustrations of (**a**) IQ imbalance and (**b**) nonlinear distortion

3.2.2 Nonlinear Distortion

Another fundamental issue, especially in low-cost communications devices, is the nonlinear distortion. It is primarily produced by the different active components, in particular the amplifiers, and can heavily distort the signal. In principle, it stems from some form of clipping, which results in the highest signal peaks being compressed. In other words, when driven sufficiently close to saturation, the gain of the amplifier is smaller when the amplitude of the input signal is higher, and thereby the relationship between the input and output signals is in fact nonlinear. In the frequency domain, nonlinear distortion can be illustrated as shown in Fig. 3.2b, where it exhibits itself as spectral regrowth. Note that, in the context of IBFD transceivers, only the nonlinear distortion falling onto the signal band needs to be considered since the distortion falling out-of-band can easily be filtered out in the receiver. The out-of-band nonlinearities are an important consideration only in terms of the transmitter spectral emission mask, which is typically defined in the system specifications to limit the interference produced on to the adjacent channels [41]. Nevertheless, since the focus of this work is only on the inband distortion, the spectral emission requirements are not explicitly considered.

In a typical case, the main source of nonlinear distortion is the TX PA [6, 7, 20, 43, 44]. The reason for this is the need for high power efficiency, while also having to amplify the signal to the required transmit power level. These two requirements mean that the PA must operate close to its saturation point, which results in the nonlinear distortion of the waveform, especially with signals having high peak-to-average-power ratio (PAPR) [44, 45]. Furthermore, even if the distortion is mild enough to fulfill the system specifications, it can still be extremely problematic for an IBFD device. For example, the strictest EVM requirement in the LTE specifications is 3.5% for the BS under the 256-QAM modulation scheme, which translates to a distortion component that is 29 dB weaker than the actual signal [46]. Making the reasonable assumption that the error component of a BS transmit signal is dominated by the PA nonlinearities [47, 48], this can be considered the highest allowed power level for the PA-induced nonlinear distortion under this modulation scheme. Hence, even with such a strict EVM requirement, the power level of the nonlinear distortion can easily be around 0 dBm at the TX output, meaning that in the RX chain it is likely orders of magnitude stronger than any signal of interest. What is more, when considering a lower-order modulation scheme, the EVM requirements are less strict, meaning that the nonlinearities may be even stronger [41, 46]. This clearly indicates that nonlinear modeling of the TX PA is necessary in IBFD transceivers.

There are various methods for modeling the PA-induced nonlinear distortion on a waveform level, such as the Volterra series or the Wiener model [49–51]. However, to limit the complexity of the model, and to ensure efficient parameter estimation, the widely-deployed parallel Hammerstein (PH) model is adopted in this chapter. In principle, a PH model refers to a system with parallel static nonlinearities, each having its own filter that models the memory effects [52]. In this chapter, the static nonlinearities are chosen to be monomials, while the memory effects are modeled using finite impulse response (FIR) filters. Denoting the baseband-equivalent PA input signal by $x_{\mathrm{PA}}^{\mathrm{in}}(n)$, its output signal can thereby be expressed with the adopted discrete-time PH model as follows [6, 7, 51, 53–55]:

$$x_{\mathrm{PA}}(n) = \sum_{\substack{p=1 \\ p \text{ odd}}}^{P} \sum_{m=0}^{M_{\mathrm{PH}}} h_p(m) \left| x_{\mathrm{PA}}^{\mathrm{in}}(n-m) \right|^{p-1} x_{\mathrm{PA}}^{\mathrm{in}}(n-m), \tag{3.1}$$

where P is the nonlinearity order of the model, M_{PH} is the memory length of the model, and $h_p(m)$ contains the coefficients of the pth-order nonlinearity. Note that it is sufficient to include only the odd-order nonlinearities in the model when analyzing the inband distortion since the even-order terms fall outside the reception bandwidth [54]. This type of a model has been shown to be accurate for modeling a wide variety of practical PAs [49, 51, 54, 55], and in Sect. 3.5 it is shown to achieve high modeling accuracy also in the context of digital SI cancellation.

In addition to the nonlinear distortion produced in the transmitter, under some circumstances the RX chain can also distort the received signal in a nonlinear manner [6, 21, 56]. This typically occurs when the amount of SI suppression before the receiver is too low, resulting in the saturation of the RX LNA. It is possible to model and attenuate also the RX-induced nonlinearities [21, 56], but the signal model becomes prohibitively complicated when also the PA is producing significant levels of distortion. Namely, then the overall coupling channel consists of the cascade of a nonlinearity, a wireless channel with memory, and another nonlinearity, resulting in an extremely large amount of nonlinear terms. For this reason, it is crucial to have sufficient RF cancellation performance since that will ensure that the power level of the receiver input signal is not high enough to produce significant nonlinear distortion in the LNA. This is also the underlying assumption in this chapter, and its validity is proven by the obtained measurement results reported in Sect. 3.5.

3.2.3 Analog-to-Digital Converter Quantization Noise

As opposed to the traditional HD systems, in IBFD transceivers also the receiver ADC plays a significant role [5, 57]. Namely, in a HD receiver, the accuracy of the quantization upon analog-to-digital conversion is rarely a bottleneck, assuming proper automatic gain control (AGC) that amplifies the overall signal to match the dynamic range of the ADC [58, p. 139]. There, the SINR is typically limited either by noise or by interference of some sort, as long as the ADC uses any reasonable amount of bits. However, the situation changes drastically when an IBFD transceiver is considered since then the ADC input signal also contains some residual SI. Even after a high amount of passive isolation and RF cancellation, the power of the

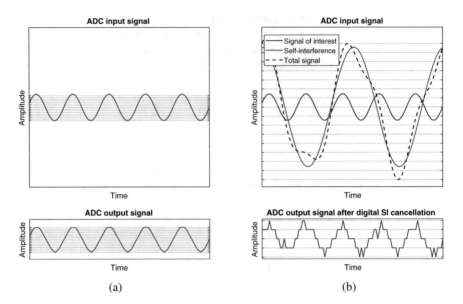

Fig. 3.3 Effect of ADC quantization (**a**) without any SI and (**b**) with a strong SI signal. The horizontal lines denote the quantization levels

residual SI can still be significantly higher than the power of the received signal of interest. This essentially means that the AGC is adjusting the gain based on the *residual SI*, not the signal of interest, to avoid clipping. Hence, upon the analog-to-digital conversion, the dynamic range of the signal of interest is well below the dynamic range of the ADC, meaning that it is quantized with fewer bits than in a corresponding HD device. Consequently, if the amount of analog SI suppression is not sufficiently high, the SINR of the signal of interest remains low due to the quantization effects, regardless of the digital cancellation performance [5, 57].

This phenomenon is illustrated in Fig. 3.3, where two scenarios are shown. In Fig. 3.3a, there is no residual SI at the ADC input, and consequently the whole dynamic range of the ADC can be used to quantize the signal of interest. As can be observed, the quantization effects are only minor and the signal quality remains good. On the other hand, in Fig. 3.3b the ADC input signal contains also some residual SI and, as a result, the AGC must use less gain in the receiver to avoid clipping in the ADC. Therefore, even after eliminating the SI with digital cancellation, the signal of interest is very noisy due to the quantization effects, as can be seen in the lower part of Fig. 3.3b. These examples show that, in the context of IBFD transceivers, the dynamic range of the ADC and the analog SI cancellation performance must be carefully considered to ensure sufficient SINR for the signal of interest in the digital domain. For a detailed analysis on the ADC dynamic range requirements, refer to [6, 10].

3.2.4 Transmitter Thermal Noise

Due to the extremely high power of the SI signal at the receiver input, even the transmitter thermal noise can result in an elevated interference floor in IBFD transceivers if not properly managed. In particular, the thermal noise present in the transmitter digital-to-analog converter (DAC) output signal will also be amplified, alongside with the actual signal, and it is further magnified by the noise figure (NF) of the transmitter. This means that the TX-induced thermal noise can reach reasonably high-power levels compared to the noise floor at the receiver input, especially with poor TX–RX isolation. In a HD transceiver, the transmitter noise can be neglected since the transmitter is either turned off during reception (cf. TDD), or it operates on a different frequency band and the noise can be filtered out (cf. FDD). However, in an IBFD transceiver, any noise included in the transmit signal will also overlap the received signal of interest in the frequency and time domains, meaning that it can potentially reduce the overall SINR. What is more, the transmitter thermal noise cannot obviously be modeled, which means that there are few options for canceling it in the receiver.

To ensure that the transmitter noise can be properly suppressed, the RF cancellation architecture illustrated in Fig. 3.1 is the preferable option. The reason for this is that in this type of an RF canceller the transmitter output signal is used as the cancellation signal, and hence it inherently includes also the transmitter noise. This means that the transmitter noise is attenuated by both the physical TX–RX isolation and the RF canceller, which together are typically enough to suppress it well below the receiver noise floor. On the other hand, if an auxiliary transmitter-based RF cancellation solution, reported, for instance, in [2], is used, the transmitter thermal noise realization is not included in the cancellation signal. This means that it is only attenuated by the passive isolation, which might not be sufficient to suppress it below the receiver noise floor. Hence, this speaks strongly for the transmitter output-based RF canceller as then also the transmitter noise is automatically taken care of.

3.2.5 Oscillator Phase Noise

The effect of phase noise has also been widely studied in the context of IBFD transceivers [59–63]. Phase noise is caused by the varying phase of the local oscillator (LO) signal during up- and downconversion, which results in a multiplicative distortion component. For the transmitter, the complex-valued baseband-equivalent phase noise model can be written as follows:

$$x_{PA}^{in}(t) = x_{IQ}^{TX}(t)e^{j\phi_{tx}(t)}, \tag{3.2}$$

where $x_{IQ}^{TX}(t)$ is the signal before upconversion, and $\phi_{tx}(t)$ is the transmitter phase noise term. The latter is a stochastic process, whose statistics depend on the quality of the LO. In the ideal case where no phase noise is produced, $\phi_{tx}(t)$ is merely a constant. In the receiver IQ mixer, the signal is downconverted, which results in the following effective phase noise realization at the ADC input:

$$y_{ADC}(t) = y_{IQ}^{RX}(t)e^{-J\phi_{rx}(t)}, \tag{3.3}$$

where $y_{IQ}^{RX}(t)$ is the signal before downconversion, and $\phi_{rx}(t)$ is the random receiver phase noise term.

Under the assumption that $\phi_{tx}(t)$ and $\phi_{rx}(t)$ are independent of each other, phase noise is indeed a serious issue in IBFD transceivers, and it results in an elevated interference floor, even after all the cancellation stages [59–61, 63]. However, using independent LO signals in an IBFD transceiver is a rather pessimistic assumption since the transmitter and receiver operate on the same center frequency. Hence, in this case, the most sensible option is to use the same LO signal for both the up- and downconversion, and consequently $\phi_{tx}(t) = \phi_{rx}(t) = \phi(t)$.

Noting that the sign of the receiver phase noise term is opposite to that of the transmitter phase noise, it can be deduced that some of the phase noise is implicitly cancelled upon downconversion when using a shared TX/RX LO [59, 63]. For the self-cancellation of the phase noise in general, the deciding factor is the delay between the TX and RX IQ mixers since that determines how well the phase noise during upconversion matches with the phase noise affecting the downconversion. Denoting the direct propagation delay between the TX and the RX IQ mixers by τ_{PN}, the effective phase noise affecting the main SI component is as follows:

$$e^{J\phi_{eff}(t)} = e^{J\phi(t)}e^{-J\phi(t+\tau_{PN})} = e^{J(\phi(t)-\phi(t+\tau_{PN}))}. \tag{3.4}$$

In particular, the common phase error (CPE), which refers to the mean value of $\phi(t)$, is perfectly cancelled upon downconversion since it can be expected to be static during the short propagation time. Also any frequency offset in the LO signal, commonly referred to as carrier frequency offset (CFO), is cancelled at this point since it is not affected by the delay. Hence, neither the CPE nor the CFO affect the overall SI waveform in an IBFD device with a shared LO, meaning that they can be omitted in this analysis.

Considering then the phase noise remaining after the self-cancellation, it can be further analyzed by making certain assumptions regarding the nature of the phase noise process. In particular, let us adopt the widely used free-running oscillator model where phase noise is modeled as a random-walk process, $\phi(t) = \sqrt{4\pi\beta_{3dB}}B(t)$, where $B(t)$ denotes Brownian motion and β_{3dB} is the 3-dB bandwidth of the phase noise [64, p. 16]. Now, the effective phase noise can be expressed as follows:

$$\phi_{eff}(t) = \sqrt{4\pi\beta_{3dB}}\left(B(t) - B(t + \tau_{PN})\right). \tag{3.5}$$

Based on the basic properties of the Brownian motion, it can easily be shown that $\phi_{eff}(t) \sim \mathcal{N}(0, 4\pi\beta_{3dB}\tau_{PN})$ [65, p. 301], where $\mathcal{N}(\mu, \sigma^2)$ denotes the normal distribution with mean μ and variance σ^2. It should also be noted that adopting this type of a model for the phase noise process can be considered a pessimistic scenario, as the phase noise performance of any real-world LO, typically utilizing some type of a phase-locked loop (PLL), is likely to be somewhat better [66–68].

In general, it can already be deduced that, with any reasonable delay, the effective phase noise of the main SI component is negligibly low [61]. It should be noted, however, that the self-cancellation of the phase noise is much weaker for the multipath SI components since their corresponding delays are longer. However, the multipath components themselves are also weaker due to the higher path loss, and hence the contribution of the effective phase noise remains negligible also in this case. These deductions are confirmed in [10], where it is be shown that, for a realistic propagation delay between the TX and RX IQ mixers, the phase noise will have practically no effect on the overall residual SI, even when assuming such a pessimistic model for the phase noise process.

3.3 Advanced Self-Interference Signal Models

In this chapter, two different advanced SI signal models are presented and used for digital cancellation. Both of the signal models are derived for a multiple-input multiple-output (MIMO) IBFD transceiver, which means that they can be readily applied to a multi-antenna system. However, the measurement-based evaluation, presented in Sect. 3.5, is done with a single-input and single-output (SISO) transceiver since that is already sufficient to show the accuracy of the different signal models. Also note that in this section it is assumed that only the SI is being received, i.e., there is no signal of interest present in the total RX signal. Nevertheless, the derivations are also valid for a case without a silent calibration period as then the signal of interest can be considered a part of the overall noise. For further analysis regarding the effect of the signal of interest on SI parameter estimation, see [69].

3.3.1 Linear Signal Model

The most basic approach in modeling the residual SI in the digital domain is to assume a perfectly linear transceiver chain. That is, all the impairments are neglected and only the different memory effects, produced by the transceiver and the wireless coupling channel, are considered. Such a linear system can be modeled as illustrated in Fig. 3.4. In this case, the baseband-equivalent output signal of the jth transmitter can be written as follows:

$$x_{j,\mathrm{PA}}(t) = f_{j,\mathrm{TX}}(t) \star x_j(t) + e_{j,\mathrm{TX}}(t), \tag{3.6}$$

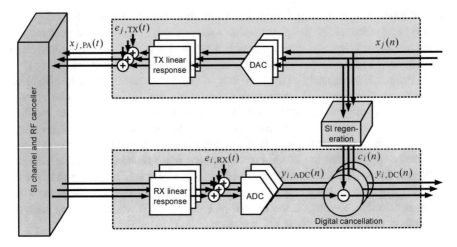

Fig. 3.4 The system model used in deriving the linear signal model for the digital canceller

where $f_{j,\mathrm{TX}}(t)$ denotes the general frequency-dependent response of the jth transmitter, including the linear gain and the possible memory effects, $e_{j,\mathrm{TX}}(t)$ represents all the unmodeled distortion components and the transmitter noise, and $\star$ denotes the convolution.

Then, the signals are transmitted and consequently received as SI, which is suppressed by the RF canceller. Now, as a MIMO transceiver is considered, the SI in the receivers consists of the sum of all the transmit signals, each with their own coupling channel. For the purposes of deriving the signal models in this section, it is assumed that the RF cancellation is performed using the PA output signals, even though that might be unfeasible if the number of transmitters and receivers is large, as is the case in, e.g., massive MIMO devices. However, as shown in [70], the essential signal model is the same also when using the auxiliary transmitter-based RF canceller reported in [2], and hence this assumption does not affect the generality of the derived signal models.

In the linear signal model, the receivers are modeled in the same way as the transmitters, i.e., only the different memory effects are considered, while all the other impairments are represented by a generic error signal. Hence, the signal before the ADC in the ith receiver can be written as:

$$y_{i,\mathrm{ADC}}(t) = f_{i,\mathrm{RX}}(t) \star \left(\sum_{j=1}^{N_t} h_{ij,\mathrm{SI}}(t) \star x_{j,\mathrm{PA}}(t) - \sum_{j=1}^{N_t} h_{ij,\mathrm{RFC}}(t) \star x_{j,\mathrm{PA}}(t) \right)$$

$$+ e_{i,\mathrm{RX}}(t)$$

$$= f_{i,\mathrm{RX}}(t) \star \sum_{j=1}^{N_t} \left(h_{ij,\mathrm{SI}}(t) - h_{ij,\mathrm{RFC}}(t) \right) \star x_{j,\mathrm{PA}}(t) + e_{i,\mathrm{RX}}(t)$$

$$= \sum_{j=1}^{N_t} f_{i,\mathrm{RX}}(t) \star h_{ij,\mathrm{RSI}}(t) \star f_{j,\mathrm{TX}}(t) \star x_j(t)$$

$$+ \sum_{j=1}^{N_t} f_{i,\mathrm{RX}}(t) \star h_{ij,\mathrm{RSI}}(t) \star e_{j,\mathrm{TX}}(t) + e_{i,\mathrm{RX}}(t), \qquad (3.7)$$

where $f_{j,\mathrm{RX}}(t)$ is the linear response of the ith receiver, $h_{ij,\mathrm{SI}}(t)$ is the SI coupling channel between the jth transmitter and the ith receiver, $h_{ij,\mathrm{RFC}}(t)$ is the response of the RF cancellation signal between the jth transmitter and the ith receiver, $h_{ij,\mathrm{RSI}}(t) = h_{ij,\mathrm{SI}}(t) - h_{ij,\mathrm{RFC}}(t)$ is the effective coupling channel after RF cancellation, and $e_{i,\mathrm{RX}}(t)$ is the total noise-plus-modeling-error signal in the ith receiver. Thus, for the purposes of a digital canceller, the RF canceller can in fact be modeled jointly with the propagation channel since it merely adds a certain amount of delayed copies of the PA output signals to the overall received signals, as is evident from (3.7) above. Consequently, the propagation channel and the RF canceller can be modeled by just a single MIMO impulse response $h_{ij,\mathrm{RSI}}(t)$. The same is also true for the auxiliary transmitter-based RF canceller, the only difference being that it utilizes the digital baseband transmit signal instead of the PA output.

Thus, based on (3.7), it can easily be observed that the overall signal model before digital cancellation in the ith receiver can be expressed as follows:

$$y_{i,\mathrm{ADC}}(n) = \sum_{j=1}^{N_t} \sum_{m=-M_1}^{M_2} h_{ij}^{\mathrm{L}}(m) x_j(n-m) + e_{i,\mathrm{tot}}(n), \qquad (3.8)$$

where $h_{ij}^{\mathrm{L}}(m)$ is the total effective linear response between the jth transmitter and the ith receiver, M_1 and M_2 are the numbers of pre-cursor and post-cursor memory taps, respectively, and $e_{i,\mathrm{tot}}(n)$ includes all the unmodeled distortion components, as well as the total noise signal. The pre-cursor taps are introduced here and in the continuation to accurately model the different memory effects occurring in a real IBFD transceiver [8, 70].

3.3.2 Nonlinear Signal Model

Another approach into modeling the residual SI signal in the digital domain is to assume that the overall distortion is dominated by the nonlinearities produced by the TX PA, as is done, for instance, in [7–9, 13, 25]. Then, the essential system model is as shown in Fig. 3.5. In particular, all the other parts of the transceiver chain are assumed to be ideal, apart from the PAs, meaning that the system is basically a parallel connection of static nonlinearities, followed by linear filters [8, 9]. This type of a signal model is referred to as a PH nonlinearity, as already discussed in some detail in Sect. 3.2.2.

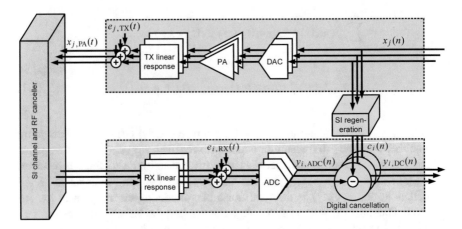

Fig. 3.5 The system model used in deriving the nonlinear signal model for the digital canceller

Now, the jth baseband-equivalent transmit signal is expressed as

$$x_{j,\mathrm{PA}}(t) = \sum_{\substack{p=1 \\ p \text{ odd}}}^{P} k_{j,p,\mathrm{TX}}(t) \star |x_j(t)|^{p-1} x_j(t) + e_{j,\mathrm{TX}}(t), \tag{3.9}$$

where P is the nonlinearity order, $e_{j,\mathrm{TX}}(t)$ represents again the modeling error and noise of the transmitter, and $k_{j,p,\mathrm{TX}}(t)$ is the response of the pth-order SI term in the jth transmitter, including also the gain. Similar to the earlier cases, the RF canceller can again be modeled jointly with the propagation channel, as also illustrated in Fig. 3.5. In addition, since all the impairments of the RX chains are omitted in this nonlinear signal model, the receivers can be modeled simply as linear filters. Hence, the signal before the analog-to-digital conversion in the ith receiver can simply be expressed as

$$y_{i,\mathrm{ADC}}(t) = f_{i,\mathrm{RX}}(t) \star \sum_{j=1}^{N_t} \left(h_{ij,\mathrm{SI}}(t) - h_{ij,\mathrm{RFC}}(t) \right) \star x_{j,\mathrm{PA}}(t) + e_{i,\mathrm{RX}}(t)$$

$$= \sum_{j=1}^{N_t} \sum_{\substack{p=1 \\ p \text{ odd}}}^{P} f_{i,\mathrm{RX}}(t) \star h_{ij,\mathrm{RSI}}(t) \star k_{j,p,\mathrm{TX}}(t) \star |x_j(t)|^{p-1} x_j(t)$$

$$+ \sum_{j=1}^{N_t} f_{i,\mathrm{RX}}(t) \star h_{ij,\mathrm{RSI}}(t) \star e_{j,\mathrm{TX}}(t) + e_{i,\mathrm{RX}}(t), \tag{3.10}$$

where the variables are as defined for the linear signal model. The final digital-domain signal can then be written as follows:

$$y_{i,\mathrm{ADC}}(n) = \sum_{j=1}^{N_t} \sum_{\substack{p=1 \\ p\ \mathrm{odd}}}^{P} \sum_{m=-M_1}^{M_2} h_{ij,p}^{\mathrm{NL}}(m)\psi_p\left(x_j(n-m)\right) + e_{i,\mathrm{tot}}(n), \qquad (3.11)$$

where $\psi_p\left(x_j(n)\right) = \left|x_j(n)\right|^{p-1}x_j(n)$ is the static pth-order nonlinear basis function, and $h_{ij,p}^{\mathrm{NL}}(m)$ is its total effective response between the jth transmitter and the ith receiver.

3.4 Parameter Estimation and Digital Self-Interference Cancellation

In order to actually cancel the SI in the digital domain, the above signal models are used to regenerate the observed residual SI signal. This obviously requires estimating the parameters of the corresponding signal model, for which reason different methods for parameter estimation are presented and discussed in this section. To facilitate more straightforward mathematical derivations, matrix-vector notations are used. Moreover, without loss of generality, the estimation and cancellation procedure is only presented for an individual receiver.

As a starting point, it is assumed that the ith receiver has an observation block of N samples of the residual SI signal $y_{i,\mathrm{ADC}}(n)$ at its disposal for performing the digital cancellation procedure. This signal can be received, for instance, during a calibration period of limited length when there are no other transmissions in the network, as discussed earlier. However, it should be noted that the signal transmitted during this calibration period can also consist of useful data since the transceiver obviously has full knowledge of its own transmit data [69, 71, 72]. Without loss of generality, the indexing of the observation block is started from zero for a more illustrative notation, and consequently it is expressed in vector form as follows:

$$\mathbf{y}_{i,\mathrm{ADC}} = \left[y_{i,\mathrm{ADC}}(0)\ y_{i,\mathrm{ADC}}(1)\ \cdots\ y_{i,\mathrm{ADC}}(N-1)\right]^T. \qquad (3.12)$$

The observation block size N is referred to as the *parameter estimation sample size*, and it determines how much data can be used for estimating the SI channel coefficients.

Moreover, the static basis functions of each transmit signal are also collected into a vector, referred to as an *instantaneous basis function vector*. The linear signal model has only one static basis function, i.e., the original transmit signal itself, and hence its instantaneous basis function vector is simply defined as follows:

$$\boldsymbol{\psi}_{\mathrm{L}}(n) = \left[x_1(n)\ x_2(n)\ \cdots\ x_{N_t}(n)\right]. \qquad (3.13)$$

Denoting the number of static basis functions for all transmit signals by K, in this case $K = N_t$. The nonlinear signal model, on the other hand, has $K = N_t \frac{P+1}{2}$ static basis functions in total and the corresponding instantaneous basis function vector is expressed as

$$\boldsymbol{\psi}_{\text{NL}}(n) = \left[\psi_1\left(x_1(n)\right) \psi_3\left(x_1(n)\right) \cdots \psi_P\left(x_1(n)\right) \psi_1\left(x_2(n)\right) \cdots \psi_P\left(x_{N_t}(n)\right)\right],$$

$$(3.14)$$

where $\psi_p\left(x_j(n)\right)$ is the static pth-order nonlinear basis function, defined in Sect. 3.3.2.

The total number of basis functions for both signal models is then simply KM, where $M = M_1 + M_2 + 1$ is the total amount of memory. That is, the term *basis function* is used to refer to all the KM entities that include also the delayed versions of the static or instantaneous basis functions. This terminology is adopted throughout the chapter.

It should also be noted here that, when applying the nonlinear basis functions to generate the nonlinear SI terms for cancellation purposes, a higher sampling frequency should be used to avoid aliasing [27]. This stems from the fact that the bandwidth of the pth-order nonlinearity is p times the bandwidth of the original signal [49]. Consequently, if the input signal of a nonlinear basis function is not properly oversampled, some of the nonlinear distortion will alias onto the original signal band, resulting in an inaccurate model of the true nonlinear process. Hence, in theory, the input signal of a 3rd-order basis function must be oversampled by a factor of 3 to avoid aliasing at the output, and so forth. In practice, however, the higher order nonlinearities have typically very little spectral content close to the edge of their theoretical bandwidth, and thus less oversampling usually suffices [27]. What is more, since in this case only the inband content of the generated nonlinear terms actually matters, some aliasing outside the actual signal band can be tolerated.

After generating the basis functions with sufficient oversampling, the resulting nonlinear signal is then decimated back to the original sampling frequency, with appropriate filtering to avoid aliasing effects. For notational simplicity, the necessary interpolation and decimation procedures are assumed to be implicitly included in the respective basis functions in the derivations of this chapter, and hence they are not explicitly considered. Equivalently, it can be assumed that the signals have already been sufficiently oversampled for the considered nonlinearity orders. For further discussion regarding this aspect, refer to [27].

Having now defined the necessary signal vectors, different methods for calculating the SI channel estimate are presented in the following sections. The presented estimation and cancellation algorithms are then evaluated in Sect. 3.5 with measurements.

3.4.1 Block Least Squares-Based Estimation and Cancellation

The least squares (LS) algorithm is a powerful and versatile parameter estimation tool since it makes no implicit or explicit assumptions regarding the statistics of the total noise signal [73, p. 219]. This is beneficial in the context of digital SI cancellation since the noise cannot in general be expected to follow any particular probability distribution due to the various RF impairments. For this reason, this section presents an LS-based digital cancellation algorithm, which operates in a block-wise nature. That is, it estimates the SI channel coefficients for a block of N received samples, after which these estimated coefficients can be used to cancel the SI until the SI channel estimation procedure is again repeated after a certain period of time. The basic operating principle of the LS-based canceller is shown on a general level in Fig. 3.6.

To lay out the LS estimation procedure in detail, let us first express the residual SI signal in the ith receiver using the defined vectors as follows:

$$y_{i,\text{ADC}}(n) = \sum_{m=-M_1}^{M_2} \boldsymbol{\psi}(n-m)\mathbf{h}_i(m) + e_{i,\text{tot}}(n), \qquad (3.15)$$

where $n = 0, 1, \ldots, N$, $\mathbf{h}_i(m)$ is the $(M_1 + M_2 + 1) \times 1$ channel coefficient vector corresponding to the basis functions with lag m, and $\boldsymbol{\psi}(n)$ is an instantaneous basis function vector without a subscript associating it with any particular signal model to present the parameter estimation procedure in a generic fashion. That is, it is either as shown in (3.13) or (3.14), depending on the utilized signal model. The vectors $\boldsymbol{\psi}(n-m)$ constitute then the complete set of basis functions when considering all the values of m. Note that (3.15) implicitly assumes that the receiver can use transmit data also from outside the given block of N samples for cancellation processing. This assumption is well justified from an estimation theoretic perspective since the

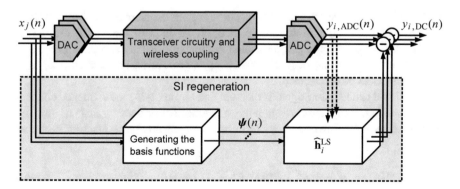

Fig. 3.6 A generic illustration of the LS-based digital SI canceller

own transmit signal is obviously known within the device and hence it does not bring any new information into the system, unlike the observed residual SI signal.

To express the whole observation block of the residual SI signal, the instantaneous basis function vectors can be collected into a single *convolution data matrix* [42], which is defined as follows:

$$
\boldsymbol{\Psi} =
\begin{bmatrix}
\boldsymbol{\psi}(M_1) & \boldsymbol{\psi}(M_1 - 1) & \cdots & \boldsymbol{\psi}(-M_2) \\
\boldsymbol{\psi}(M_1 + 1) & \boldsymbol{\psi}(M_1) & \cdots & \boldsymbol{\psi}(-M_2 + 1) \\
\vdots & \vdots & \ddots & \vdots \\
\boldsymbol{\psi}(N + M_1 - 1) & \boldsymbol{\psi}(N + M_1 - 2) & \cdots & \boldsymbol{\psi}(N - M_2 - 1)
\end{bmatrix}. \tag{3.16}
$$

The corresponding complete channel vector for the ith receiver is then simply

$$
\mathbf{h}_i = \left[\mathbf{h}_i^T(-M_1)\, \mathbf{h}_i^T(-M_1 + 1) \cdots \mathbf{h}_i^T(M_2) \right]^T, \tag{3.17}
$$

which contains the coefficients of all the basis functions. With these, the residual SI vector in the ith receiver can be written as follows:

$$
\mathbf{y}_{i,\text{ADC}} = \boldsymbol{\Psi}\mathbf{h}_i + \mathbf{e}_i, \tag{3.18}
$$

where $\mathbf{e}_i$ is the $N \times 1$ noise-plus-modeling-error vector.

Since the overall signal model in (3.18) is in fact linear in parameters with both the received signal vector $\mathbf{y}_{i,\text{ADC}}$ and the matrix $\boldsymbol{\Psi}$ being obviously known, linear LS can be used to estimate the channel coefficients. Assuming that $\boldsymbol{\Psi}$ has full column rank, the LS solution to the SI channel estimation problem is given by [7, 42] and [73, p. 223]

$$
\widehat{\mathbf{h}}_i^{\text{LS}} = \min_{\mathbf{h}_i} \left\| \mathbf{y}_{i,\text{ADC}} - \boldsymbol{\Psi}\mathbf{h}_i \right\|_2^2 = \left(\boldsymbol{\Psi}^H \boldsymbol{\Psi} \right)^{-1} \boldsymbol{\Psi}^H \mathbf{y}_{i,\text{ADC}}, \tag{3.19}
$$

where $\|\cdot\|_2$ denotes the L^2-norm. The accuracy of this estimate is largely determined by the parameter estimation sample size N, as is well known in estimation theory [73]. However, the computational complexity of the estimation procedure is higher for larger values of N, and hence a proper trade-off between accuracy and complexity must be determined.

The digital cancellation signal is then obtained by applying the estimated channel coefficients in $\widehat{\mathbf{h}}_i^{\text{LS}}$ into the corresponding SI terms. This is done by first stacking the instantaneous basis function vectors to form the complete basis function vector as follows:

$$
\boldsymbol{\Psi}(n) = \left[\boldsymbol{\psi}(n + M_1)\, \boldsymbol{\psi}(n + M_1 - 1) \cdots \boldsymbol{\psi}(n - M_2) \right], \tag{3.20}
$$

which is of the same form as the nth row of the convolution data matrix Ψ. The corresponding digital cancellation signal in the ith receiver is then simply

$$c_i(n) = \Psi(n)\widehat{\mathbf{h}}_i^{\mathrm{LS}}. \tag{3.21}$$

Thus, the signal after the LS-based digital canceller is as follows:

$$y_{i,\mathrm{DC}}(n) = y_{i,\mathrm{ADC}}(n) - c_i(n) = y_{i,\mathrm{ADC}}(n) - \Psi(n)\widehat{\mathbf{h}}_i^{\mathrm{LS}} \approx e_{i,\mathrm{tot}}(n), \tag{3.22}$$

where the final approximation is obviously valid only when the SI channel estimate is sufficiently accurate.

An important aspect of the LS estimation scheme is its block-wise nature. Namely, the SI channel estimate is always calculated for a certain piece of the observed signal, whose length in this case is determined by the parameter estimation sample size N. Thus, LS is well suited for scenarios where the SI channel is estimated during a dedicated training period when there are no other transmissions in the network [69, 71]. The same estimate must then be used until the next calibration period, during which the LS estimate is recalculated. Moreover, it should be noted that obtaining the LS estimate requires solving the pseudoinverse of the convolution data matrix Ψ, as can be observed in (3.19). This can be computationally intensive, especially if the number of basis functions is large. Hence, especially in mobile-scale devices, simpler and less computationally demanding estimation methods might be preferable.

3.4.2 Least Mean Squares-Based Adaptive Estimation and Cancellation

One approach to decreasing the computational requirements of the digital SI cancellation procedure is to utilize adaptive algorithms for estimating the channel coefficients. A widely used solution for this type of an estimation problem is the least means squares (LMS) parameter learning algorithm, which has been described and evaluated in [8, 9, 25]. The basic operating principle of the LMS-based digital canceller is illustrated in Fig. 3.7. Essentially, the LMS canceller aims at minimizing the power of its output signal [74, p. 150], i.e., $y_{i,\mathrm{DC}}(n)$, which it does by utilizing a predetermined signal model for the residual SI.

However, as also shown in Fig. 3.7, before the actual parameter learning, the different static basis functions must be orthogonalized. The reason for this is the poor convergence performance of the LMS algorithm if the elements of the input vector are highly correlated, caused by the large eigenvalue spread of the input signal covariance matrix [75, p. 417], [23, 25]. Since in this case the static basis functions can indeed be expected to be correlated as they are all dependent on the original transmit signal, orthogonalizing them is necessary to ensure efficient

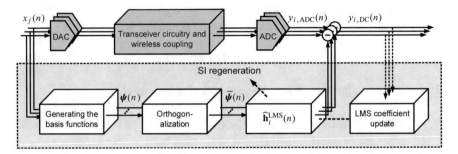

Fig. 3.7 A generic illustration of the LMS-based digital SI canceller

parameter learning by the LMS algorithm. Although there are various alternative methods for performing the orthogonalization, in this chapter it is done with an orthogonalization matrix that can be obtained starting from the covariance matrix of the instantaneous basis functions, defined as follows [75, p. 100]:

$$\mathbf{R}_{\boldsymbol{\psi}} = \mathbb{E}\left[\boldsymbol{\psi}^H(n)\boldsymbol{\psi}(n)\right], \tag{3.23}$$

where the subscript of the instantaneous basis function vector has again been omitted to present the LMS-based solution in a generic fashion. The next step in deriving the orthogonalization matrix is calculating the eigen decomposition of the above covariance matrix:

$$\mathbf{R}_{\boldsymbol{\psi}} = \mathbf{U}_{\boldsymbol{\psi}}\boldsymbol{\Lambda}_{\boldsymbol{\psi}}\mathbf{U}_{\boldsymbol{\psi}}^H, \tag{3.24}$$

where $\boldsymbol{\Lambda}_{\boldsymbol{\psi}}$ is a diagonal matrix containing the eigenvalues of $\mathbf{R}_{\boldsymbol{\psi}}$, and $\mathbf{U}_{\boldsymbol{\psi}}$ is a unitary matrix containing the corresponding eigenvectors [76, p. 21]. The orthogonalization matrix is then simply given by [25] and [76, p. 140]

$$\mathbf{S}_{\boldsymbol{\psi}} = \mathbf{U}_{\boldsymbol{\psi}}\boldsymbol{\Lambda}_{\boldsymbol{\psi}}^{-1/2}, \tag{3.25}$$

where $\boldsymbol{\Lambda}_{\boldsymbol{\psi}}^{-1/2}$ denotes an element-wise square root and inverse of the diagonal elements. The orthogonalization matrix is written here in a slightly different form than in [25] since now the instantaneous basis functions are expressed as a row vector, whereas in [25] they are collected into a column vector. Nevertheless, the orthogonalization principle is still identical, apart from these notational differences. The static basis functions can then be orthogonalized simply by

$$\tilde{\boldsymbol{\psi}}(n) = \boldsymbol{\psi}(n)\mathbf{S}_{\boldsymbol{\psi}}, \tag{3.26}$$

which can easily be shown to be orthogonal as follows:

$$\mathbb{E}\left[\tilde{\boldsymbol{\psi}}^H(n)\tilde{\boldsymbol{\psi}}(n)\right] = \mathbf{S}_{\boldsymbol{\psi}}^H\mathbb{E}\left[\boldsymbol{\psi}^H(n)\boldsymbol{\psi}(n)\right]\mathbf{S}_{\boldsymbol{\psi}} = \boldsymbol{\Lambda}_{\boldsymbol{\psi}}^{-1/2}\mathbf{U}_{\boldsymbol{\psi}}^H\mathbf{U}_{\boldsymbol{\psi}}\boldsymbol{\Lambda}_{\boldsymbol{\psi}}\mathbf{U}_{\boldsymbol{\psi}}^H\mathbf{U}_{\boldsymbol{\psi}}\boldsymbol{\Lambda}_{\boldsymbol{\psi}}^{-1/2} = \mathbf{I}_K,$$

$$(3.27)$$

where $\mathbf{I}_K$ is a $K \times K$ identity matrix. Note that the orthogonalization matrix $\mathbf{S}_{\boldsymbol{\psi}}$ only depends on the statistical properties of the original transmit signal (via the covariance matrix $\mathbf{R}_{\boldsymbol{\psi}}$) and hence it does not change with respect to time, as long as the transmit waveforms remains the same. This means that the orthogonalization matrix can be precomputed offline, and only the actual orthogonalization in (3.26) must be performed in real time. However, it is also possible to calculate the orthogonalization matrix adaptively during the actual digital cancellation procedure, as shown in [8].

Having orthogonalized the basis functions, they can then be used for learning the SI channel coefficients with the LMS algorithm. Now, the input vector of the LMS filter, containing all the orthogonalized basis functions, is defined as follows:

$$\tilde{\boldsymbol{\Psi}}(n) = \left[\tilde{\boldsymbol{\psi}}(n + M_1) \ \tilde{\boldsymbol{\psi}}(n + M_1 - 1) \cdots \tilde{\boldsymbol{\psi}}(n - M_2)\right]. \tag{3.28}$$

Then, denoting the LMS SI channel estimate in the ith receiver after n iterations by $\hat{\mathbf{h}}_i^{\mathrm{LMS}}(n)$ (where $n = 0, 1, \ldots, N$), the cancelled signal is given by

$$y_{i,\mathrm{DC}}(n) = y_{i,\mathrm{ADC}}(n) - \tilde{\boldsymbol{\Psi}}(n)\hat{\mathbf{h}}_i^{\mathrm{LMS}}(n), \tag{3.29}$$

after which the LMS algorithm updates the SI channel estimate using the following rule [25]:

$$\hat{\mathbf{h}}_i^{\mathrm{LMS}}(n + 1) = \hat{\mathbf{h}}_i^{\mathrm{LMS}}(n) + \mathbf{M}y_{i,\mathrm{DC}}(n)\tilde{\boldsymbol{\Psi}}^H(n), \tag{3.30}$$

where $\mathbf{M}$ is a diagonal matrix containing the step sizes for the different orthogonalized basis functions on its diagonal. If no further side information is available, the channel estimate is initialized as $\hat{\mathbf{h}}_i^{\mathrm{LMS}}(0) = \mathbf{0}$.

Comparing the LS-based and LMS-based channel estimation procedures, it can be observed that the LMS rule in (3.30) requires only additions and multiplications, whereas the LS estimation involves a costly matrix inversion, among other matrix operations. Hence, as shown in more detail below, the LMS-based digital cancellation algorithm is computationally more efficient than performing the SI parameter estimation with LS. Moreover, the LMS-based digital canceller is also capable of tracking the SI channel under time-varying conditions, unlike the LS estimator which assumes the SI channel to be static during the whole observation period of N samples. However, as is well known in estimation theory, with a sufficient amount of learning data and Gaussian-distributed noise, the variance of the channel estimate given by the LMS algorithm is higher than that given by the LS estimator [75, p. 397]. This is an inherent cost for the many upsides of the LMS canceller. Consequently, in some cases a suitable compromise might be to utilize a parameter estimation solution that falls somewhere in between LS and LMS in terms of accuracy and complexity, such as the RLS algorithm [75, p. 562].

3.4.3 Computational Complexity of Digital Cancellation

Let us then briefly analyze the computational complexities of the two alternative parameter estimation algorithms for an individual receiver. Here, the so-called *Big O notation*, denoted by $O(\cdot)$, is used to characterize their asymptotic complexities for large data sets, written with respect to complex arithmetic operations [77, p. 107]. Such analysis describes how the number of arithmetic operations required for large data sets is related to the dimensions of the input matrices and/or vectors, which is a common approach for comparing the computational complexities of different algorithms [78–80].

3.4.3.1 Least Squares

Starting from the LS-based solution, it consists of first estimating the SI channel coefficients with (3.19), after which the SI signal is regenerated and cancelled. The former consists of the following parts:

- calculating the matrix product $\boldsymbol{\Psi}^H \boldsymbol{\Psi}$;
- calculating the matrix-vector product $\boldsymbol{\Psi}^H \mathbf{y}_{i,\text{ADC}}$;
- inverting the matrix $\boldsymbol{\Psi}^H \boldsymbol{\Psi}$;
- calculating the matrix-vector product between the inverse of $\boldsymbol{\Psi}^H \boldsymbol{\Psi}$ and $\boldsymbol{\Psi}^H \mathbf{y}_{i,\text{ADC}}$.

The consecutive SI regeneration and cancellation simply consists of first calculating the matrix-vector product $\boldsymbol{\Psi}(n)\widehat{\mathbf{h}}_i^{\text{LS}}$ and then subtracting it from the corresponding input sample $y_{i,\text{ADC}}(n)$. In order to quantify then the overall arithmetic complexity, recall that $\mathbf{y}_{i,\text{ADC}}$ is a $N \times 1$ vector, $\widehat{\mathbf{h}}_i^{\text{LS}}$ is a $KM \times 1$ vector, and $\boldsymbol{\Psi}$ is a $N \times KM$ matrix, where N is the parameter estimation sample size, K is the number of static basis functions, and M is the total number of memory taps. To facilitate a straightforward comparison with the LMS-based solution, the cancellation is assumed to be performed over N samples, meaning that the regeneration and cancellation must be repeated N times.

It is easy to show that the arithmetic complexity of calculating the matrix product $\boldsymbol{\Psi}^H \boldsymbol{\Psi}$ between the $KM \times N$ matrix and the $N \times KM$ matrix is $O(K^2M^2N)$, while the corresponding complexity of inverting the resulting $KM \times KM$ matrix is $O(K^3M^3)$. The latter assumes that a LUP decomposition-based approach is used to find the inverse [81, p. 828]. Determining then the arithmetic complexities of the rest of the operations in a similar manner, the total complexity of the LS-based digital cancellation procedure is

$$O\left(K^2 M^2 N\right) + O\left(KMN\right) + O\left(K^3 M^3\right) + O\left(K^2 M^2\right) + O\left(KMN\right)$$

$$= O\left(K^3 M^3\right) + O\left(K^2 M^2 N\right). \tag{3.31}$$

Furthermore, in any practical system $N \gg KM$, which means that the term $O\left(K^2 M^2 N\right)$ typically dominates the arithmetic complexity asymptotically.

3.4.3.2   Least Mean Squares

Analyzing then the adaptive LMS-based digital cancellation procedure, now each iteration involves the following computations:

- calculating the vector-matrix product $\boldsymbol{\psi}(n)\mathbf{S}_{\boldsymbol{\psi}}$;
- calculating the dot product $\widetilde{\boldsymbol{\Psi}}(n)\widehat{\mathbf{h}}_i^{\mathrm{LMS}}(n)$, and subtracting it from $y_{i,\mathrm{ADC}}(n)$;
- calculating the matrix-vector product $\mathbf{M} y_{i,\mathrm{DC}}(n)\widetilde{\boldsymbol{\Psi}}^H(n)$ and adding it to $\widehat{\mathbf{h}}_i^{\mathrm{LMS}}(n)$.

Since the complete signal model has K static basis functions and M memory taps, it is easy to show that the overall asymptotic arithmetic complexity over N iterations is now

$$O\left(K^2 N\right) + O\left(KMN\right) + O\left(KMN\right) = O\left(K^2 N\right) + O\left(KMN\right). \tag{3.32}$$

Noting then that typically the number of memory taps is higher than the number of static basis functions, i.e., $M > K$, it can be concluded that the term $O\left(KMN\right)$ usually dominates asymptotically.

Comparing then the arithmetic complexities of the LS-based and LMS-based digital cancellers, it can be observed that the complexity of the former is asymptotically relative to the square of the number of parameters KM, while the complexity of the latter is only linearly related to it. Hence, especially when a large number of taps and/or static basis functions are used, the LMS-based solution can be expected to be a more computationally efficient method for obtaining the SI channel estimate than the LS algorithm. However, as mentioned earlier, the cost of this is the lower accuracy of the LMS estimate.

3.5 Measurement-Based Self-Interference Cancellation Performance Evaluation

Let us then finally study measurement results where the proposed digital SI cancellation solutions are evaluated under different circumstances. Especially, in the measurements, the digital canceller complements an IBFD transceiver prototype, which suppresses the SI also in the analog/RF domain. This allows for evaluating the total SI cancellation performance, illustrating that true wireless IBFD operation is indeed possible.

3.5.1 Measured Self-Interference Cancellation Performance of a Generic Inband Full-Duplex Device

Let us first consider a prototype implementation designed to be a generic SISO radio device. The hereby obtained results are reported also in [9, 10, 25], wherein further details regarding the prototype and the measurements can be found. The actual measurement setup is as shown in Fig. 3.8a, while Fig. 3.8b illustrates the basic structure of the prototype. Moreover, the relevant parameters of the measurement setup and the used digital cancellers are listed in Table 3.2. Similar to the assumption in Sect. 3.3, no received signal of interest is present in the forthcoming measurement results to concentrate on evaluating the actual SI cancellation performance. In this implementation, the National Instruments (NI) PXIe-5645R vector signal

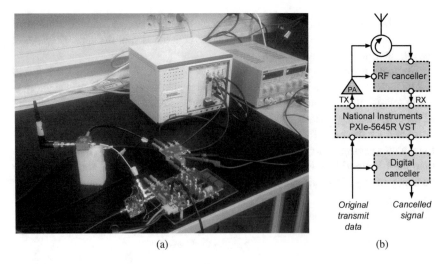

(a)   (b)

Fig. 3.8 (a) The measurement setup used in evaluating the digital cancellation performance of the generic IBFD transceiver prototype, and (b) the basic structure of the evaluated prototype

Table 3.2 The essential parameters of the measured generic IBFD transceiver prototype

Parameter	Value
Center frequency	2.46 GHz
Transmit waveform	OFDM
Bandwidth	20/40/80 MHz
Transmit power	6–8 dBm
TX/RX sampling rate	120 MHz
PA gain	24 dB
Parameter estimation sample size (N)	500,000
Number of pre-cursor taps (M_1)	10
Number of post-cursor taps (M_2)	20
Order of the nonlinear canceller (P)	11

transceiver (VST) is used both as the transmitter and the receiver, the TX output signal being further amplified by a low-cost PA (Texas Instruments CC2595 [82]). The transmit power is in the order of 6–8 dBm due to the limited output power of the VST, as well as due to the losses introduced by the RF canceller [83]. The PA output is then connected to the antenna via a circulator, which allows the usage of a single antenna while still isolating the TX and RX chains to some extent, as discussed earlier.

After the circulator, the received signal is fed to the first active SI cancellation stage, a multi-tap RF canceller, which is discussed in detail in [83], and also illustrated on a general level in Fig. 3.1. Having performed RF cancellation, the signal is fed to the RX input of the VST and digitized for further offline post-processing. Digital cancellation is then performed on the digitized and recorded signal, using Matlab.

Figure 3.9 shows the PSDs of the SI signal at different interfaces of the IBFD transceiver for the three considered bandwidths, the signal powers being referred to the RX input. Investigating first the 20-MHz case in Fig. 3.9a, here the residual SI power after RF cancellation is −61 dBm, which is clearly above the receiver noise floor and hence calls for further cancellation in the digital domain. It can firstly be observed that, due to the highly nonlinear low-cost PA, the linear digital canceller is not capable of fully suppressing the residual SI, achieving roughly 18 dB of cancellation. On the other hand, the nonlinear canceller manages to suppress the SI to the level of the receiver noise floor, canceling it by over 25 dB with both the LMS-based and LS-based parameter learning solutions. Hence, in total, the amount of obtained SI cancellation is 94 dB, of which 20 dB is physical isolation provided by the antenna and the circulator, while 47 dB of the overall suppression is contributed by the RF canceller. The small difference in the residual SI powers after the LMS-based and LS-based cancellers can likely be attributed to some very slight temporal changes in the SI coupling channel, which the LS algorithm is incapable of tracking.

The nonlinear digital canceller copes well also with the wider bandwidths, since the residual SI is still cancelled close to the receiver noise floor in the 40-MHz

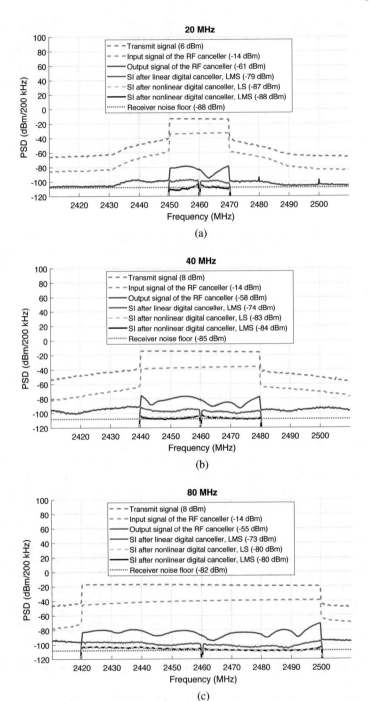

Fig. 3.9 The PSDs after the different SI cancellation stages in the generic IBFD device for (**a**) 20-MHz bandwidth, (**b**) 40-MHz bandwidth, and (**c**) 80-MHz bandwidth. The absolute powers are measured over the useful signal bandwidth

and 80-MHz cases, as can be observed from Fig. 3.9b, c. Moreover, similar to the 20-MHz scenario, the residual SI after linear digital cancellation is roughly 10 dB above the noise floor also with these wider bandwidths, indicating that nonlinear modeling is indeed necessary. With a bandwidth of 40 MHz, the total amount of SI cancellation is therefore 92 dB, of which 44 dB is provided by the RF canceller, while the nonlinear digital canceller suppresses the SI by 26 dB. Correspondingly, in the 80-MHz scenario, the SI is suppressed by 88 dB in total, consisting of 41 dB of RF cancellation and 25 dB of nonlinear digital cancellation. Thus, the overall SI cancellation performance remains on a high level even when using such wide instantaneous bandwidths. Moreover, the chosen parameter estimation scheme does not seem to affect the cancellation accuracy, indicating that the LMS-based solution is capable of obtaining sufficiently accurate coefficient estimates, regardless of its lower computational complexity in comparison to LS estimation.

In conclusion, with the considered transmit powers of 6–8 dBm and instanta-neous bandwidths of 20–80 MHz, the implemented IBFD prototype is capable of canceling the residual SI to the level of the receiver noise floor when using the developed nonlinear digital canceller with LMS-based parameter learning. Even though the utilized RF canceller is introducing rather significant levels of noise into the digital canceller input signal [10], the performance of the digital canceller is still sufficient to efficiently suppress the residual SI. Hence, by further improving the RF canceller design, a fully functional IBFD transceiver can be implemented by utilizing the proposed nonlinear digital canceller in conjunction with the analog SI suppression techniques.

3.5.2 Measured Self-Interference Cancellation Performance of an Inband Full-Duplex Relay

Another IBFD prototype implementation utilizing the digital cancellers described in this chapter is reported in [8], where a relay-type scenario is considered. Namely, in this prototype, a compact high-isolation back-to-back relay antenna, designed for a center frequency of 2.56 GHz, is used as the radiating element. The antenna element is designed such that the transmit and receive directions are on the opposite sides of the structure, which provides a high level of inherent physical isolation, especially when complemented with the so-called wavetraps that further decrease the SI coupling between the TX and RX sides [84].

The high physical isolation of the antenna structure also means that no active RF canceller is actually needed in the prototype. In particular, the power of the SI signal coupling to the receiver is suppressed sufficiently low already by the antenna structure, meaning that it is within the dynamic range of the RX chain without any additional analog cancellation. Hence, the residual SI can be suppressed by performing active cancellation only in the digital domain. This obviously decreases the complexity of the overall SI cancellation architecture as no

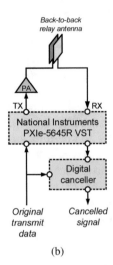

(a)   (b)

Fig. 3.10 (**a**) The measurement setup used in evaluating the digital cancellation performance of the IBFD relay prototype, and (**b**) the basic structure of the relay prototype

Table 3.3 The essential parameters of the measured IBFD relay prototype

Parameter	Value
Center frequency	2.56 GHz
Transmit waveform	OFDM
Bandwidth	20/40/80 MHz
Transmit power	24 dBm
TX/RX sampling rate	120 MHz
PA gain	36 dB
RX losses	4 dB
Parameter estimation sample size (N)	1,000,000
Number of pre-cursor taps (M_1)	25
Number of post-cursor taps (M_2)	50
Order of the nonlinear canceller (P)	11

additional RF hardware is required. However, the burden on the digital canceller is correspondingly greater, as it must reduce the SI level by a somewhat larger amount.

The measurement setup is now as shown in Fig. 3.10a, while the general structure of the prototype is illustrated in Fig. 3.10b. Moreover, all the relevant parameters of the measurement setup and the digital cancellers are listed in Table 3.3. The used measurement location is in the Kampusareena library, at Tampere University, meaning that it represents a realistic indoor deployment scenario with various reflecting surfaces. Again, the NI PXIe-5645R VST is used both as the transmitter and the receiver, the TX output signal being also amplified by an external PA (Mini-Circuits ZVE-8G+ [85]). The final transmit power of this prototype is approximately

24 dBm, owing to the high-gain PA. For further technical details regarding the prototype, please refer to [10].

Now, the PA output signal is directly fed to the TX port of the high-isolation antenna, the SI coupling to the RX port via the surrounding reflections and the direct leakage through or around the structure (although the latter is greatly weakened by the wavetraps). As there is no RF canceller in this prototype, the RX port of the antenna structure is then directly connected to the receiver (PXIe-5645R VST) where the received signal is digitized for offline post-processing. Similar to the results in Sect. 3.5.1, the cancellation performance of both the linear and nonlinear digital cancellers is evaluated when receiving only the SI from the own transmitter, i.e., no signal of interest is present.

The PSDs of the signal after the different SI cancellation stages are shown in Fig. 3.11 for all the considered bandwidths, using again the RX input as the reference point for the different power levels. Investigating first the isolation provided by the antenna, it can be observed that it is in the order of 60 dB for each bandwidth when considering also the cable losses in the RX path (4 dB). Hence, the power of the residual SI signal entering the digital domain is roughly −40 dBm in all the cases, which is already sufficiently low for the RX chain to operate without saturation or excessive distortion. Such a power level is also within the dynamic range of the ADC, meaning that the quantization noise remains well below the receiver noise floor. Consequently, no active cancellation is required before the ADC.

Investigating then the digital cancellation performance, it is clear that also now the linear digital canceller is incapable of fully suppressing the residual SI. Namely, the residual SI power after linear cancellation is still 10–15 dB above the receiver noise floor, rendering the IBFD operation infeasible. On the other hand, the nonlinear canceller with LMS-based parameter learning is capable of very efficient SI cancellation with all the considered bandwidths, reducing the overall noise-plus-interference power practically to the level of the receiver noise floor. Hence, the amount of digital SI cancellation is beyond 40 dB for each considered bandwidth and, taking into account the RX cable losses, the overall amounts of SI suppression are 106 dB, 103 dB, and 100 dB over 20 MHz, 40 MHz, and 80 MHz, respectively.

However, as opposed to the results in Sect. 3.5.1, now the accuracy of the LS parameter estimates is observably lower than those obtained with the LMS algorithm. This likely stems from the more time-variant nature of the measurement environment, resulting in the SI channel changing even during the rather short recorded signal sequence. Consequently, tracking of the SI channel coefficients is required, and for this reason the LMS is better suited for the parameter estimation in this case, thanks to its adaptive nature.

In general, these measurement results obtained with an IBFD relay prototype demonstrate the flexibility of the developed LMS-based nonlinear digital cancellation algorithm, as it is shown to be capable of efficient SI cancellation even without an RF cancellation stage. This greatly reduces the overall complexity of the IBFD device as only one active cancellation stage is required, albeit the digital canceller requires now more memory taps due to the higher cancellation demands.

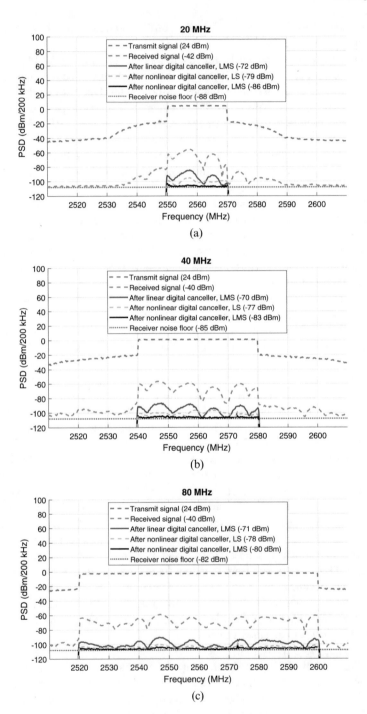

Fig. 3.11 The PSDs after the different SI cancellation stages in the IBFD relay for (**a**) 20-MHz bandwidth, (**b**) 40-MHz bandwidth, and (**c**) 80-MHz bandwidth. The absolute powers are measured over the useful signal bandwidth.

Nevertheless, no additional RF hardware is required, which likely results in a lower overall cost for the transceiver.

3.6   Conclusions

In this chapter, state-of-the-art digital cancellation solutions were presented and also experimentally verified. It was shown that accurate digital cancellation requires the modeling of the power amplifier-induced nonlinear distortion. To this end, a nonlinear signal model was proposed, which can be used to reconstruct an accurate replica of the observed self-interference signal even under a very nonlinear power amplifier. Moreover, two alternative parameter estimation solutions were presented, one based on least squares and another based on least mean squares. The former is a block-wise procedure while the latter is an iterative algorithm for adaptively estimating the SI channel coefficients.

The proposed digital cancellation solution was then evaluated with real-life RF measurements, using two alternative full-duplex prototype architectures. In the first prototype, which utilized a shared TX/RX antenna and an active RF canceller, the nonlinear digital canceller could suppress the SI to the level of the receiver noise floor with bandwidths ranging from 20 to 80 MHz. With the used transmit power levels, this translates to roughly 90 dB of SI suppression. The performance of the nonlinear digital canceller was also successfully evaluated in an IBFD relay with a high-isolation back-to-back antenna. There, even when using only a digital canceller without any active cancellation in the analog domain, the SI was cancelled to the level of the receiver noise floor, again for bandwidths between 20 and 80 MHz. This means that the total amount of SI suppression, obtained with only physical antenna isolation and active digital cancellation, was in the order of 100–110 dB. These are some of the highest reported overall SI cancellation performances to date.

Considering then potential future work items, efficient SI cancellation with transmit powers beyond 30 dBm is an area that still requires further research. Especially, the results reported this chapter have been achieved using transmit powers of 24 dBm or lower, which is insufficient for applications that require a wide area of coverage, such as macro cells with IBFD-capable base stations. Employing such high transmit powers is likely to require even more advanced signal models for sufficiently accurate SI regeneration, which also emphasizes the role of the different complexity reduction schemes to maintain the computational requirements of the digital cancellation procedure on a reasonable level.

Acknowledgements We would like to acknowledge the financial support received from the Tampere University of Technology Graduate School, Nokia Foundation, Tuula and Yrjö Neuvo Research Fund, Emil Aaltonen Foundation, and Pekka Ahonen Fund. In addition, we also wish to acknowledge the funding received from Academy of Finland (under the projects #259915 "In-Band Full-Duplex MIMO Transmission: A Breakthrough to High-Speed Low-Latency Mobile Networks", #301820 "Competitive Funding to Strengthen University Research Profiles", and #304147 "In-Band Full-Duplex Radio Technology: Realizing Next Generation Wireless Transmission"),

Finnish Funding Agency for Technology and Innovation (Tekes, under the projects "Full-Duplex Cognitive Radio" and "TAKE-5"), and Intel Corporation.

References

1. D. W. Bliss, P. A. Parker, and A. R. Margetts, "Simultaneous transmission and reception for improved wireless network performance," in *Proc. 14th IEEE/SP Workshop on Statistical Signal Processing (SSP)*, Aug. 2007, pp. 478–482.
2. M. Duarte and A. Sabharwal, "Full-duplex wireless communications using off-the-shelf radios: Feasibility and first results," in *Proc. 44th Asilomar Conference on Signals, Systems, and Computers (ASILOMAR)*, Nov. 2010, pp. 1558–1562.
3. J. I. Choi, M. Jain, K. Srinivasan, P. Levis, and S. Katti, "Achieving single channel full duplex wireless communication," in *Proc. 16th Annual International Conference on Mobile Computing and Networking (MobiCom)*, Sep. 2010, pp. 1–12.
4. Z. Zhang, K. Long, A. V. Vasilakos, and L. Hanzo, "Full-duplex wireless communications: Challenges, solutions, and future research directions," *Proceedings of the IEEE*, vol. 104, no. 7, pp. 1369–1409, Jul. 2016.
5. A. Sabharwal, P. Schniter, D. Guo, D. W. Bliss, S. Rangarajan, and R. Wichman, "In-band full-duplex wireless: Challenges and opportunities," *IEEE Journal on Selected Areas in Communications*, vol. 32, no. 9, pp. 1637–1652, Sep. 2014.
6. D. Korpi, T. Riihonen, V. Syrjälä, L. Anttila, M. Valkama, and R. Wichman, "Full-duplex transceiver system calculations: analysis of ADC and linearity challenges," *IEEE Transactions on Wireless Communications*, vol. 13, no. 7, pp. 3821–3836, Jul. 2014.
7. L. Anttila, D. Korpi, V. Syrjälä, and M. Valkama, "Cancellation of power amplifier induced nonlinear self-interference in full-duplex transceivers," in *Proc. 47th Asilomar Conference on Signals, Systems and Computers (ASILOMAR)*, Nov. 2013, pp. 1193–1198.
8. D. Korpi, M. Heino, C. Icheln, K. Haneda, and M. Valkama, "Compact inband full-duplex relays with beyond 100 dB self-interference suppression: Enabling techniques and field measurements," *IEEE Transactions on Antennas and Propagation*, vol. 65, pp. 960–965, Feb. 2017.
9. D. Korpi, J. Tamminen, M. Turunen, T. Huusari, Y.-S. Choi, L. Anttila, S. Talwar, and M. Valkama, "Full-duplex mobile device: Pushing the limits," *IEEE Communications Magazine*, vol. 54, no. 9, pp. 80–87, Sep. 2016.
10. D. Korpi, "Full-duplex wireless: Self-interference modeling, digital cancellation, and system studies," Ph.D. dissertation, Tampere University of Technology, Dec. 2017.
11. D. Pozar, *Microwave Engineering*. Wiley, 2012.
12. M. Jain, J. I. Choi, T. Kim, D. Bharadia, S. Seth, K. Srinivasan, P. Levis, S. Katti, and P. Sinha, "Practical, real-time, full duplex wireless," in *Proc. 17th Annual International Conference on Mobile computing and Networking (MobiCom)*, Sep. 2011, pp. 301–312.
13. D. Bharadia, E. McMilin, and S. Katti, "Full duplex radios," in *Proc. SIGCOMM'13*, Aug. 2013, pp. 375–386.
14. V. Tapio, M. Juntti, A. Pärssinen, and K. Rikkinen, "Real time adaptive RF and digital self-interference cancellation for full-duplex transceivers," in *Proc. 50th Asilomar Conference on Signals, Systems and Computers (ASILOMAR)*, Nov. 2016, pp. 1558–1562.
15. M. S. Amjad and O. Gurbuz, "Linear digital cancellation with reduced computational complexity for full-duplex radios," in *Proc. IEEE Wireless Communications and Networking Conference (WCNC)*, Mar. 2017.
16. W. Chung, D. Hong, R. Wichman, and T. Riihonen, "Interference cancellation architecture for full-duplex system with GFDM signaling," in *Proc. 24th European Signal Processing Conference (EUSIPCO)*, Aug. 2016, pp. 788–792.

17. M. Chung, M. S. Sim, J. Kim, D. K. Kim, and C. b. Chae, "Prototyping real-time full duplex radios," *IEEE Communications Magazine*, vol. 53, no. 9, pp. 56–63, Sep. 2015.
18. D. Wu, C. Zhang, S. Gao, and D. Chen, "A digital self-interference cancellation method for practical full-duplex radio," in *Proc. IEEE International Conference on Signal Processing, Communications and Computing (ICSPCC)*, Aug. 2014, pp. 74–79.
19. M. S. Sim, M. Chung, D. K. Kim, and C. B. Chae, "Low-complexity nonlinear self-interference cancellation for full-duplex radios," in *Proc. IEEE Globecom Workshops*, Dec. 2016.
20. A. C. M. Austin, A. Balatsoukas-Stimming, and A. Burg, "Digital predistortion of power amplifier non-linearities for full-duplex transceivers," in *Proc. 17th IEEE International Workshop on Signal Processing Advances in Wireless Communications (SPAWC)*, Jul. 2016.
21. E. Ahmed, A. M. Eltawil, and A. Sabharwal, "Self-interference cancellation with nonlinear distortion suppression for full-duplex systems," in *Proc. 47th Asilomar Conference on Signals, Systems and Computers (ASILOMAR)*, Nov. 2013, pp. 1199–1203.
22. D. W. Bliss and Y. Rong, "Full-duplex self-interference mitigation performance in nonlinear channels," in *Proc. 48th Asilomar Conference on Signals, Systems and Computers (ASILO-MAR)*, Nov. 2014, pp. 1696–1700.
23. M. Emara, M. Faerber, L. G. Baltar, J. Nossek, and K. Roth, "Nonlinear digital self-interference cancellation with reduced complexity for full duplex systems," in *Proc. International ITG Workshop on Smart Antennas (WSA)*, Mar. 2017.
24. Z. Luan, H. Qu, J. Zhao, and B. Chen, "Robust digital non-linear self-interference cancellation in full duplex radios with maximum correntropy criterion," *China Communications*, vol. 13, no. 9, pp. 53–59, Sep. 2016.
25. D. Korpi, Y.-S. Choi, T. Huusari, S. Anttila, L. Talwar, and M. Valkama, "Adaptive nonlinear digital self-interference cancellation for mobile inband full-duplex radio: algorithms and RF measurements," in *Proc. IEEE Global Communications Conference (GLOBECOM)*, Dec. 2015.
26. L. Anttila, D. Korpi, E. Antonio-Rodríguez, R. Wichman, and M. Valkama, "Modeling and efficient cancellation of nonlinear self-interference in MIMO full-duplex transceivers," in *Proc. IEEE Globecom Workshops*, Dec. 2014, pp. 862–868.
27. D. Korpi, L. Anttila, and M. Valkama, "Asymmetric full-duplex with contiguous downlink carrier aggregation," in *Proc. 17th International Workshop on Signal Processing Advances in Wireless Communications (SPAWC)*, Jul. 2016.
28. M. Heino, D. Korpi, T. Huusari, E. Antonio-Rodríguez, S. Venkatasubramanian, T. Riihonen, L. Anttila, C. Icheln, K. Haneda, R. Wichman, and M. Valkama, "Recent advances in antenna design and interference cancellation algorithms for in-band full-duplex relays," *IEEE Communications Magazine*, vol. 53, no. 5, pp. 91–101, May 2015.
29. W. Zhao, C. Feng, F. Liu, C. Guo, and Y. Nie, "Polarization mismatch based self-interference cancellation against power amplifier nonlinear distortion in full duplex systems," in *Proc. 26th Annual IEEE International Symposium on Personal, Indoor, and Mobile Radio Communications (PIMRC)*, Aug. 2015, pp. 256–260.
30. A. Balatsoukas-Stimming, "Non-linear digital self-interference cancellation for in-band full-duplex radios using neural networks," in *Proc. IEEE 19th International Workshop on Signal Processing Advances in Wireless Communications (SPAWC)*, Jun. 2018.
31. Y. Kurzo, A. Burg, and A. Balatsoukas-Stimming, "Design and implementation of a neural network aided self-interference cancellation scheme for full-duplex radios," in *Proc. 52nd Asilomar Conference on Signals, Systems, and Computers*, Oct. 2018, pp. 589–593.
32. H. Guo, J. Xu, S. Zhu, and S. Wu, "Realtime software defined self-interference cancellation based on machine learning for in-band full duplex wireless communications," in *Proc. International Conference on Computing, Networking and Communications (ICNC)*, Mar. 2018, pp. 779–783.
33. M. Duarte, C. Dick, and A. Sabharwal, "Experiment-driven characterization of full-duplex wireless systems," *IEEE Transactions on Wireless Communications*, vol. 11, no. 12, pp. 4296–4307, Dec. 2012.

34. M. Duarte, A. Sabharwal, V. Aggarwal, R. Jana, K. Ramakrishnan, C. Rice, and N. Shankara-narayanan, "Design and characterization of a full-duplex multiantenna system for WiFi networks," *IEEE Transactions on Vehicular Technology*, vol. 63, no. 3, pp. 1160–1177, Mar. 2014.

35. M. Duarte, "Full-duplex wireless: Design, implementation and characterization," Ph.D. dissertation, Rice University, 2012.

36. E. Everett, A. Sahai, and A. Sabharwal, "Passive self-interference suppression for full-duplex infrastructure nodes," *IEEE Transactions on Wireless Communications*, vol. 13, no. 2, pp. 680–694, Feb. 2014.

37. D. Bharadia and S. Katti, "Full duplex MIMO radios," in *Proc. 11th USENIX Conference on Networked Systems Design and Implementation (NSDI)*, Apr. 2014, pp. 359–372.

38. B. Debaillie, D. J. van den Broek, C. Lavín, B. van Liempd, E. A. M. Klumperink, C. Palacios, J. Craninckx, B. Nauta, and A. Pärssinen, "Analog/RF solutions enabling compact full-duplex radios," *IEEE Journal on Selected Areas in Communications*, vol. 32, no. 9, pp. 1662–1673, Sep. 2014.

39. M. Valkama, M. Renfors, and V. Koivunen, "Advanced methods for I/Q imbalance compensation in communication receivers," *IEEE Transactions on Signal Processing*, vol. 49, no. 10, pp. 2335–2344, Oct. 2001.

40. J. K. Cavers and M. W. Liao, "Adaptive compensation for imbalance and offset losses in direct conversion transceivers," *IEEE Transactions on Vehicular Technology*, vol. 42, no. 4, pp. 581–588, Nov. 1993.

41. ETSI, "LTE; evolved universal terrestrial radio access (E-UTRA); user equipment (UE) radio transmission and reception (3GPP TS 36.101 version 14.3.0 release 14)," Sophia Antipolis Cedex, France, Jan. 2017.

42. D. Korpi, L. Anttila, V. Syrjälä, and M. Valkama, "Widely linear digital self-interference cancellation in direct-conversion full-duplex transceiver," *IEEE Journal on Selected Areas in Communications*, vol. 32, no. 9, pp. 1674–1687, Sep. 2014.

43. S. Li and R. D. Murch, "An investigation into baseband techniques for single-channel full-duplex wireless communication systems," *IEEE Transactions on Wireless Communications*, vol. 13, no. 9, pp. 4794–4806, Sep. 2014.

44. Y. Rahmatallah and S. Mohan, "Peak-to-average power ratio reduction in OFDM systems: A survey and taxonomy," *IEEE Communications Surveys Tutorials*, vol. 15, no. 4, pp. 1567–1592, Fourth Quarter 2013.

45. T. Jiang and Y. Wu, "An overview: Peak-to-average power ratio reduction techniques for OFDM signals," *IEEE Transactions on Broadcasting*, vol. 54, no. 2, pp. 257–268, Jun. 2008.

46. ETSI, "LTE; evolved universal terrestrial radio access (E-UTRA); base station (BS) radio transmission and reception (3GPP TS 36.104, version 14.3.0, release 14)," Sophia Antipolis Cedex, France, Mar. 2017.

47. F. M. Ghannouchi and O. Hammi, "Behavioral modeling and predistortion," *IEEE Microwave Magazine*, vol. 10, no. 7, pp. 52–64, Dec. 2009.

48. F. M. Ghannouchi, "Power amplifier and transmitter architectures for software defined radio systems," *IEEE Circuits and Systems Magazine*, vol. 10, no. 4, pp. 56–63, Fourth Quarter 2010.

49. D. Morgan, Z. Ma, J. Kim, M. Zierdt, and J. Pastalan, "A generalized memory polynomial model for digital predistortion of RF power amplifiers," *IEEE Transactions on Signal Processing*, vol. 54, no. 10, pp. 3852–3860, Oct. 2006.

50. A. Abdelhafiz, A. Kwan, O. Hammi, and F. M. Ghannouchi, "Digital predistortion of LTE-A power amplifiers using compressed-sampling-based unstructured pruning of Volterra series," *IEEE Transactions on Microwave Theory and Techniques*, vol. 62, no. 11, pp. 2583–2593, Nov. 2014.

51. A. S. Tehrani, H. Cao, S. Afsardoost, T. Eriksson, M. Isaksson, and C. Fager, "A comparative analysis of the complexity/accuracy tradeoff in power amplifier behavioral models," *IEEE Transactions on Microwave Theory and Techniques*, vol. 58, no. 6, pp. 1510–1520, Jun. 2010.

52. M. Schoukens, R. Pintelon, and Y. Rolain, "Parametric identification of parallel Hammerstein systems," *IEEE Transactions on Instrumentation and Measurement*, vol. 60, no. 12, pp. 3931–3938, Dec. 2011.
53. L. Ding, G. T. Zhou, D. R. Morgan, Z. Ma, J. S. Kenney, J. Kim, and C. R. Giardina, "A robust digital baseband predistorter constructed using memory polynomials," *IEEE Transactions on Communications*, vol. 52, no. 1, pp. 159–165, Jan. 2004.
54. M. Isaksson, D. Wisell, and D. Rönnow, "A comparative analysis of behavioral models for RF power amplifiers," *IEEE Transactions on Microwave Theory and Techniques*, vol. 54, no. 1, pp. 348–359, Jan. 2006.
55. L. Anttila, P. Händel, and M. Valkama, "Joint mitigation of power amplifier and I/Q modulator impairments in broadband direct-conversion transmitters," *IEEE Transactions on Microwave Theory and Techniques*, vol. 58, no. 4, pp. 730–739, Apr. 2010.
56. D. Korpi, M. Valkama, T. Riihonen, and R. Wichman, "Implementation challenges in full-duplex radio transceiver," in *Proc. XXXIII Finnish URSI Convention on Radio Science*, Apr. 2013, pp. 181–184.
57. B. P. Day, A. R. Margetts, D. W. Bliss, and P. Schniter, "Full-duplex bidirectional MIMO: Achievable rates under limited dynamic range," *IEEE Transactions on Signal Processing*, vol. 60, no. 7, pp. 3702–3713, Jul. 2012.
58. Q. Gu, *RF System Design of Transceivers for Wireless Communications*. Springer, 2006.
59. V. Syrjälä, M. Valkama, L. Anttila, T. Riihonen, and D. Korpi, "Analysis of oscillator phase-noise effects on self-interference cancellation in full-duplex OFDM radio transceivers," *IEEE Transactions on Wireless Communications*, vol. 13, no. 6, pp. 2977–2990, Jun. 2014.
60. A. Sahai, G. Patel, C. Dick, and A. Sabharwal, "On the impact of phase noise on active cancelation in wireless full-duplex," *IEEE Transactions on Vehicular Technology*, vol. 62, no. 9, pp. 4494–4510, Nov. 2013.
61. T. Riihonen, P. Mathecken, and R. Wichman, "Effect of oscillator phase noise and processing delay in full-duplex OFDM repeaters," in *Proc. 46th Asilomar Conference on Signals, Systems and Computers (ASILOMAR)*, Nov. 2012, pp. 1947–1951.
62. X. Quan, Y. Liu, S. Shao, C. Huang, and Y. Tang, "Impacts of phase noise on digital self-interference cancellation in full-duplex communications," *IEEE Transactions on Signal Processing*, vol. 65, no. 7, pp. 1881–1893, Apr. 2017.
63. A. Masmoudi and T. Le-Ngoc, "A maximum-likelihood channel estimator for self-interference cancelation in full-duplex systems," *IEEE Transactions on Vehicular Technology*, vol. 65, no. 7, pp. 5122–5132, Jul. 2016.
64. V. Syrjälä, "Analysis and mitigation of oscillator impairments in modern receiver architectures," Ph.D. dissertation, Tampere University of Technology, 2012.
65. R. Durrett, *Probability: Theory and Examples*, 4th ed. Cambridge University Press, 2010.
66. Q. Zou, A. Tarighat, and A. H. Sayed, "Compensation of phase noise in OFDM wireless systems," *IEEE Transactions on Signal Processing*, vol. 55, no. 11, pp. 5407–5424, Nov. 2007.
67. D. Petrovic, W. Rave, and G. Fettweis, "Effects of phase noise on OFDM systems with and without PLL: Characterization and compensation," *IEEE Transactions on Communications*, vol. 55, no. 8, pp. 1607–1616, Aug. 2007.
68. L. Tomba, "On the effect of Wiener phase noise in OFDM systems," *IEEE Transactions on Communications*, vol. 46, no. 5, pp. 580–583, May 1998.
69. D. Korpi, L. Anttila, and M. Valkama, "Impact of received signal on self-interference channel estimation and achievable rates in in-band full-duplex transceivers," in *Proc. 48th Asilomar Conference on Signals, Systems and Computers (ASILOMAR)*, Nov. 2014, pp. 975–982.
70. ——, "Nonlinear self-interference cancellation in MIMO full-duplex transceivers under crosstalk," *EURASIP Journal on Wireless Communications and Networking*, vol. 2017, no. 1, p. 24, Feb. 2017.
71. D. Korpi, T. Riihonen, and M. Valkama, "Achievable rate regions and self-interference channel estimation in hybrid full-duplex/half-duplex radio links," in *Proc. 49th Annual Conference on Information Sciences and Systems (CISS)*, Mar. 2015.

72. D. Korpi, T. Riihonen, K. Haneda, K. Yamamoto, and M. Valkama, "Achievable transmission rates and self-interference channel estimation in hybrid full-duplex/half-duplex MIMO relaying," in *Proc. 82nd IEEE Vehicular Technology Conference (VTC Fall)*, Sep. 2015.

73. S. M. Kay, *Fundamentals of Statistical Signal Processing*. Prentice Hall, 1993.

74. S. Haykin, *Neural Networks: A Comprehensive Foundation*, 2nd ed. Prentice Hall, 1999.

75. ——, *Adaptive Filter Theory*, 3rd ed. Prentice Hall, 1996.

76. A. Hyvärinen, J. Karhunen, and E. Oja, *Independent Component Analysis*. Wiley, 2001.

77. D. Knuth, *The Art of Computer Programming*, 3rd ed. Addison–Wesley, 1997, vol. 1, Fundamental Algorithms.

78. K. Deb, A. Pratap, S. Agarwal, and T. Meyarivan, "A fast and elitist multiobjective genetic algorithm: NSGA-II," *IEEE Transactions on Evolutionary Computation*, vol. 6, no. 2, pp. 182–197, Apr. 2002.

79. S. Verdu, "Minimum probability of error for asynchronous Gaussian multiple-access channels," *IEEE Transactions on Information Theory*, vol. 32, no. 1, pp. 85–96, Jan. 1986.

80. D. Kivanc, G. Li, and H. Liu, "Computationally efficient bandwidth allocation and power control for OFDMA," *IEEE Transactions on Wireless Communications*, vol. 2, no. 6, pp. 1150–1158, Nov. 2003.

81. T. H. Cormen, C. E. Leiserson, R. L. Rivest, and C. Stein, *Introduction to Algorithms*, 3rd ed. MIT Press, 2009.

82. Texas Instruments Incorporated, "CC2595 RF front-end transmit power amplifier for 2.4 GHz ISM band systems," Dallas, Texas, USA.

83. J. Tamminen, M. Turunen, D. Korpi, T. Huusari, Y.-S. Choi, S. Talwar, and M. Valkama, "Digitally-controlled RF self-interference canceller for full-duplex radios," in *Proc. 24th European Signal Processing Conference (EUSIPCO)*, Aug. 2016, pp. 783–787.

84. M. Heino, S. Venkatasubramanian, C. Icheln, and K. Haneda, "Design of wavetraps for isolation improvement in compact in-band full-duplex relay antennas," *IEEE Transactions on Antennas and Propagation*, vol. 64, no. 3, pp. 1061–1070, Mar. 2016.

85. Mini-Circuits, "ZVE-8G+ coaxial amplifier," Brooklyn, New York, USA.

Chapter 4
Filter Design for Self-Interference Cancellation

Risto Wichman

Abstract Mitigation of self-interference is the prime challenge in making full-duplex technology feasible in wireless communications. In this chapter, we first present system model of a wireless communication link including a source, full-duplex transceiver, and destination, and discuss different approaches and assumptions when building the signal model. Then we will present frequency, time, and spatial signal processing techniques to mitigate self-interference in digital baseband. We concentrate on time-domain filtering instead of frequency-domain filtering on a subcarrier basis, because time-domain filtering need not assume OFDM waveforms or synchronization between transmitted and received signals. We complement the time-domain filtering with spatial filters that together are able to mitigate the effects of non-linear transmitter distortions using only linear operations.

4.1 Motivation

In ideal case, full-duplex link doubles the spectral efficiency of half-duplex link being able to transmit and receive simultaneously within the same radio resource. To compensate the loss half-duplex transmitter should approximately double the transmit power (assuming high signal-to-noise (SNR)) regime when compared to the full-duplex transmitter. Conversely, assuming that the transmitted signal powers are fixed, full-duplex link outperforms half-duplex link if the power of noise and self-interference in the full-duplex receiver is less than double of the noise power in the half-duplex receiver. Therefore, to make full-duplex technology useful, self-interference cancellation and suppression should push the residual self-interference below noise level.

R. Wichman (✉)
Aalto University School of Electrical Engineering, Espoo, Finland
e-mail: risto.wichman@aalto.fi

© Springer Nature Singapore Pte Ltd. 2020   99
H. Alves et al. (eds.), *Full-Duplex Communications for Future Wireless Networks*,
https://doi.org/10.1007/978-981-15-2969-6_4

Self-interference cancellation techniques can be classified as passive and active. The former refers to spatial separation of transmit and receive antennas, and antenna design techniques to minimize over-the-air leakage of the transmitted signal to the receiver chain. Active cancellation techniques can be further divided into radio frequency (RF), or analog cancellation, and in digital baseband cancellation.

Isolation and analog interference cancellation must be applied before digital cancellation, because the dynamic range of any front-end circuitry is finite and large difference in power levels may saturate the receiver. By partitioning transmit and receive antennas into separate arrays, physical isolation arises from the path-loss between the transmit and receive antenna arrays. Furthermore, polarization, directivity, antenna design, and geometry of transmit and receive antenna arrays can be used to improve the attenuation between the transmitter and the receiver chains. In case of sharing the antenna between the transmitter and the receiver, circulators or electrical balance duplexers are used to provide the isolation between receiver and transmitter chains.

Next, self-interference should be mitigated further in radio frequency (RF) domain to avoid saturation of analog-to-digital converter (ADC) by the self-interference signal. The principle of the self-interference cancellation in analog and digital domains is the same: a replica of the interfering signal is created and then subtracted from the received signal. In RF domain the replica is generated by an auxiliary transmitter chain, or by splitting the transmitted RF signal.

As long as the self-interference signals fit to the dynamic range of the ADC, cancellation of self-interference in digital baseband is agnostic to physical isolation and RF cancellation techniques. In practice, the effect of the cancellation techniques is not strictly additive, though. If the interfering signal is already weak when converted into digital baseband, the estimation of the self-interference channel becomes more difficult and the estimate of the replica signal becomes less accurate, due to the noisy channel estimate.

The received signal of a full-duplex transceiver may be more than 100 dB weaker than the transmitted signal in cellular systems. For example, in LTE, the maximum transmit power of the user equipment is 23 dBm while the receiver's noise floor is −97 dBm assuming 10 MHz received signal bandwidth and 7 dB noise figure. Thus, there is a 120 dB difference between the transmitted signal level and the noise floor. Empirical results suggest that physical isolation between transmitter and receiver chains may be insufficient for high-rate full-duplex transmission with high power and long range. A straightforward way to reduce the self-interference is to decrease the power of the transmitted signal or bring the end points of the communications links closer to each other. This would compromise transmission range and throughput and push full-duplex transceivers into margin in cellular systems. This gives a good motivation to develop analog and digital interference cancellation schemes for full-duplex transceivers.

4.2   System Model

Figure 4.1 depicts a generic communication link in digital baseband involving a full-duplex multiantenna transceiver. The source S and destination D have N_S transmit and N_D receive antennas, respectively, and the full-duplex transceiver T is equipped with N_{rx} receive and N_{tx} transmit antennas. In practice, full-duplex transceivers are likely to be implemented with spatially separated receive and transmit arrays which yields physical isolation. This is not viable for handsets due to their small form factor, but is more amenable when transceiver is a part of a wireless infrastructure, i.e., it is a relay or an access point. However, the following signal model is also applicable for full-duplex transmission with a shared antenna array by setting $N_{rx} = N_{tx}$ and assuming high gain for the self-interference channel $\mathbf{H}_i[n]$.

Matrices $\mathbf{H}_{rx}[n] \in \mathbb{C}^{N_{rx} \times N_S}$ and $\mathbf{H}_{tx}[n] \in \mathbb{C}^{N_D \times N_{tx}}$ may represent combined MIMO channels in digital baseband from all sources to the transceiver, and from the transceiver to all destinations. In what follows, we assume a single link where the destination and the source are equipped with multiple antennas, but the model can be interpreted to comprise multiple source and destination nodes as well. All channels may vary between the transmitted symbols.

At time instant (subcarrier index) n, the source transmits signal vector $\mathbf{x}_S[n] \in \mathbb{C}^{N_S}$, and the transceiver transmits signal vector $\mathbf{x}_T[n] \in \mathbb{C}^{N_{tx}}$ while simultaneously receiving signal vector $\mathbf{y}_T[n] \in \mathbb{C}^{N_{rx}}$. This creates an unavoidable feedback loop from the transceiver output to the transceiver input through channel $\mathbf{H}_i[n] \in \mathbb{C}^{N_{rx} \times N_{tx}}$. For example, if the full-duplex transceiver is assumed to be an access point in a wireless network, $\mathbf{H}_{rx}[n]$ and $\mathbf{H}_{tx}[n]$ correspond to uplink and downlink channel, respectively. Finally, the respective received signals in the full-duplex transceiver and in a destination node in digital baseband can be expressed as

$$\mathbf{y}_T[n] = \mathbf{H}_{rx}[n]\,\mathbf{x}_S[n] + \mathbf{H}_i[n]\,\mathbf{x}_T[n] + \mathbf{n}_T[n],$$

$$\mathbf{y}_D[n] = \mathbf{H}_{tx}[n]\,\mathbf{x}_T[n] + \mathbf{n}_D[n] \tag{4.1}$$

where $\mathbf{n}_T[n] \in \mathbb{C}^{N_{rx}}$ and $\mathbf{n}_D[n] \in \mathbb{C}^{N_D}$ are additive noise vectors in the transceiver and in the destination, respectively. They can be further decomposed to different noise components due to hardware imperfections.

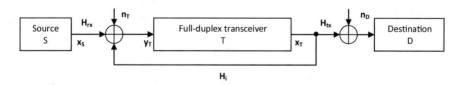

Fig. 4.1 System model of a full-duplex communication link. Subcarrier (or time) index n has been excluded for simplicity of the notation

This model is viable in frequency-selective channels when subcarriers remain approximately orthogonal in the receiver. In OFDM, this requires that phase noise is small enough, and carrier frequency offset has been estimated and compensated. As a rule of thumb, bandwidth of the phase noise and carrier frequency offset should be less than 10% of the subcarrier width for the assumption to hold. In addition, channel coherence time should exceed the length of the OFDM symbol such that channel does not change within an OFDM symbol. If the channel must be equalized, the benefits of the OFDM design and simple equalization are lost. In the same vein, the length of cyclic prefix (CP) should exceed the length of the multipath channel.

The assumptions are plenty, but they are commonly accepted in engineering literature. Several channel measurement and simulation campaigns in projected usage scenarios ensure that system parameters are selected in a way that the assumptions hold in commercial wireless systems. Deviations from the ideal case can be included into implementation margin among other imperfections in receiver implementation. In system level studies the implementation margin may be, say, 3 dB which gives the penalty factor for the received SINR when compared to the ideal case.

The model applies also in time domain in case of flat fading, single path, channels. In this case, the channel matrices and transmitted symbols are simply interpreted as time-domain variables. A signal model for multipath channels in time domain is presented in Sect. 4.2.6.

4.2.1   Dynamic Range

Basic building blocks of a full-duplex transceiver without detailing self-interference cancellation are depicted in Fig. 4.2. For simplicity of presentation, the transceiver employs separate transmit and receive antennas, one for each direction.

First of all, the signal of interest $x_S[n]$ (part of the received baseband signal) together with the self-interference should fit to the dynamic range of the receiver chain. If the analog cancellation signal is injected into the receiver chain after low-

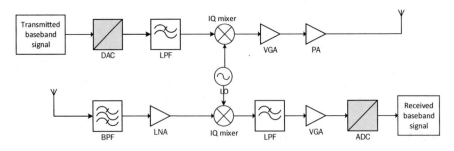

Fig. 4.2 Full-duplex transceiver without self-interference cancellation

noise amplifier (LNA), the self-interference signal should fit to the dynamic range of the LNA. Alternatively, cancellation signal can be injected before LNA, which relaxes the requirements of the dynamic range of the LNA, but increases the noise figure of the receiver. Noise figure is caused by receiver electronics and it describes the degradation of the signal-to-noise ratio in a real receiver when compared to ideal receiver.

If self-interference saturates the ADC, the signal of interest is lost and self-interference cancellation in digital baseband is not able to recover the signal of interest. The dynamic range of ADC is typically smaller than that of LNA and therefore it presents the limiting factor in the receiver chain. The task of the variable gain amplifier (VGA) is to control the received signal such that it stays within limits. The dynamic range of ADC is given by Rice [1]

$$\gamma = 6.02b + 4.76 - 10\log_{10}\frac{x_{max}}{\mathcal{E}\{x[n]\}} \ [\text{dB}]$$

where b is the number of bits in ADC, and the last term refers to the estimated peak-to-average power ratio (PAPR) of the input signal $x[n]$, be it due to the self-interference or due to the signal of interest. This assumes that the input signal is not clipped, and additional headroom has been reserved to accommodate signal peaks. Typically one bit is reserved for this, although PAPR depends on the modulation and waveform used. In case of OFDM signal that can be assumed Gaussian when the number of subcarriers is large, optimized dynamic range in terms of quantization range and clipping noise can be approximated as [2]

$$\gamma = 5.54b - 3.26 \ [\text{dB}]$$

In addition to reserving headroom to avoid excessive clipping, another quantization bit is typically reserved to avoid the system from being quantization limited but noise limited. For example, if ADC has 14 bits, effective number of bits being 11, the dynamic range of ADC becomes 54 dB. Therefore, passive isolation together with RF cancellation should attenuate the transmitted signal at least by 66 dB when assuming 120 dB difference between receiver noise floor and the transmit power.

When designing the algorithms in digital baseband it is usually assumed that the signal of interest and the self-interference fit to the digital baseband such that the clipping noise can be ignored in the design. However, quantization noise $\delta\mathbf{y}_T[n]$ can be included in the signal model in addition to receiver noise term as $\mathbf{n}_T[n]$ in (4.1).

4.2.1.1 Quantization Noise

Uniform distribution is generally considered a good model for quantization noise [3]. Assuming that the signal is not saturated and its dynamic range equals $2x_{max}$, a uniform quantizer with b bits partitions the signal in 2^b levels, and the width of each level becomes $\Delta = 2x_{max}/2^b$. Uniformly distributed quantization error is bounded in $[-\Delta/2, \Delta/2)$ and the variance of the error is given by $\sigma^2 = \Delta/12$.

For analysis purposes it is often more convenient to model the quantization noise as Gaussian distributed [4, 5]. This has been justified by assuming that in addition to quantization, the transceiver is subject to several other impairments making their joint contribution Gaussian. When the effect of quantization noise is studied separately from the thermal noise its contribution can be included in $\delta\mathbf{y}_T[n]$ such that the system model (4.1) becomes

$$\mathbf{y}_T[n] = \mathbf{H}_{rx}[n]\mathbf{x}_S[n] + +\mathbf{H}_i[n]\mathbf{x}_T[n] + \delta\mathbf{y}_T[n] + \mathbf{n}_T[n],$$

$$\mathbf{y}_D[n] = \mathbf{H}_{tx}[n]\mathbf{x}_T[n] + \mathbf{n}_D[n] \tag{4.2}$$

The covariance of the quantization noise is defined by

$$\mathbf{R}_{\delta\mathbf{y}_T} \triangleq \mathcal{E}\{\delta\mathbf{y}_T[n]\delta\mathbf{y}_T[n]^\dagger\} = \epsilon_\mathbf{y}\frac{\mathcal{E}\left\{\|\mathbf{y}_T(t)\|^2\right\}}{N_{rx}}\mathbf{I} \tag{4.3}$$

where $\mathbf{y}_T(t)$ is the received signal before analog-to-digital conversion.

4.2.2 Transmit Signal Noise

The transceiver is obviously able to know the digital baseband signal, denoted by $\tilde{\mathbf{x}}_T[n]$, it has generated, but the corresponding transmitted analog passband signal $\mathbf{x}_T(t)$ is not perfectly known any more, because signal conversion between baseband and radio frequencies is prone to various distortion effects. In (4.1) signal $\mathbf{x}_T[n]$ represents this analog signal in digital baseband after down-conversion and sampling in the receiver. Therefore, in addition to $\tilde{\mathbf{x}}_T[n]$, it contains the unknown noise term due to RF imperfections as well. As for the analog signal $\mathbf{x}_T(t)$, it is digital-to-analog converted, up-converted and amplified version of the transmitted digital baseband signal $\tilde{\mathbf{x}}_T[n]$. As shown in Fig. 4.2 the received signal is first bandpass filtered, amplified with LNA, and down-converted directly to the baseband and sampled. The in-phase (I) and quadrature (Q) components of the complex signal are then separately low-pass filtered, amplified, and sampled. This kind of direct-conversion architecture is simple, but it suffers from some characteristic impairments like direct-current (DC) offset, I/Q imbalance, and second order intermodulation. In addition, DAC adds quantization noise to the signal at digital baseband. However, power amplifier nonlinearity is typically the main source of the distortion in the transmitted signal. Together these components make up the unknown part of self-interference, denoted by $\delta\mathbf{x}_T[n]$, that makes cancellation of the self-interference challenging.

In addition to in-band distortion causing self-interference in full-duplex transceivers, non-linear distortion of the transmitted signal causes spectral regrowth and therefore adjacent channel interference. However, this out-of-band distortion is the same for half-duplex and full-duplex transceivers. The only exception to this is full-duplex amplify-and-forward relay, which amplifies the residual non-linear distortion after self-interference cancellation. Therefore, the transmitted signal is different when compared to half-duplex amplify-and-forward relay.

There are two approaches to cope with the RF distortion in the design and analysis of self-interference cancellation algorithms in digital baseband: One option is to model transmit imperfections in detail using behavioral models [6] in digital baseband. After the imperfections have been modeled their effect is included in the replica of the self-interference signal and subtracted from the received signal. This reduces the power of the remaining $\delta x_T[n]$ but increases the complexity of signal processing. Moreover, one can argue that if the PA nonlinearity can be reliably identified in digital domain, it would be more useful to predistort the transmitted signal based on the identified nonlinearity [7]. In addition, to improving the SINR in the full-duplex receiver, predistortion will reduce the out-of-band interference and improve SNR in the destination as well. Using the behavioral models for self-interference cancellation only improves the performance of the receiver but not the transmitter. In any case, there is a large body of literature on behavioral modeling for self-interference cancellation [8–10] and several prototypes implementing non-linear cancellation have been built and demonstrated as well [11–14].

Another option is to model the joint effect of RF impairments in digital baseband $\delta x_T[n]$ as an additional noise source. The transmitted signal is simply expressed as

$$\mathbf{x}_T[n] = \tilde{\mathbf{x}}_T[n] + \delta \mathbf{x}_T[n] \tag{4.4}$$

and the elements of $\delta \mathbf{x}_T[n]$ are modeled as independent identically distributed (i.i.d.) circularly symmetric complex Gaussian random variables. Consequently, the covariance matrix of transmit signal noise can be expressed as

$$\mathbf{R}_{\delta \mathbf{x}_T} \triangleq \mathcal{E}\{\delta \mathbf{x}_T[n] \delta \mathbf{x}_T[n]^\dagger\} = \epsilon_{\mathbf{x}} \frac{\text{tr}\{\mathbf{R}_{\tilde{\mathbf{x}}_T}\}}{N_{\text{tx}}} \mathbf{I} \tag{4.5}$$

in which $\mathbf{R}_{\tilde{\mathbf{x}}_T} \triangleq \mathcal{E}\{\tilde{\mathbf{x}}_T[n]\tilde{\mathbf{x}}_T[n]^\dagger\}$ and the variance of $\delta \mathbf{x}_T[n]$ is defined with relative distortion level $\epsilon_{\mathbf{x}}$ and normalized by the number of transmit antennas N_{tx}. In addition, $\tilde{\mathbf{x}}_T[n]$ and $\delta \mathbf{x}_T[n]$ are assumed to be uncorrelated which implies that $\mathbf{R}_{\mathbf{x}_T} = \mathbf{R}_{\tilde{\mathbf{x}}_T} + \mathbf{R}_{\delta \mathbf{x}_T}$. The level of transmit distortion is typically well below the actual data signal in typical wireless communication systems, i.e., $\epsilon_{\mathbf{x}} \ll 1$.

4.2.2.1   Error Vector Magnitude

In conformance tests of wireless transceivers, distortion in the transmitted signal $\delta \mathbf{x}_T[n]$ is typically quantified as error-vector magnitude (EVM) that sums up the effects of different impairments. The EVM is experimentally measured by

$$\text{EVM} = \frac{\sum_{n=0}^{N-1} \sqrt{|x[n] - y[n]|^2}}{\sum_{n=0}^{N-1} |y[n]|^2} \tag{4.6}$$

where $x[n]$ refers to transmitted symbol sequence in digital baseband, $y[n]$ denotes the symbol sequence after transmission and demodulation, and N is the number of samples used in the test. Specifications define maximum EVM limits for certified devices depending on modulation, because higher order modulation requires smaller EVM levels. For example, in LTE, maximum allowed EVM levels for base station transmitter are 17.5%, 12.5%, and 8% for QPSK, 16 QAM, and 64 QAM, respectively [15]. These figures correspond to -22 to $-15\,dB$ power levels for the distortion w.r.t. the transmitted baseband symbols, i.e., $x[n]$. LTE small cell enhancement targeting home, local area, and medium range base stations specifies even 256 QAM requiring better than 3.5% EVM level, i.e., better than $-30\,dB$. As a reference high-end spectral analyzers achieve EVM levels of the order of $-45\,dB$. Therefore, even if thermal noise in the receiver is assumed to be zero, EVM level in the transmitter defines the maximum SNR that receiver can ever achieve. In addition, EVM in the receiver degrades the received SNR even more.

When transmitter and interferers are located within comparable distance from the receiver, EVM levels as specified in, e.g., LTE are tolerable, and EVM value is inversely proportional to the maximum SINR value the receiver can experience. However, when the interfering signal is much stronger than the signal of interest as may happen within full-duplex transceivers, these kind of EVM levels may seriously degrade the performance of a full-duplex transceiver.

4.2.3 Channel Estimation Error

A full-duplex transceiver may use any regular channel estimation technique or techniques developed specifically for full-duplex transceivers to obtain respective channel estimates $\hat{\mathbf{H}}_i[n]$ and $\hat{\mathbf{H}}_{rx}[n]$ of the channels $\mathbf{H}_i[n]$ and $\mathbf{H}_{rx}[n]$. However, strong self-interference component $\mathbf{H}_i[n]\mathbf{x}_T[n]$ may create additional problems when estimating the signal of interest $\mathbf{H}_{rx}[n]\mathbf{x}_S[n]$. Treating self-interference as additional noise would result in a large estimator variance. The same happens when estimating self-interference channel $\mathbf{H}_i[n]$ in the presence of the signal of interest. The worse the estimate of the self-interference channel is, the more residual self-interference remains after cancellation and suppression.

Channel estimation can be modeled with additive error coefficients $\delta\mathbf{H}_i[n]$ and $\delta\mathbf{H}_{rx}[n]$ so that the estimates $\hat{\mathbf{H}}_i[n]$ and $\hat{\mathbf{H}}_{rx}[n]$ differ from physical channels as

$$\hat{\mathbf{H}}_i[n] = \mathbf{H}_i[n] + \delta\mathbf{H}_i[n]$$

$$\hat{\mathbf{H}}_{rx}[n] = \mathbf{H}_{rx}[n] + \delta\mathbf{H}_{rx}[n] \tag{4.7}$$

All elements of $\delta\mathbf{H}_i[n]$ and $\delta\mathbf{H}_{rx}[n]$ are assumed to be mutually independent circularly symmetric complex Gaussian random variables. In case of MMSE channel estimation, the channel estimates and the errors are uncorrelated conditioned on received signals and pilot sequences.

Joint estimation of the two channels is a reasonable choice if timing between the two signals is unknown, or if the signal of interest does not have any control signal structure resulting in blind channel estimation [16]. It is reasonable to assume that the transmitted and the received signals are synchronized at least in symbol level, because the transmitter and the receiver should employ the same local oscillator to reduce distortion and residual self-interference to mitigate the detrimental effect of phase noise [17, 18]. Even this assumption may not always hold, though, e.g., when baseband processing is done in software running on a general purpose computer. In this case operating system can cause unexpected delays driving input and output streams out of synch. This causes problems to self-interference cancellation and may lead to a sudden increase in self-interference levels.

When baseband is implemented in hardware, synchronization in symbol level can be enforced. Typically, wireless communication systems adopt a frame and slot structure such that the received and the transmitted frames can be synchronized within full-duplex transceiver. Frame and symbol synchronization make it possible to employ orthogonal pilot sequences between the received signal of interest and the transmitted signal for channel estimation. This kind of arrangement is known to be optimal in MIMO systems [19]. Most wireless systems employ a pilot signal structure for channel estimation, so orthogonal pilot sequences should be applied in full-duplex communications as well whenever possible. Large self-interference levels still deteriorate the channel estimation, because even when the orthogonality of the pilot signals is easily achieved, transmitter distortion is not orthogonal to received pilot sequence.

Given the noise due to transmitter imperfections it is better to employ orthogonal pilot sequences in time and/or frequency domain instead of in code domain. In the latter case, transmitter distortions deteriorate channel estimation of the signal of interest even when the pilot sequences themselves are orthogonal. In the former case, self-interference signal is not present when estimating the channel of the signal of interest so residual self-interference does not pose a problem for channel estimation. However, as elaborated in the next subsection, in case of different modulation parameters or non-ideal hardware, self-interference leaks from neighbor subcarriers into the pilot subcarrier. Therefore, in practice LTE-like pilot structure, where pilot symbols are time-frequency multiplexed among subcarriers, is more prone to self-interference than time-multiplex pilot structure.

Radio resources reserved for orthogonal pilot signals decrease the throughput in uplink and downlink, but this overhead is the same in all domains, time, frequency, and code, when pilot sequences are orthogonal. The throughput of the link can be further improved when taking into account the self-interference in the power allocation between pilot and data symbols [20].

4.2.4 Self-Interference Channel

Relative channel estimation errors depend on channel statistics. In a realistic environment, transmitted signals from a full-duplex transceiver are reflected back to the receiver chain through different paths which are each characterized by an attenuation and a delay giving rise to a frequency-selective channel. Early channel measurement results on full-duplex MIMO transmission have been presented, e.g., in [21–23]. The study reported in [22] builds one of the first prototypes of antenna arrays for full-duplex MIMO transceiver and measures self-interference channels and physical isolation achieved with the prototypes. Measurement results presented [23] showed that physical isolation has a larger spread than exponential random variable. Since there was no interference cancellation besides antenna isolation, the physical isolation is the same as the power of self-interference. Recently, the same antenna geometry was used to measure and characterize self-interference channels in outdoor scenarios at street level [24]. The first two channel taps experience Rician characteristics while the rest of the channels are Rayleigh distributed. Even when the transceiver is static, the self-interference channel is changing due to movement in the vicinity, i.e., due to pedestrians and vehicles.

It was shown in [12] that analog cancellation changes the Rician K-factor (measuring the relative strength of the line-of-sight component) of the self-interference channel. The measurement campaign assumed a line-of-sight (LOS) path between transmitter and receiver antennas, and when the strongest LOS path of the self-interference was cancelled, the multipath channel profile was changed accordingly. Since the strongest path is largely suppressed, the K-factor becomes smaller.

Antenna design and analog cancellation change the statistics of the self-interference channels, but when developing algorithms in digital baseband, the effect of isolation and analog cancellation is merely visible in the power of the self-interference channel. The number of channel taps can vary, and each tap is modeled as a random variable characterized, e.g., a Rayleigh, Nakagami-m AWGN, or Rician distribution. This far there is no standardized model of self-interference that could be used to benchmark different algorithms in wireless communication systems.

4.2.5 Self-Interference Signal

Coming back to OFDM signaling, the signal model (4.1) when interpreted in frequency domain assumes that the transmitted and received signals are synchronized in time and frequency so that inter-carrier interference (ICI) can be ignored. Since the full-duplex transceiver can control the signal it is transmitting this may be look a reasonable assumption. However, ICI cannot be always avoided in a wireless communication system the full-duplex transceiver is a part of. The transmitter and receiver links may well have different parameters such as: time offset (TO), carrier frequency offset (CFO), number or subcarriers, cyclic prefix (CP) length, etc.

A full-duplex transceiver can be attached to two different cells having different synchronization parameters. Even within one cell the transmitted (uplink) signal can be subject to timing advantage correction giving rise to time offset between the transmitted and the received signal. Uplink and downlink signals can also use different waveforms like in LTE where uplink uses single carrier and downlink OFDM. In self-backbone networks, the backbone network may operate with different modulation parameters than the access network, like in the high-speed train (HST) relay link [25]. To make the relay—train link more robust against the low channel coherence time, the inter-carrier spacing is increased. When channel changes within an OFDM symbol, the subcarriers do not remain orthogonal, but when the subcarriers are more widely spaced, ICI is better localized. On the other hand, in the backhaul link between the base station and the relay, the inter-carrier spacing remains the same, resulting in different modulation parameters in the transmitter and the receiver.

In these cases, the signal model (4.1) is not sufficient in frequency domain, and a more general model is required:

$$\mathbf{y}_T[n] = \mathbf{H}_{rx}[n]\mathbf{x}_S[n] + \mathbf{H}_i[n]\,\mathbf{x}_T[n] + \sum_{k \neq n} \mathbf{H}_i[k]\,\mathbf{x}_T[k] + \mathbf{n}_T[n] \tag{4.8}$$

Here the term $\sum_{k \neq n} \mathbf{H}_i[k]\,\mathbf{x}_T[k]$ represents additional self-interference leaking from other OFDM subcarriers due to different modulation parameters. Thus, even if the subcarrier n is switched off in the transmitter, self-interference falling on the received subcarrier n still remains [26].

Figure 4.3 shows an example of the leakage γ (in dB) of the self-interference power to adjacent subcarriers due to time offset between transmitted and received OFDM waveforms in a full-duplex relay. Both transmit and receive symbols lengths

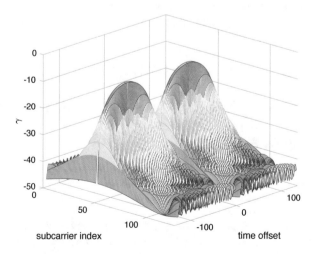

Fig. 4.3 Coupling interference γ (dB) in the full-duplex receiver in case of time offset between the transmitted and the received OFDM waveforms

are the same, but the number of subcarriers is 64 and 128 in the transmitter and the receiver, respectively. Thus, the subcarrier spacing in the transmitter is double from that in the receiver. Time offset in the figure is defined such that the offset ± 128 corresponds to the length of one OFDM symbol in time.

Only the subcarrier no. 50 is active while the other subcarriers are unmodulated. Therefore, there is no additional leakage to subcarrier 50 in any time offset according to the definition. There is no leakage either around time offset zero when the offset falls within the cyclic prefix. However, for other time offsets, self-interference leaks to neighboring subcarriers. Similar study can be presented for frequency offset as well, and it can be shown that it has a significant effect to performance in case of full-duplex relays serving high-speed trains [26].

4.2.6 Time-Domain Signal Model

In case of OFDM, mitigation in the frequency domain may process each subcarrier signal individually, when the assumptions of the signal model (4.1) are satisfied. Non-orthogonal multicarrier modulation schemes like filterbank multicarrier (FBMC) and generalized frequency division multiplexing (GFDM) inherently cause inter-carrier interference that must be taken into account when subtracting interference in frequency domain as outlined in Sect. 4.3.1.1. Thus, single carrier model does not apply to non-orthogonal multicarrier modulation even when the effect of time offset and frequency offset can be ignored.

Time-domain self-interference cancellation is agnostic to modulation technique used and it can be applied to OFDM, FBMC, and GFDM alike. Thus, it applies also to the case when the transmitted signal and the signal of interest use different modulation. Frequency flat channel model is not valid in general, and multipath channel should be included in the system model instead. In the received signal model multiplications in (4.1) are substituted for convolutions and the received signal in the full-duplex transceiver becomes

$$\mathbf{y}_T[n] = \mathbf{H}_{rx}[n] \star \mathbf{x}_S[n] + \mathbf{H}_i[n] \star \mathbf{x}_T[n] + \mathbf{n}_T[n]$$

$$= \sum_{k=0}^{L_{rx}} \mathbf{H}_{rx}[k-n]\mathbf{x}_S[n] + \sum_{k=0}^{L_i} \mathbf{H}_i[k-n]\mathbf{x}_T[n] + \mathbf{n}_T[n]$$

$$\mathbf{y}_D[n] = \mathbf{H}_{tx}[n] \star \mathbf{x}_T[n] + \mathbf{n}_D[n]$$

$$= \sum_{k=0}^{L_{tx}} \mathbf{H}_{tx}[k-n]\mathbf{x}_T[n] + \mathbf{n}_D[n] \tag{4.9}$$

where $\star$ refers to convolution. As before, the dimensions of the channel matrices are $\mathbf{H}_{rx}[n] \in \mathbb{C}^{N_{rx} \times N_S}$, $\mathbf{H}_i[n] \in \mathbb{C}^{N_{tx} \times N_{tx}}$, and $\mathbf{H}_{tx}[n] \in \mathbb{C}^{N_D \times N_{tx}}$, and L_{rx}, L_i, and L_{tx} refer to the order of the channels. Thus, the number of multipath components is

$L_{rx} + 1$, $L_i + 1$, and $L_{tx} + 1$, respectively. The signal model can be further cast in the form

$$y_T[n] = \mathcal{H}_{rx}[n]\chi_S[n] + \mathcal{H}_i[n]\chi_T[n] + n_T[n]$$

$$y_D[n] = \mathcal{H}_{tx}[n]\chi_T[n] + n_D[n]$$

(4.10)

where $\chi_S[n] = [x_S^T[n], \ldots, x_S^T[n-L_{rx}]]^T$ is L_{rx}th-order column expansion of $x_S[n]$, or in other words, $\chi_S[n] = \mathrm{vec}(x_S[n], \ldots, x_S[n-L_{rx}])$ is the vectorization of $x_S[n], \ldots, x_S[n-L_{rx}]$. In the same vein, $\chi_T[n]$, $n_T[n]$, $y_D[n]$, and $n_D[n]$ are vectorizations of $x_T[n]$, $n_T[n]$ $y_D[n]$, and $n_D[n]$, respectively. Moreover, $\mathcal{H}_{rx}[n] = [\mathbf{H}_{rx}[n], \ldots, \mathbf{H}_{rx}[n-L_{rx}]]$, $\mathcal{H}_i[n] = [\mathbf{H}_i[n], \ldots, \mathbf{H}_i[n-L_i]]$, and $\mathcal{H}_{tx}[n] = [\mathbf{H}_{tx}[n], \ldots, \mathbf{H}_{tx}[n-L_{tx}]]$ refer to row-expansion matrices of $\mathbf{H}_{rx}[n]$, $\mathbf{H}_i[n]$, and $\mathbf{H}_{tx}[n]$. The formulation (4.10) of the time-domain signal model makes it possible to apply standard linear algebra tools when developing adaptive interference cancellation techniques as shown in Sect. 4.4.1. For simplicity we assumed in (4.10) that $L_{rx} = L_i = L_{tx}$ to keep the dimensions compatible. This assumption can be easily relaxed by padding signals with zeros appropriately.

4.3 Mitigation of Self-Interference

The basic approaches to mitigate self-interference are cancellation and suppression and their combinations. In case of subtracting the replica signal, its fidelity will determine the level of residual self-interference after cancellation. In order to achieve a nearly interference-free system, choosing the proper cancellation method for each application is important. In this chapter, we first discuss on self-interference cancellation in frequency domain in case of orthogonal and non-orthogonal multicarrier modulation. Linear filtering in frequency or time domain is not able to suppress non-linear transmitter distortions, and therefore, we will add spatial transmit and receive filters to the transceiver model.

4.3.1 Frequency-Domain Cancellation

In signal model at digital baseband (4.1) self-interference is modeled as the sum of transmitted information signal and RF distortion. The latter is assumed to comprise a single noise term in the transceiver input. Assuming that the transmitted signal in digital baseband is known and channel estimation is ideal, the known information signal can be subtracted from the received signal.

Under the assumptions that inter-carrier interference and inter-symbol interference can be ignored, the frequency-domain self-interference cancellation is simply done on a subcarrier basis as

$$\mathbf{y}_T[n] = \mathbf{H}_{rx}[n]\,\mathbf{x}_S[n]\ + \mathbf{H}_i[n]\,\mathbf{x}_T[n] - \hat{\mathbf{H}}_i[n]\bar{\mathbf{x}}_T[n] + \mathbf{n}_T[n] \qquad (4.11)$$

where $\hat{\mathbf{H}}_i[n]$ and $\bar{\mathbf{x}}_T[n]$ refer to the estimate of the self-interference channel and the known part of the transmitted signal, respectively. Thus, self-interference cancellation involves estimating the self-interference channel to generate the replica of the transmitted signal in the receiver chain. The residual self-interference term becomes $\mathbf{H}[n]\boldsymbol{\delta}\mathbf{x}_T[n] + \hat{\mathbf{H}}_i[n]\boldsymbol{\delta}\mathbf{x}_T[n]$. This term cannot be cancelled with linear filter in time or frequency domain, but assuming a sufficient number of antennas, it can be suppressed in spatial domain as discussed in Sect. 4.3.2.

In case of orthogonal multicarrier modulation like OFDM and frequency-domain cancellation, the argument n refers to the subcarrier n. Assuming that interference between subcarriers can be ignored, the argument can be dropped to simplify the notation. The same model can be used in time domain when assuming single path, frequency flat MIMO channels and interpreting the variables as time-domain variables instead of frequency-domain ones.

4.3.1.1   Non-Orthogonal Multicarrier Modulation

The promise of non-orthogonal multicarrier waveforms in place of OFDM is to relax the overhead due to cyclic prefix and facilitate better localization in frequency domain than the OFDM waveform. This is achieved by sophisticated filter design that controls the distribution of energy in time and frequency, and good localization in frequency is achieved by extending and overlapping the multicarrier symbols in time. Thus, multicarrier modulation schemes trade off time and frequency localization, and this trade-off is controlled by prototype filter design. Eventually, the trade-off in time and frequency cannot be avoided because of well-known Heisenberg's uncertainty principle from quantum mechanics [27]. Exploiting the trade-off requires more complicated signal processing than OFDM waveforms, and mitigation of the interference from neighboring subcarriers in the receiver becomes necessary. Several different non-orthogonal multicarrier techniques have been promoted as alternatives to OFDM for 5G systems, e.g., FBMC [28] and GFDM [29]. Properties of waveforms have been compared and their suitability to 5G have been evaluated, e.g., in [30, 31].

Let us use GFDM as an example, how a non-orthogonal modulation scheme effects self-interference cancellation in frequency domain. Self-interference cancellation should be applied before channel equalization of the desired signal to avoid amplifying the self-interference. The cancellation process is executed in frequency domain after GFDM demodulator by subtracting a replica of the transmitted signal as usual. However, since the modulation is non-orthogonal it is necessary to extend the interference cancellation over the whole GFDM symbol block in frequency domain instead of an individual subcarrier n.

The generated GFDM symbol block at the full-duplex transmitter before adding cyclic prefix is given by

$$\mathbf{x_T} = \sum_{m=0}^{M-1} \mathbf{G}_m \mathbf{P} \mathbf{W} \mathbf{s}_m = \mathbf{U}_m \mathbf{s}_{T,m} \tag{4.12}$$

where M is the number of subsymbols $\mathbf{s}_{T,m}$ each containing N subcarriers, $\mathbf{W}$ is the $N \times N$ Fourier transform matrix, and $\mathbf{P}$ is an M-fold row expansion of $N \times N$ identity matrices given by $\mathbf{P} = [\mathbf{I}, \dots, \mathbf{I}]^T$. Diagonal matrix $\mathbf{G}_m$ is of size $MN \times MN$ window function having real coefficients, and its diagonal elements are cyclic shifts of the elements of $\mathbf{G}_0$ by Nm elements. Thus, $\mathbf{x_T}$ is a vector of size $MN \times 1$, and OFDM can be considered as a special case of GFDM when $M = 1$. Since cyclic prefix is only added after the M subsymbols, the subsymbols do not remain orthogonal when transmitted over multipath channel.

After removing the cyclic prefix the subsymbol $\mathbf{y}_{T,m}$ in the full-duplex receiver is obtained from the received signal $\mathbf{y_T}$ by

$$\mathbf{y}_{T,m} = \mathbf{U}_m^\dagger \mathbf{y_T} = \mathbf{P}^T \mathbf{W}^\dagger \mathbf{G}_m \mathbf{y_T} \tag{4.13}$$

When taking into account ISI and ICI components, self-interference cancellation becomes

$$\mathbf{z}_{T,m} = \mathbf{y_{T,m}} - \sum_{k=0}^{M-1} \mathbf{C}_{m,k} \mathbf{s}_{T,k} \tag{4.14}$$

where $\mathbf{C}_{m,k}$ is the coefficient matrix for replica generation.

The main task of the replica generation process is to build the coefficient matrices $\mathbf{C}_{m,k}$ based on the GFDM signaling and the estimated self-interference channel. To reproduce the subsymbol interference of GFDM signaling, the replica generation sums up all data symbols with its own coefficient matrix. The coefficient matrix is constructed by convolving the estimated self-interference channel with the ICI pattern. The coefficient matrix for replica generation is given by

$$\mathbf{C}_{m,k} = \bar{\mathbf{G}}_m \mathbf{H_i} \bar{\mathbf{G}}_k \tag{4.15}$$

where $\bar{\mathbf{G}}_m = (\mathbf{PW})^\dagger \mathbf{G}_m \mathbf{PW}$ is the frequency-domain filter matrix, and $\mathbf{H_i}$ refers now to frequency-domain self-interference channel matrix interpolated by M. The channel matrix is interpolated, because the self-interference is passed through the channel being upsampled in frequency domain. By using the interpolated channel, the canceller is able to duplicate the upsampled self-interference signal. After cancellation, the receiver performs channel equalization.

Due to non-orthogonal signaling, self-interference cancellation in GFDM system has larger computational complexity than that of orthogonal signaling like OFDM. Even when the filter response $\bar{\mathbf{G}}_m$ is deterministic, calculating full coefficient matrix requires high computational complexity. This is because the coefficient matrix $\mathbf{C}_{m,k}$ should be updated whenever the self-interference channel is changing.

Significant self-interference terms can be determined by examining cross correlation terms of transmission pulse responses $\mathbf{U}_m^\dagger \mathbf{U}_k$. When $m = k$, the diagonal part indicates the self-interference due to transmitted symbol vector $\mathbf{x}_{\mathrm{T},m}$. When $m \neq k$ the diagonal part indicates the ISI of self-interference pattern. The non-diagonal parts of $\mathbf{U}_m^\dagger \mathbf{U}_k$ indicate the ICI from its own subsymbol or different subsymbols. The diagonal part of matrices cannot be ignored but off-diagonal elements are small compared to the diagonal elements, and interference power becomes smaller the further away from the diagonal of the matrix. This observation is valid for GFDM signaling in general regardless of the channel. Thus, the computational complexity of the replica generation can be reduced by using only the coefficients of $\mathbf{C}_{m,k}$ near the diagonal and ignoring the rest of the off-diagonal part [32]. The same kind of approximation has been used to simplify the equalization of OFDM signal over doubly dispersive channels [33, 34] when the channel is changing within one OFDM symbol destroying the orthogonality of the subcarriers.

4.3.2 Spatial Suppression

Only the subtraction of the estimated self-interference is possible in SISO transceivers, but a new dimension opens up for mitigation when considering MIMO transceivers. In addition to subtracting the replica signal the full-duplex MIMO transceiver may also apply spatial suppression in case transmitter and receiver antennas are spatially separated. Precoder at the transmitter side and beamformer at the receiver can be designed to simultaneously attenuate the self-interference signal and receive the signal of interest. The beamformer is able to attenuate the effect of RF imperfections as well, because the imperfections pass through the same channel together with the known part of the self-interference.

The price of spatial suppression is the loss of degrees of freedom for information transfer. Spatial suppression techniques transform the physical $N_{\mathrm{rx}} \times N_{\mathrm{tx}}$ transceiver to an equivalent "interference-free" $\bar{N}_{\mathrm{rx}} \times \bar{N}_{\mathrm{tx}}$ transceiver. Here $\bar{N}_{\mathrm{rx}}$ and $\bar{N}_{\mathrm{tx}}$ represent the number of spatial streams reserved for spatial multiplexing. Without loss of generality, we assume that $\bar{N}_{\mathrm{rx}} \leq N_{\mathrm{rx}}$ and $\bar{N}_{\mathrm{tx}} \leq N_{\mathrm{tx}}$.

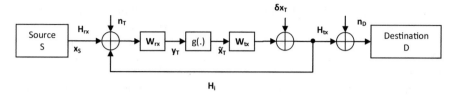

Fig. 4.4 Full-duplex communication link with spatial suppression. The function $g(\cdot)$ refers to other baseband signal processing tasks besides spatial filtering

Again, the target is to push residual self-interference to receiver noise level or below. The transceiver applies MIMO receive filter $\mathbf{W}_{rx}[n] \in \mathbb{C}^{\bar{N}_{rx} \times N_{rx}}$ and MIMO transmit filter $\mathbf{W}_{tx}[n] \in \mathbb{C}^{N_{tx} \times \bar{N}_{tx}}$ for spatial suppression as illustrated in Fig. 4.4. The signal model (4.1) becomes

$$\bar{\mathbf{y}}_T[n] = \mathbf{W}_{rx}[n](\mathbf{H}_{tx}[n]\mathbf{x}_S[n] + \mathbf{H}_i[n](\mathbf{W}_{tx}[n]\tilde{\mathbf{x}}_T[n] + \delta\mathbf{x}_T[n]) + \mathbf{n}_T[n])$$

$$\mathbf{y}_D[n] = \mathbf{H}_{tx}(\mathbf{W}_{tx}[n]\tilde{\mathbf{x}}_T[n] + \delta\mathbf{x}_T[n]) + \mathbf{n}_D[n]$$

$$(4.16)$$

where $\bar{\mathbf{y}}_T \in \mathbb{C}^{\bar{N}_{rx}}$ and $\bar{\mathbf{x}}_T \in \mathbb{C}^{\bar{N}_{tx}}$ are the respective receive and transmit signal vectors of the equivalent interference-free transceiver. The residual self-interference power becomes as

$$P_I = \mathcal{E}\left\{\|\mathbf{W}_{rx}[n]\mathbf{H}_i[n]\mathbf{W}_{tx}[n]\bar{\mathbf{x}}_T[n] + \mathbf{W}_{rx}[n]\mathbf{H}_i[n]\delta\mathbf{x}_T[n]\|_2^2\right\} \qquad (4.17)$$

Residual self-interference can be now mitigated by appropriately designing spatial filters $\mathbf{W}_{rx}[n]$ and $\mathbf{W}_{tx}[n]$ to minimize the first term of P_I and/or by designing only $\mathbf{W}_{rx}[n]$ to minimize the second term that is due to transmit signal noise. The cross term vanishes, because the transmitted signal and transmitter noise are uncorrelated. The spatial filters can be designed, independently, separately, or jointly, where the last option is supposed to provide the best performance. Different approaches to precoder design are, e.g., antenna selection, general eigenbeam selection, null-space projection, and minimum mean squared error filtering [35]. In practice, only the estimate $\hat{\mathbf{H}}_i[n]$ of the self-interference channel is available, and spatial suppression requires a sufficiently accurate channel estimation to be feasible. Spatial suppression techniques may also embrace the forward channel [36] when it can be assumed known by channel reciprocity or channel state feedback. Here we do not make any assumptions on antenna geometry and the structure of $\hat{\mathbf{H}}_i[n]$. Experiments on spatial suppression have been conducted in [37] using two-dimensional planar antenna array and in [38] using one-dimensional linear antenna array.

Null-space projection is particularly attractive, because it may suppress the self-interference completely (assuming $\mathbf{H}_i[n]$ is known) with the price of reducing the degrees of freedom for data transmission. In null-space projection, the spatial filters $\mathbf{W}_{rx}[n]$ and $\mathbf{W}_{tx}[n]$ are selected such that the transceiver receives and transmits in different subspaces, i.e., transmit beams are projected to the null-space of the self-interference channel combined with the receive filter and vice versa. Such a condition can be formalized for joint or separate filter design as

$$\mathbf{W}_{rx}[n]\mathbf{H}_i[n]\mathbf{W}_{tx}[n] = \mathbf{0} \tag{4.18}$$

to eliminate the first term in (4.17), i.e., the known part of the transmitter signal. Similarly, for suppressing the transmit signal noise, the condition becomes $\mathbf{W}_{rx}[n]\mathbf{H}_i[n] = \mathbf{0}$ to eliminate the second term in (4.17).

In particular, the number of data streams is constrained by

$$\bar{N}_{rx} + \bar{N}_{tx} + \text{rank}(\mathbf{H}_i[n]) \leq N_{rx} + N_{tx} \tag{4.19}$$

This condition defines also the general existence of joint null-space projection, if $\mathbf{W}_{rx}[n]$ and $\mathbf{W}_{tx}[n]$ are additionally constrained to have full rank. Even if $\mathbf{H}_i[n]$ may be rank-deficient, $\hat{\mathbf{H}}_i[n]$ is of full rank in practice due to channel estimation errors which also cause residual self-interference.

In the case of $\bar{N}_{rx} = \bar{N}_{tx}$, the total number of antennas ($N_{rx} + N_{tx}$) is minimized for null-space projection by choosing $N_{rx} = 2\bar{N}_{rx} = 2N_{tx} = 2\bar{N}_{tx}$ or $N_{tx} = 2\bar{N}_{tx} = 2N_{rx} = 2\bar{N}_{rx}$ when $\text{rank}(\mathbf{H}_i[n]) = \min\{N_{rx}, N_{tx}\}$ meaning that $\mathbf{H}_i[n]$ has a full rank. Thus, self-interference can be completely suppressed if the number of transmit antennas is two times larger than the number of receive antennas, or vice versa. Selecting $N_{rx} > N_{tx}$ may be preferable due to transmit signal noise as discussed later.

When $\mathbf{W}_{rx}[n]$ is designed separately given $\mathbf{W}_{tx}[n]$, null-space projection can be implemented by applying projection matrix

$$\mathbf{W}_{rx}[n] = \sqrt{\frac{N_{rx}}{N_{rx} - \text{rank}(\mathbf{H}_i[n]\mathbf{W}_{tx}[n])}} \left(\mathbf{I} - \mathbf{H}_i[n]\mathbf{W}_{tx}[n]\left(\mathbf{H}_i[n]\mathbf{W}_{tx}[n]\right)^{\#}\right) \tag{4.20}$$

where # denotes Moore–Penrose pseudoinverse [39], and it is assumed that $\text{rank}(\mathbf{H}_i[n]\mathbf{W}_{tx}[n]) < N_{rx}$. Separate design for $\mathbf{W}_{tx}[n]$ is given by a similar projection matrix which is obtained by replacing $\mathbf{H}_i[n]\mathbf{W}_{tx}[n]$ above with $\mathbf{W}_{rx}[n]\mathbf{H}_i[n]$. In case of having full rank for $\mathbf{H}_i[n]\mathbf{W}_{tx}[n]$ or $\mathbf{W}_{rx}[n]\mathbf{H}_i[n]$ in above leads to $\mathbf{W}_{rx}[n] = \mathbf{0}$ or $\mathbf{W}_{tx}[n] = \mathbf{0}$, respectively.

4.3.3   Spatial Suppression and Frequency-Domain Cancellation

Frequency-domain and time-domain cancellation suffer from residual interference that is mainly due to the transmit signal noise. On the other hand, spatial-domain suppression requires more antennas than data streams for effective mitigation of self-interference. These two approaches are not mutually exclusive, and a question arises, how cancellation and suppression should be combined together.

Four options to combine frequency or time-domain cancellation and spatial domain suppression are shown in Fig. 4.5. With independent or separate filter design, the design (and performance) of these options differs according to the

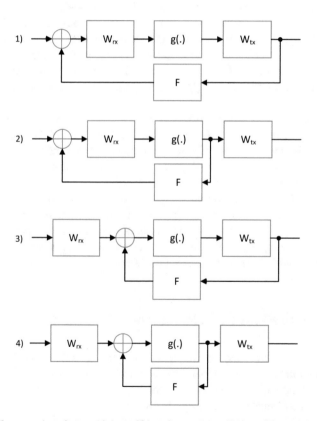

Fig. 4.5 Different options for combining self-interference cancellation with spatial suppression

residual self-interference channels:

$$\hat{\mathbf{H}}_{\mathrm{i}}[n] = \begin{cases} \mathbf{W}_{\mathrm{rx}}[n]\mathbf{H}_{\mathrm{i}}[n]\mathbf{W}_{\mathrm{tx}}[n] + \mathbf{F}[n], & \text{option 1)} \\ (\mathbf{W}_{\mathrm{rx}}[n]\mathbf{H}_{\mathrm{i}}[n] + \mathbf{F}[n])\mathbf{W}_{\mathrm{tx}}[n], & \text{option 2)} \\ \mathbf{W}_{\mathrm{rx}}[n](\mathbf{H}_{\mathrm{i}}[n]\mathbf{W}_{\mathrm{tx}}[n] + \mathbf{F}[n]), & \text{option 3)} \\ \mathbf{W}_{\mathrm{rx}}[n](\mathbf{H}_{\mathrm{i}}[n] + \mathbf{F}[n])\mathbf{W}_{\mathrm{tx}}[n], & \text{option 4)} \end{cases} \qquad (4.21)$$

However, these become equivalent if all three filters are designed jointly.

Options (1) and (4) allow the most straightforward implementation, because the two layers of mitigation can be decoupled and joint design is still feasible for $\mathbf{W}_{\mathrm{rx}}[n]$ and $\mathbf{W}_{\mathrm{tx}}[n]$: in option (1), spatial suppression is first applied, and the residual interference is mitigated with cancellation while option (4) is the converse case. In options (2) and (3), all three filters must be implemented sequentially leading to suboptimal separate design. Moreover, these two options cannot exploit the full potential of the spatial transmit filter, because cancellation has already eliminated the known part of interference. Thus, option (1) turns out to be the best choice from spatial suppression point of view: As cancellation is blind to the spatial domain, it is better to first design jointly $\mathbf{W}_{\mathrm{rx}}[n]$ and $\mathbf{W}_{\mathrm{tx}}[n]$ for exploiting all possible degrees of freedom.

The combination of cancellation and spatial suppression can offer better performance than either alone. However, when the rank-deficient self-interference channel enables the usage of null-space projection, adding cancellation on top of suppression may reduce the overall performance due to transmit signal noise. This is because null-space projection can completely eliminate transmit signal noise together with the actual interference signal, while cancellation effectively acts as an extra interference signal due to channel estimation error. In general, link-level simulation [40] results show that spatial-domain suppression is better than time-domain (or frequency domain) cancellation whenever there is a sufficient number of antennas compared to the number of spatial streams.

4.4 Algorithms for Self-Interference Cancellation

Mitigating self-interference in frequency domain on subcarrier basis requires several assumptions on the system: Transmit and receive signals are synchronized, multicarrier modulation is orthogonal, modulation parameters are the same, and inter-carrier interference and inter-symbol interference are small enough. Moreover, complexity is related to the number of subcarriers instead of the channel order.

In this chapter, we first present time-domain self-interference cancellation algorithms that do not require synchronization or pilot symbols for channel estimation. Furthermore, the algorithms do not make any assumptions on waveforms used. The time-domain algorithms apply a linear filter that is unable to mitigate the effect of transmitter distortion. Therefore, we complement the time-domain filtering with spatial filtering. Together these linear filters are able to mitigate transmitter distortion as well at the expense of the degrees of freedom available for spatial multiplexing.

We assume that transmit and receive signals $\mathbf{x}_S[n]$ and $\mathbf{x}_T[n]$ are independent. This is trivially satisfied when full-duplex transceiver is an access point serving simultaneously uplink and downlink users, or when the transceiver works as a decode-and-forward relay. We exclude amplify-and-forward relays from the study, although they are very popular in scientific literature. In practical networks, amplify-and-forward relays or repeaters complicate network planning, because they amplify noise and interference as well, and their usage is not recommended by vendors. Furthermore, we ignore the effect of traffic models to system performance, and assume that the source and the full-duplex transceiver are continuously transmitting. In any case, full-duplex operation is most useful when both transmit and receive directions are active.

4.4.1 Adaptive Algorithms

Let us first review two well-known adaptive filters: least mean squares (LMS) and recursive least squares (RLS) filters. A linear adaptive filter consists of a linear filter whose coefficients are controlled by an adaptive algorithm aiming to minimize a suitable cost function. In case of single-input single-output system, the problem can be formulated as

$$\mathbf{w} = \arg\min_{\mathbf{w}[n]} \mathcal{E}\left\{ |e[n]|^2 \right\} \tag{4.22}$$

where $\mathbf{w}[n]$ is a vector containing the coefficients of the adaptive filter, and $e[n] = d[n] - \mathbf{w}^\dagger[n]\mathbf{x}[n]$ is the scalar error signal between the desired signal $d[n]$ and the filter's output signal.

LMS algorithm [41, 42] approximates the mean square error $\mathcal{E}\{|e[n]|^2\}$ as the instantaneous error $|e[n]|^2$. The adaptation step is based on the stochastic approximation of the gradient as

$$\mathbf{w}[n] = \mathbf{w}[n-1] + \mu e[n] \tag{4.23}$$

where μ is the adaptation step size. Thus, the LMS algorithm is one example of stochastic gradient descent SGD algorithms. A proper step size μ and thereby the convergence speed depends on the autocorrelation of the input signal.

The RLS criterion approximates the minimum mean square solution by minimizing the sum of weighted squared errors as

$$\mathbf{w} = \arg\min_{\mathbf{w}[n]} \sum_{k=0}^{n} \lambda^{n-k} |e[k]|^2 \tag{4.24}$$

where $0 < \lambda < 1$ denotes the forgetting factor. The filter coefficients at time instant n become

$$\mathbf{w}[n] = \mathbf{R}^{-1}[n]\mathbf{p}[n] \tag{4.25}$$

where $\mathbf{R}[n]$ and $\mathbf{p}[n]$ are referred to as deterministic correlation matrix of the input signal and cross correlation between the desired signal and the output, respectively, given by

$$\mathbf{R}[n] = \sum_{k=0}^{n} \lambda^{n-k}\mathbf{x}[k]\mathbf{x}^{\dagger}[k]$$

$$\mathbf{p}[n] = \sum_{k=0}^{n} \lambda^{n-k}\mathbf{x}[k]d^{*}[k] \tag{4.26}$$

Matrix inverse of $\mathbf{R}[n]$ is a computationally intensive operation of the complexity $\mathcal{O}(L^3)$, where L refers to the length of the filter $\mathbf{w}$. Instead of calculating the inverse for every time step, it can be updated according to the matrix inverse lemma as

$$\mathbf{S}[n] = \mathbf{R}^{-1}[n] = \frac{1}{\lambda}\left(\mathbf{S}[n-1] - \frac{\mathbf{S}[n-1]\mathbf{x}[n]\mathbf{x}^{\dagger}[n]\mathbf{S}[n-1]}{\lambda + \mathbf{x}[n]^{\dagger}\mathbf{S}[n-1]\mathbf{x}[n]}\right) \tag{4.27}$$

while $\mathbf{p}[n] = \lambda\mathbf{p}[n-1] + \mathbf{x}[n]d^{*}[n]$. RLS algorithm has larger computation complexity than the LMS algorithm. On the other hand, RLS converges faster and the speed of convergence is independent of the spectral characteristics of the input signal. Several other adaptive algorithms and modifications to LMS and RLS algorithms have been proposed in the literature [41, 42]. In the following we will extend LMS and RLS algorithms for the self-interference cancellation in case of multiple transmit and receive antennas.

4.4.2 Stochastic Gradient Descent Algorithm

The building blocks for adaptive digital baseband cancellation in a MIMO full-duplex transceiver according to option (1) in Fig. 4.5 are depicted in Fig. 4.6. Cancellation is done in timedomain following the signal model (4.9). Linear time-domain cancellation aims at estimating the self-interference channel $\mathbf{H}_i[n], n = 0, \ldots, L_i$, where L_i refers to the order of the channel. Then the canceller subtracts $\mathbf{F}[n] \star \tilde{\mathbf{x}}_T[n] = \hat{\mathbf{H}}_i[n] \star \tilde{\mathbf{x}}_T[n]$ from the received signal, where $\hat{\mathbf{H}}_i[n]$ refers to the estimate of the self-interference channel.

Let us arrange the $L_F + 1$ MIMO matrices $\mathbf{F}[n]$ of the interference canceller into the $N_{rx} \times N_{rx}(L_F + 1)$ row-expansion matrix as $\mathcal{F}[n] = [\mathbf{F}[n], \ldots, \mathbf{F}[n-L_F]]$ in the similar manner as in (4.10). In order to obtain a sufficient estimate of $\mathbf{H}_i[n]$, the

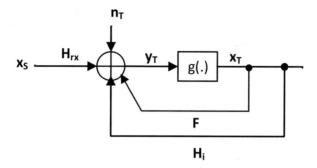

Fig. 4.6 Adaptive baseband interference cancellation. The notation is simplified by excluding the time index n

order of $\mathbf{F}[n]$ should be larger than that of $\mathbf{H}_i[n]$; $L_F \geq L_i$. We may write the signal after interference cancellation as

$$y_T[n] = \mathcal{H}_{rx}[n] x_S[n] + \mathcal{H}_i[n] x_T[n] + \mathcal{F}[n-1] x_T[n-1] + n_T[n] \qquad (4.28)$$

where $x_T[n]$, $y_T[n]$, $n_T[n]$, $\mathcal{H}_{rx}[n]$, and $\mathcal{H}_{rx}[n]$ are defined as in (4.10).

If the source and full-duplex transceiver employ orthogonal pilot sequences for channel estimation, and the transmit and receive signals are synchronized to maintain orthogonality, then the estimation process for $\mathcal{F}[n]$ is similar to conventional multiuser channel estimation. However, here we assume that the transmitted and received signals are not necessarily synchronized and therefore we oversample the received signal. Due to the uncorrelated transmit and receive signals, we may use a power minimization algorithm without resorting to pilot signals for the estimation of the self-interference channel. This kind of architecture makes it possible to add the digital cancellation block to an existing transceiver when, e.g., more attenuation is needed after antenna isolation and RF cancellation.

The power of the error signal $||\mathbf{e}[n]||^2 = ||\mathbf{H}_i[n] \star \mathbf{x}_T[n] + \mathbf{F}[n] \star \mathbf{x}_T[n]||^2$ is obviously minimized when $\mathbf{F}[n] = -\mathbf{H}_i[n]$. To this end, we minimize the cost function $\mathcal{E}\left\{||y_T[n]||^2\right\}$ with respect to $\mathcal{F}[n]$ and approximate the gradient descent by using the SGD algorithm. The adaptation rule becomes

$$\mathcal{F}[n] = \mathcal{F}[n-1] - \mu \nabla_{\mathcal{F}^*}\left\{||y_T[n]||^2\right\}$$

$$= \mathcal{F}[n-1] - \mu \nabla_{\mathcal{F}^*}$$

$$\times \left\{||(\mathcal{H}_{rx}[n] x_S[n] + \mathcal{H}_i[n] x_T[n] + \mathcal{F}[n-1] x_T[n-1] + n_T[n])^{\dagger} y_T[n]||^2\right\}$$

$$= \mathcal{F}[n-1] - \mu y_T[n] x_T[n-1]^{\dagger} \qquad (4.29)$$

where $\mu > 0$ is the adaptation step size controlling stability, convergence speed, and misadjustment. When computing the gradient, we treat $\mathcal{F}[n]$ and $\mathcal{F}^{\dagger}[n]$ as independent entities and apply the rule $\nabla_{\mathbf{X}^*}\{\mathbf{u}^{\dagger}\mathbf{X}^{\dagger}\mathbf{v}\} = \mathbf{v}\mathbf{u}^{\dagger}$.

It is important to ensure that $\mathbf{x}_T[n]$ and $\mathbf{y}_T[n]$ are uncorrelated so that the signal-of-interest is not accidentally suppressed in the process. This condition is satisfied when the full-duplex transceiver is an access point, because the uplink and downlink data are different. The condition is also true when the full-duplex transceiver is a decode-and-forward relay, because decoding and re-encoding the received signal necessarily imposes a delay at least the length of the code block. This is typically longer than the delay due to the self-interference channel. In case of full-duplex amplify-and-forward relay, the transmitted signal can be deliberately delayed to ensure the condition is met.

The algorithm operates in time domain and is independent of data modulation and pilot symbols. In contrast, the operation domain of interference cancellation using channel estimation and pilot symbols is determined by the structure of the pilot symbols. For example, scattered pilot symbols in a time-frequency grid as in LTE would imply frequency-domain processing. In time-domain, data symbols would appear as noise in the channel estimation while in frequency-domain data and pilot symbols do not interfere.

It can be shown [43] that the stationary points of the algorithm provide perfect cancellation of the self-interference signal (in the sense that $\mathbf{F}[n]$ converges to $\mathbf{H}_i[n]$) if the order of $\mathcal{F}[n]$ is sufficient and $\mathbf{x}_T[n]$ is persistently exciting. A stationary process $\mathbf{x}_T[n]$ is persistently exciting iff its autocorrelation matrix is of full rank. The SGD algorithm can be combined with spatial suppression such that after the convergence of the algorithm, the spatial filters are alternatively optimized as discussed in Sect. 4.4.4.

4.4.3 RLS Stochastic Gradient Descent Algorithm

The recursive least squares (RLS) version of the adaptive self-interference cancellation algorithm minimizes the weighted cost function as

$$\mathcal{F}[n] = \arg \min_{\mathcal{F}[n]} \sum_{k=0}^{n} \lambda^{n-k} ||y_T[k]||^2 \tag{4.30}$$

Differentiating the cost function w.r.t. $\mathcal{F}[n]$ yields

$$\mathcal{F}[n]\mathbf{R}[n] = -\mathbf{P}[n] \tag{4.31}$$

where

$$\mathbf{R}[n] = \sum_{k=0}^{n} \lambda^{n-k} x_T[k] x_T[k]^\dagger$$

$$\mathbf{P}[n] = \sum_{k=0}^{n} \lambda^{n-k} y_T'[k] x_T[k-1]^\dagger \tag{4.32}$$

and $y_T'[k]$ presents the received signal before the self-interference cancellation. Continuing,

$$\mathcal{F}[n]\mathbf{R}[n] = -\lambda\mathbf{P}[n-1] - y_T'[n]x_T[n-1]^\dagger$$

$$= \lambda\mathcal{F}[n-1]\mathbf{R}[n-1] - y_T'[n]x_T[n-1]^\dagger$$

$$= \mathcal{F}[n-1]\mathbf{R}[n] - y_T[n]x_T[n-1]^\dagger \qquad (4.33)$$

This leads to the update rule

$$\mathcal{F}[n] = \mathcal{F}[n-1] + y_T[n]x_T[n-1]^\dagger\mathbf{R}^{-1}[n] \qquad (4.34)$$

Matrix $\mathbf{R}[n]$ is updated by applying matrix inversion lemma similar to the RLS algorithm. Comparing (4.34) and (4.29) shows that $\mathbf{R}^{-1}[n]$ in (4.34) has the same role as the step size μ in (4.29).

Figure 4.7 shows an example of the performance of the SGD and RLS-SGD algorithms in a MIMO-OFDM transceiver. The figure depicts mean estimation error of the self-interference channel per iteration for SGD and RLS-SGD adaptation algorithms. Parameter κ defines the power ratio between information signal transmitted by the source node and self-interference, i.e., when $\kappa = 0$, the power of the self-interference and information signal are the same, and when $\kappa = \infty$ there is no self-interference. The antenna configuration of the relay is of size 2×2 for transmitting two data streams simultaneously. The self-interference channel has three taps and SNR is 15 dB. The number of subcarriers is 8192 and the length of the cyclic prefix is 8192/4 samples so the algorithms converge within few symbols

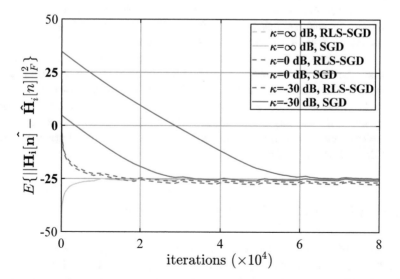

Fig. 4.7 Convergence of SGD and RLS-SGD adaptive self-interference cancellation algorithms toward the steady-state residual self-interference

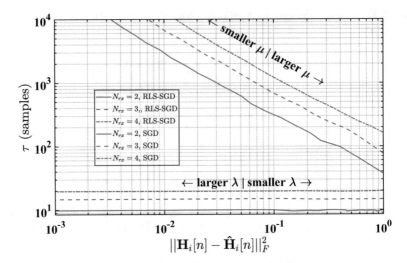

Fig. 4.8 Convergence speed vs. adaptation step size μ in SGD and forgetting factor λ in RLS-SGD algorithms

depending on κ. The step size of SGD $\mu = 0.0005$ and the forgetting factor of RLS-SGD $\lambda = 0.9999$ and $\epsilon = 0.0001$ is used to initialize the autocorrelation matrix in RLS-SGD algorithm; $\mathbf{R}[0] = \epsilon\mathbf{I}$. The algorithms are able to suppress the self-interference below noise level. The residual interference depends also on the up-sampling factor, i.e., the proportion of the signal w.r.t. the bandwidth [43]. Here the up-sampling factor is 1.15.

Adaptive algorithms typically trade off the convergence speed and the residual error. Figure 4.8 shows the trade-off between convergence speed and self-interference channel estimation error for SGD and RLS-SGD algorithms for different fixed adaptation step sizes μ and forgetting factors λ, and for different number of antennas in the full-duplex transceiver: 2×2, 3×3, and 4×4. The OFDM symbol has 8192 subcarriers (like in 8K mode of DVB-T, for example) so that the algorithms converge fast w.r.t symbol length. When the number of antennas in the relay increases, there are more parameters to estimate resulting in longer convergence time and larger residual self-interference. Here the received power ratio between the signal transmitted from the source and self-interference signal is $-25\,\text{dB}$ and SNR is $15\,\text{dB}$.

4.4.4 Adaptive Cancellation and Spatial Suppression

In practice only the transmitted signal $\tilde{\mathbf{x}}_T[n]$ is known in digital baseband, so $\mathbf{x}_T[n]$ in (4.29) and (4.34) should be substituted for when $\mathbf{x}_T[n] = \tilde{\mathbf{x}}_T[n] + \delta\mathbf{x}_T[n]$. The linear filter $\mathbf{F}[n]$ is not able to mitigate the effects of transmitter impairments $\delta\mathbf{x}_T[n]$, because $\delta\mathbf{x}_T[n]$ is not included in the adaptation.

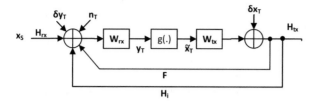

Fig. 4.9 Adaptive spatio-temporal self-interference cancellation in digital baseband

Let us add spatial transmit and receive filters in the system model as in Fig. 4.9 to be able to mitigate non-linear transmitter distortion as well using linear filters. The total receiver noise is modeled as $\mathbf{n}_T[n] = \mathbf{n}_I[n] + \delta\mathbf{y}_T[n] + \mathbf{H}_i[n] \star \delta\mathbf{x}_T[n]$, where $\delta\mathbf{x}_T[n]$ models the RF imperfections as Gaussian noise, whose power depends on EVM value, $\delta\mathbf{y}_T[n]$ models the quantization noise, and $\mathbf{n}_I[n]$ denotes thermal noise in the receiver. By modifying $\delta\mathbf{y}_T[n]$ it is possible to model the effects of limited dynamic range of analog-to-digital converter. The aim is to maximize received SINR given by

$$\gamma = \frac{\mathcal{E}\left\{||\mathbf{W}_{rx}[n] \star \mathbf{H}_{rx}[n] \star \mathbf{x}_S[n]||^2\right\}}{\mathcal{E}\left\{||\mathbf{W}_{rx}[n] \star (\mathbf{H}_i[n] + \mathbf{F}[n]) \star \mathbf{W}_{tx}[n] \star \tilde{\mathbf{x}}_T[n] + \mathbf{W}_{rx}[n] \star +\mathbf{z}[n] + \mathbf{n}_I[n]||^2\right\}}$$
(4.35)

where $\mathbf{z}[n] = \mathbf{H}_i[n] \star \delta\mathbf{x}_T[n] + \delta\mathbf{y}_T[n]$. This may not be the performance optimal criterion, because that depends on the MIMO transmission scheme used. Without specifying a MIMO transmission scheme, this is a reasonable criterion to use, though. The optimization problem becomes

$$\max_{\mathbf{W}_{tx}[n], \mathbf{W}_{rx}[n], \mathbf{F}[n]} \gamma$$

$$\text{subject to } \mathcal{E}\left\{||\mathbf{W}_{tx}[n] \star \tilde{\mathbf{x}}_T[n]||^2\right\} \leq P_{\max}$$

with the necessary condition to keep transmit power in reasonable limits. Assuming that the time-domain SGD algorithm has converged, the term containing $\tilde{\mathbf{x}}_T[n]$ in (4.35) vanishes. The optimization problem becomes

$$\max_{\mathbf{W}_{tx}[n], \mathbf{W}_{rx}[n]} \frac{\mathcal{E}\left\{||\mathbf{W}_{rx}[n] \star \mathbf{H}_{rx}[n] \star \mathbf{x}_S[n]||^2\right\}}{\mathcal{E}\left\{||\mathbf{W}_{rx}[n] \star (\mathbf{H}_i[n] \star \delta\mathbf{x}_T[n] + \delta\mathbf{y}_T[n])||^2\right\}}$$

$$\text{subject to } \mathcal{E}\left\{||\mathbf{W}_{tx}[n] \star \tilde{\mathbf{x}}_T[n]||^2\right\} \leq P_{\max}$$
(4.36)

where the thermal noise term $\mathbf{n}_I[n]$ has been dropped as well, because it is independent of the spatial and temporal filters. The remaining noise term in the denominator depends both on the transmit and receive spatial filters, and therefore the filters should be designed jointly. Moreover, this coupling makes the optimization problem non-convex as well.

Assuming i.i.d. transmit noise as in (4.5) the power due to transmit noise becomes $||\epsilon_{\mathbf{x}}^2/N_T \mathbf{W}_{tx}[n]\mathbf{W}_{tx}^{\dagger}[n]||_F^2$. Setting $\mathbf{W}_{tx}[n] = \mathbf{0}$ would trivially minimize the noise term in the denominator. At the same time the transmitted information signal would be completely suppressed, so the problem requires additional constraints. Possible linear constraints are, e.g., shortening the convolution of the spatial transmit filter and the forward channel $\mathbf{W}_{tx}[n] \star \mathbf{H}_{tx}[n]$ [5] or setting the convolution to a predefined value [44]. This requires channel state information of the forward channel $\mathbf{H}_{tx}$ in the full-duplex transmitter. Alternatively the SINR of the received signal could be added to the objective, but this would make the optimization problem even more difficult to solve.

Once appropriate linear constraints are set, the problem (4.36) can be solved by alternative optimization as follows [5]: Let us first fix $\mathbf{W}_{tx}[n]$ and solve for $\mathbf{W}_{rx}[n]$. Then we fix $\mathbf{W}_{rx}[n]$ we just solved and solve for $\mathbf{W}_{tx}[n]$. This procedure is repeated until the filters converge. This does not guarantee a global optimum of the non-convex problem, but the algorithm can be shown to converge [45].

4.4.4.1 Non-Iterative Design of Spatial Filters

Alternative optimization converges, but the iterations may take too long time if radio propagation environments is changing and the filters must be updated frequently. To get around this and to simplify the optimization problem we choose to minimize the self-interference to the receiver chain as

$$\min_{\mathbf{W}_{tx}[n]} ||\mathbf{H}_i[n] \star (\mathbf{W}_{tx}[n] \star \tilde{\mathbf{x}}_T[n] + \delta\mathbf{x}_T[n])||^2$$

$$\text{subject to } \mathcal{E}\left\{||\mathbf{W}_{tx}[n] \star \mathbf{x}_T[n]||^2\right\} \leq P_{\max}$$

$$\mathbf{H}_{tx}[n] \star \mathbf{W}_{tx}[n] = \mathbf{H}_{tx}^{eq}[n] \qquad (4.37)$$

The motivation behind this formulation is to decouple the design of $\mathbf{W}_{rx}[n]$ and $\mathbf{W}_{tx}[n]$ and to minimize the total amount of self-interference leaking to the receiver chain. This is useful when the signal of interest and the self-interference must fit to the dynamic range of ADC. Now the linear constraint prevents the trivial solution $\mathbf{W}_{tx}[n] = \mathbf{0}$ and guarantees a reasonable communication quality between the full-duplex transceiver and the destination. This constraint can be set as, for example,

$$\mathbf{H}_{tx}^{eq}[n] = \begin{cases} \mathbf{I}, & n = 0 \\ \mathbf{0}, & \text{otherwise} \end{cases} \qquad (4.38)$$

so that the equalization in spatial or in time domain is not needed in the destination. Other option for the linear constraint is to enforce channel shortening by requiring

$$\mathbf{H}_{tx}^{eq}[n] = \mathbf{0}, \ n = L, \cdots L_i + L_{tx}$$

This is useful when an OFDM waveform employs cyclic prefix of L samples, since equalization can be efficiently performed by discrete Fourier transform as long as the delay spread does not exceed the length of the cyclic prefix.

To solve (4.37), let us first stack the columns of $\mathbf{W}_{tx}[n], \cdots \mathbf{W}_{tx}[n-L_i]$ to $\bar{N}_{tx}N_{tx}(L_i + 1) \times 1$ vector $\mathbf{w}_{tx}$ as $\mathbf{w}_{tx} = \text{vec}(\mathbf{W}_{tx}[n], \cdots \mathbf{W}_{tx}[n-L_i])$ where we dropped the time index n from $\mathbf{w}_{tx}$ to simplify the notation. From now on, the vectors and matrices without time index n refer to variables that are obtained by shuffling the elements of matrices and vectors according to $\mathbf{w}_{tx}$. At the same time matrix-vector multiplications substitute convolutions. The optimization problem (4.37) can be cast to the form

$$\min_{\mathbf{w}_{tx}} \ \mathbf{w}_{tx}^{\dagger}(\tilde{\mathbf{R}} + \mathbf{R}_\delta)\mathbf{w}_{tx}$$

$$\text{subject to} \ \mathbf{w}_{tx}^{\dagger}\mathbf{R}\mathbf{w}_{tx} \le P_{max}$$

$$\mathbf{H}_{tx}\mathbf{w}_{tx} = \mathbf{h}_{tx}^{eq} \tag{4.39}$$

where $\tilde{\mathbf{R}}$ is obtained from $\mathcal{E}\left\{||\mathbf{H}_i[n] \star \mathbf{W}_{tx}[n] \star \tilde{\mathbf{x}}_T[n]\}||^2$ and $\mathbf{R}_\delta$ is obtained from $\mathcal{E}\left\{||\mathbf{H}_i[n] \star \delta\mathbf{x}_T[n]\}||^2$ by shuffling the elements of the matrices to match $\mathbf{w}_{tx}$. In the same vein, $\mathbf{R}$ is obtained from $\mathcal{E}\left\{||\mathbf{W}_{tx}[n] \star \tilde{\mathbf{x}}_T[n]||^2\right\}$. Transmit noise is assumed to be uncorrelated with the information signal such that the cross terms containing $\tilde{\mathbf{x}}_T$ and $\delta\mathbf{x}_T$ in the objective (4.37) vanish. The dimension of $\mathbf{w}_{tx}$ should be large enough to satisfy the linear constraint, i.e., $\bar{N}_{tx}N_{tx}(L_i + 1) > \text{rank}(\mathbf{H}_{tx})$. In general, this implies that $N_{tx} > N_D$ and the required order of $\mathbf{W}_{tx}[n]$ grows linearly with the order of $\mathbf{H}_{tx}[n]$.

All candidate vectors $\mathbf{w}_{tx}$ satisfying the second constraint are given by

$$\mathbf{w}_{tx} = \mathbf{H}_{tx}^{\#}\mathbf{h}_{tx}^{eq} + \mathbf{N}\mathbf{v} \tag{4.40}$$

where # refers to Moore–Penrose pseudoinverse, $\mathbf{v}$ is a vector of size, and $\mathbf{N}$ forms the basis of the null-space of $\mathbf{H}_{tx}$. Now the problem is to find $\mathbf{v}$ that consequently gives $\mathbf{w}_{tx}$ as a solution to (4.36). This can be expressed as a standard least squares problem with quadratic constraints as follows:

$$\min_{\mathbf{v}} \ ||\mathbf{Q}\mathbf{N}\mathbf{v} + \mathbf{Q}\mathbf{H}_{tx}^{\#}\mathbf{h}_{tx}^{eq}||^2$$

$$\text{subject to} \ ||\mathbf{L}\mathbf{N}\mathbf{v} + \mathbf{L}\mathbf{H}_{tx}^{\#}\mathbf{h}_{tx}^{eq}||^2 \le P_{max} \tag{4.41}$$

where we applied square-root factorizations $\mathbf{Q}^{\dagger}\mathbf{Q} = \tilde{\mathbf{R}} + \mathbf{R}_\delta$ and $\mathbf{L}^{\dagger}\mathbf{L} = \mathbf{R}$. This can be solved by decomposing $\mathbf{Q}\mathbf{N}$ and $\mathbf{L}\mathbf{N}$ using generalized singular value decomposition [46]. Given $\mathbf{v}$, the corresponding $\mathbf{w}_{tx}$ is obtained from (4.40), and $\mathbf{W}_{tx}[n]$ is reconstructed from $\mathbf{w}_{tx}$. Then we fix $\mathbf{W}_{tx}[n]$ and solve $\mathbf{W}_{rx}[n]$ from (4.36). This is done by first defining $\mathbf{w}_{rx} = \mathrm{vec}(\mathbf{W}_{rx}[n], \ldots, \mathbf{W}_{rx}[n-L_{rx}])$ and modifying (4.36) in the similar manner as solving for $\mathbf{w}_{rx}$. The problem becomes a general eigenvalue problem [46] and solved accordingly.

The original problem of maximizing the received SINR in the full-duplex transceiver was modified to be able to find the transmit and receive spatial filters using generalized singular value and eigenvalue decompositions. The solution is sub-optimum but it is faster than the alternative optimization, because multiple iterations are not needed. Naturally, the transmit and receive spatial filters must be solved again, or tracked, whenever the underlying channels change.

Instead of maximizing SINR as in (4.35) it is possible to minimize post-processing MSE given by

$$\mathcal{E}\left\{||\mathbf{W}_{rx}[n] \star (\mathbf{H}_{rx}[n] \star \mathbf{x}_S[n] + \mathbf{H}_i[n] \star \delta\mathbf{x}_T[n] + \delta\mathbf{y}_T[n]) + \mathbf{n}_I[n] - \mathbf{x}_S[n - \tau]||^2\right\} \tag{4.42}$$

assuming again that the time-domain filter has converged and the known part of the self-interference has been suppressed. The delay τ is a design parameter that can be adjusted based on filter lengths. This approach together with alternating optimization has been pursued in [47]. Transmitter distortions and imperfect channel estimation can be also used to formulate a robust filter design problem [48].

The following two figures illustrate the performance of the spatial filter design. The source and the destination have two antennas, and the source sends two independent data streams from two antennas. Transmitted OFDM waveform from each antenna has 8192 subcarriers and 64-QAM modulation. The oversampling factor in the full-duplex receiver is two. The number of taps is two in all channels and filters, and the channels are drawn from complex Gaussian distribution. The self-interference channel is 30 dB stronger than the other two channels, and the maximum transmit power is 20 dB. The linear constraint is set according to (4.38). The destination has two receive antennas, and the number of transmit and receive antennas in the full-duplex transceiver and the transmit and receive noise levels is varying.

Figure 4.10 shows the additional isolation by the spatial self-interference suppression w.r.t. reference system with time-domain cancellation only (the ratio between the respective residual interference powers) as a function of transmitter distortion (4.5). The capability of suppressing self-interference strongly depends on the degrees of freedom available, i.e., on the number of transmit antennas w.r.t. the number of receive antennas. When the number of antennas is the same, the gain from spatial suppression is close to 0 dB. The isolation can be improved also by increasing the order of $\mathbf{W}_{tx}$ [44].

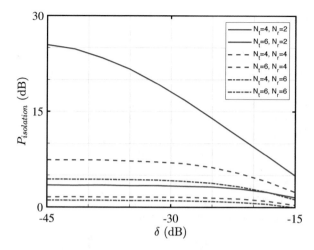

Fig. 4.10 Suppression of self-interference for different number of transmit and receive antennas as a function of transmit noise level

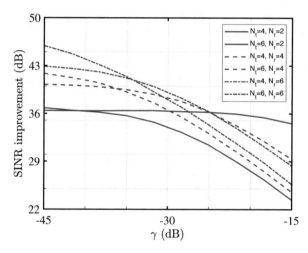

Fig. 4.11 SINR improvement as a function of dynamic range in the receiver (quantization noise level)

Figure 4.11 shows the SINR improvement as a function of receiver dynamic range w.r.t. to the reference system using $\epsilon_y = -30$ dB in (4.3). When the receiver has large dynamic range ($\epsilon_y < -30$) the noise due to self-interference is rather slow, and the best performance is obtained when the number of receive antennas is large, and when there are more transmit than receive antennas. With narrow dynamic range $\epsilon_y > -30$, a larger number of transmit antennas is preferable.

4.5 Summary

Passive isolation and active RF cancellation are usually not enough, and full-duplex transceivers need to cancel the self-interference also in digital baseband to push the residual interference below noise level. Cancellation and filtering in frequency domain on subcarrier basis works well as long as the subcarriers in the received multicarrier waveform remain sufficiently orthogonal. Here we presented adaptive filtering and cancellation in time domain that is agnostic to transmitted waveform and does not require neither pilot symbols for channel estimation nor synchronization between transmitted and received waveforms. The adaptive algorithms aim to minimize the power of the received signal and they are able to suppress the self-interference below noise level.

Linear finite-impulse response filters for self-interference cancellation are inherently stable and they can get rid of the known interfering signal, but they are not able to cancel unknown transmitter distortions. Unfortunately, transmitter distortions are often so large that suppressing only the known transmitted signal in digital baseband is not enough. A well-known solution is to model the analog distortion terms by behavioral models in digital baseband and include the effect of RF imperfections in the cancellation process. The drawback of these non-linear cancellation techniques is that their performance is more difficult to analyze than that of the linear ones.

Instead of applying non-linear filters, we complemented the time-domain interference cancellation with spatial linear filters in the transmitter and receiver chains. Together these linear filters are able to suppress transmitter and receiver distortion effects as long as there are enough degrees of freedom in terms of filter order, and transmitter and receiver antennas.

Acknowledgements The author wants to thank Dr. Gustavo J. González for providing the full-duplex interference figure, Dr. Emilio Antonio Rodríguez for providing the simulation results of the adaptive filters, and Prof. Taneli Riihonen for the years spent with full-duplex research.

References

1. M. Rice, *Digital Communications: A Discrete-Time Approach*, Prentice Hall, 2009
2. T. Riihonen and R. Wichman, "Analog and digital self-interference cancellation in full-duplex MIMO-OFDM transceivers with limited resolution in A/D conversion," in *Proc. 46th Asilomar Conference on Signals, Systems and Computers*, Nov 2012, pp. 45–49.
3. B. Widrow and I. Kollár, *Quantization Noise: Roundoff Error in Digital Computation, Signal Processing, Control, and Communications*. Cambridge, UK: Cambridge University Press, 2008.
4. B. P. Day, A. R. Margetts, D. W. Bliss, and P. Schniter, "Full-Duplex MIMO Relaying: Achievable Rates under Limited Dynamic Range," *IEEE Journal on Selected Areas in Communications*, vol. 30, no. 8, pp. 1541–1553, sep 2012.
5. E. Antonio-Rodriguez, R. Lopez-Valcarce, T. Riihonen, S. Werner, and R. Wichman, "Subspace-constrained SINR optimization in MIMO full-duplex relays under limited dynamic range," in *IEEE Workshop on Signal Processing Advances in Wireless Communications, SPAWC*, 2015.

6. T. R. Turlington, *Behavioral Modeling of Nonlinear RF and Microwave Devices*. Artech House, 2000.
7. M. Cheong, S. Werner, M. Bruno, J. Figueroa, J. Cousseau, and R. Wichman, "Adaptive piecewise linear predistorters for nonlinear power amplifiers with memory," *IEEE Transactions on Circuits and Systems I: Regular Papers*, vol. 59, no. 7, 2012.
8. G. Karam and H. Sari, "Analysis of predistortion, equalization, and ISI cancellation techniques in digital radio systems with nonlinear transmit amplifiers," *IEEE Transactions on Communications*, vol. 37, no. 12, pp. 1245–1253, 1989.
9. L. Ding, R. Raich, and G. Zhou, "A Hammerstein predistortion linearization design based on the indirect learning architecture," in *International Conference on Acoustics, Speech, and Signal Processing (ICASSP)*, vol. 3, 2002.
10. F. Gregorio, J. Cousseau, S. Werner, T. Riihonen, and R. Wichman, "Compensation of IQ imbalance and transmitter nonlinearities in broadband MIMO-OFDM," in *Proceedings - IEEE International Symposium on Circuits and Systems*, 2011.
11. J. Choi, M. Jain, K. Srinivasan, P. Levis, and S. Katti, "Achieving single channel, full duplex wireless communication," in *Proceedings of the 16th Annual International Conference on Mobile Computing and Networking (Mobicom)*, 2010.
12. M. Duarte, C. Dick, and A. Sabharwal, "Experiment-driven characterization of full-duplex wireless systems," *IEEE Transactions on Wireless Communications*, vol. 11, no. 12, pp. 4296–4307, Dec 2012.
13. M. Chung, M. S. Sim, J. Kim, D. K. Kim, and C. Chae, "Prototyping real-time full duplex radios," *IEEE Communications Magazine*, vol. 53, no. 9, pp. 56–63, 2015.
14. D. Korpi, M. Heino, C. Icheln, K. Haneda, and M. Valkama, "Compact Inband Full-Duplex Relays with Beyond 100 dB Self-Interference Suppression: Enabling Techniques and Field Measurements," *IEEE Transactions on Antennas and Propagation*, vol. 65, no. 2, 2017.
15. 3GPP, "Evolved Universal Terrestrial Radio Access (E-UTRA); Base Station (BS) radio transmission and reception," 3GPP TS 36.104 version 12.7.0 Release 12), sep 2014.
16. A. Koohian, H. Mehrpouyan, M. Ahmadian, and M. Azarbad, "Bandwidth efficient channel estimation for full duplex communication systems," in *2015 IEEE International Conference on Communications (ICC)*, 2015, pp. 4710–4714.
17. T. Riihonen, P. Mathecken, and R. Wichman, "Effect of oscillator phase noise and processing delay in full-duplex OFDM repeaters," in *Asilomar Conference on Signals, Systems and Computers*, 2012.
18. A. Sahai, G. Patel, and A. Sabharwal, "Phase Noise: Understanding the Bottleneck in Full-duplex Designs," in *Asilomar Conference on Signals, Systems and Computers*, 2012.
19. B. Hassibi and B. M. Hochwald, "How much training is needed in multiple-antenna wireless links?" *IEEE Transactions on Information Theory*, vol. 49, no. 4, pp. 951–963, 2003.
20. M. Vehkaperä, T. Riihonen, R. Wichman, and B. Xu, "Power allocation for balancing the effects of channel estimation error and pilot overhead in full-duplex decode-and-forward relaying," in *Signal Processing Advances in Wireless Communications, SPAWC*, 2016.
21. P. Persson, M. Coldrey, A. Wolfgang, and P. Bohlin, "Design and Evaluation of a 2 x 2 MIMO Repeater," in *Proc. 3rd European Conference on Antennas and Propagation*, mar 2009.
22. K. Haneda, E. Kahra, S. Wyne, C. Icheln, and P. Vainikainen, "Measurement of Loop-Back Interference Channels for Outdoor-to-Indoor Full-Duplex Radio Relays," in *4th European Conference on Antennas and Propagation*, Apr 2010.
23. T. Riihonen, A. Balakrishnan, K. Haneda, S. Wyne, S. Werner, and R. Wichman, "Optimal eigenbeamforming for suppressing self-interference in full-duplex MIMO relays," in *45th Annual Conference on Information Sciences and Systems (CISS)*. IEEE, Mar 2011, pp. 1–6.
24. S. N. Venkatasubramanian, C. Zhang, L. Laughlin, K. Haneda, and M. A. Beach, "Geometry-based modelling of self-interference channels for outdoor scenarios," *IEEE Transactions on Antennas and Propagation*, 2019.
25. T. Levanen, J. Talvitie, R. Wichman, V. Syrjala, M. Renfors, and M. Valkama, "Location-aware 5G communications and Doppler compensation for high-speed train networks," in *EuCNC 2017 - European Conference on Networks and Communications*, 2017.

26. G. González, F. Gregorio, J. Cousseau, T. Riihonen, and R. Wichman, "Generalized Self-Interference Model for Full-Duplex Multicarrier Transceivers," *IEEE Trans. Comm.*, 2019.

27. G. B. Folland and A. Sitaram, "The uncertainty principle: A mathematical survey," *J. Fourier Anal. and Appl.*, vol. 3, no. 3, pp. 207–238, 1997.

28. F. Schaich, "Filterbank based multi carrier transmission (FBMC)—evolving OFDM: FBMC in the context of WiMAX," in *European Wireless Conference (EW)*, Apr 2010, pp. 1051–1058.

29. N. Michailow, I. Gaspar, S. Krone, M. Lentmaier, and G. Fettweis, "Generalized frequency division multiplexing: Analysis of an alternative multi-carrier technique for next generation cellular systems," in *International Symposium on Wireless Communication Systems (ISWCS)*, Aug 2012, pp. 171–175.

30. G. Berardinelli, K. Pajukoski, E. Lahetkangas, R. Wichman, O. Tirkkonen, and P. Mogensen, "On the Potential of OFDM Enhancements as 5G Waveforms," in *IEEE Vehicular Technology Conference (VTC Spring)*, May 2014, pp. 1–5.

31. C. Boyd, R.-A. Pitaval, O. Tirkkonen, and R. Wichman, "On the time-frequency localisation of 5G candidate waveforms," in *IEEE Workshop on Signal Processing Advances in Wireless Communications, SPAWC*, 2015.

32. W. Chung, D. Hong, T. Riihonen, and R. Wichman, "Interference Cancellation Architecture for Full-Duplex System with GFDM Signaling," in *European signal processing conference (EUSIPCO)*, aug 2016.

33. P. Schniter, "Low-complexity equalization of OFDM in doubly selective channels," *IEEE Transactions on Signal Processing*, vol. 52, no. 4, pp. 1002–1011, 2004.

34. G. González, F. Gregorio, J. Cousseau, R. Wichman, and S. Werner, "Uplink CFO compensation for FBMC multiple access and OFDMA in a high mobility scenario," *Physical Communication*, vol. 11, 2014.

35. T. Riihonen, S. Werner, and R. Wichman, "Mitigation of Loopback Self-Interference in Full-Duplex MIMO Relays," *IEEE Transactions on Signal Processing*, vol. 59, no. 12, pp. 5983–5993, Dec 2011.

36. U. Ugurlu, T. Riihonen, and R. Wichman, "Optimized in-band full-duplex MIMO relay under single-stream transmission," *IEEE Transactions on Vehicular Technology*, vol. 65, no. 1, 2016.

37. E. Everett, C. Shepard, L. Zhong, and A. Sabharwal, "SoftNull: Many-Antenna Full-Duplex Wireless via Digital Beamforming," *IEEE Transactions on Wireless Communications*, vol. 15, no. 12, pp. 8077–8092, Dec 2016.

38. J. P. Doane, K. E. Kolodziej, and B. T. Perry, "Simultaneous transmit and receive performance of an 8-channel digital phased array," in *2017 IEEE International Symposium on Antennas and Propagation & USNC/URSI National Radio Science Meeting*, Jul 2017, pp. 1043–1044.

39. R. Penrose, "A Generalized Inverse for Matrices," *Mathematical Proceedings of the Cambridge Philosophical Society*, vol. 51, no. 3, pp. 406–413, Jul 1955.

40. T. Riihonen, S. Werner, and R. Wichman, "Mitigation of loopback self-interference in full-duplex MIMO relays," *IEEE Transactions on Signal Processing*, vol. 59, no. 12, 2011.

41. P. Diniz, *Adaptive Filtering: Algorithms and Practical Implementation*, 4th ed. Springer, 2013.

42. S. S. Haykin and Simon, *Adaptive filter theory*. Prentice Hall, 1996.

43. E. Antonio-Rodríguez, S. Werner, R. López-Valcarce, T. Riihonen, and R. Wichman, "Wideband full-duplex MIMO relays with blind adaptive self-interference cancellation," *Signal Processing*, vol. 130, 2017.

44. E. Antonio-Rodriguez, R. Lopez-Valcarce, T. Riihonen, S. Werner, and R. Wichman, "SINR optimization in wideband full-duplex MIMO relays under limited dynamic range," in *Proceedings of the IEEE Sensor Array and Multichannel Signal Processing Workshop*, 2014.

45. J. C. Bezdek and R. J. Hathaway, "Convergence of Alternating Optimization," *Neural, Parallel Sci. Comput.*, vol. 11, no. 4, pp. 351–368, Dec 2003.

46. G. H. Golub and C. F. Van Loan, *Matrix Computations (3rd Ed.)*. Baltimore, MD, USA: Johns Hopkins University Press, 1996.

47. E. Antonio-Rodríguez, S. Werner, R. López-Valcarce, and R. Wichman, "MMSE Filter Design for Full-duplex Filter-and-forward MIMO Relays under Limited Dynamic Range," *Signal Processing*, vol. 156, pp. 208–219, 2019.
48. E. Antonio-Rodriguez, S. Werner, T. Riihonen, and R. Wichman, "Robust filter design for full-duplex relay links under limited dynamic range," in *IEEE Workshop on Signal Processing Advances in Wireless Communications, SPAWC*, 2017.

Part II
Future Trends and Applications

Chapter 5
Interference Management in Full-Duplex Cellular Networks

Ali Cagatay Cirik and Yingbo Hua

Abstract Despite its promising potential to double the throughput of a point-to-point radio link, a full-duplex (FD) radio has not been comprehensively analyzed in current cellular systems due to the high levels of interference it generates, which significantly degrades its performance. If not carefully planned and managed, an FD operation might lead to much higher interference in both uplink and downlink than existing half-duplex (HD) operation, limiting its potential gains greatly. This chapter provides an overview of the challenges caused by FD radio links in cellular systems as well as the techniques to overcome those challenges to unlock the full potential of FD wireless communications.

5.1 Introduction

The current cellular networks are no longer voice-centric but data-centric, especially with smart-phones leading to a proliferation of data-hungry applications. For example, data traffic in mobile communication systems is expected to increase by around 1000 times by 2020 and by more than $40,000$ times by 2030 [1] from 2010. Mobile broadband has been identified as essential to creating jobs, growing economy and benefiting the society because advances in mobile networks will create new opportunities for innovative applications, including devices and services for health, transportation, manufacturing. To support the exponentially growing demand of mobile services to provide high data rates and low latencies, key emerging technologies are being developed in mobile communication systems,

A. C. Cirik (✉)
Ofinno Technologies, Reston, VA, USA
e-mail: acirik@ofinno.com

Y. Hua
Department of Electrical and Computer Engineering, University of California, Riverside, CA, USA
e-mail: yhua@ece.ucr.edu

© Springer Nature Singapore Pte Ltd. 2020
H. Alves et al. (eds.), *Full-Duplex Communications for Future Wireless Networks*,
https://doi.org/10.1007/978-981-15-2969-6_5

such as full-duplex (FD) radio, massive multiple-input multiple-output (MIMO), non-orthogonal multiple access (NOMA), etc.

Although great progress has been achieved in designing and implementing FD wireless nodes with the capability of suppressing the self-interference to a sufficiently low level [2–6], the impact of FD transmissions in a wireless network has not been extensively analyzed, especially in a cellular environment, which has a more complicated interference environment. In particular, how to best utilize an FD radio in a wireless cellular network needs sufficient attention. Given the complexity of radio link configurations in a typical cellular network, how to minimize the mutual interferences between FD radio links and/or the performance improvement obtained by the FD communication over the half-duplex (HD) communication are not straightforward for cellular networks.

At the network level, employing FD nodes, such as FD base stations (BSs) creates additional challenges to the existing challenges associated with the HD systems. As seen in Fig. 5.1, compared to HD networks, a single-cell FD cellular network with a BS operating in FD mode and user equipments (UEs) operating in HD mode newly experiences intra-cell UE-to-UE interference (or called as co-channel interference (CCI)) from the uplink users to the downlink users, as well as the self-interference at the BS. Moreover, a multi-cell FD network suffers from even worse interference situation (e.g., inter-cell interference from the BSs and the uplink users in the neighboring cells), which might lead to a severe performance deterioration, even eating up the potential gain of FD networks. Therefore, unless proper interference mitigation techniques are applied, the gain of having FD radios in cellular networks can be severely limited, and thus the use of FD BS in each cell could have a negative impact on the network throughput.

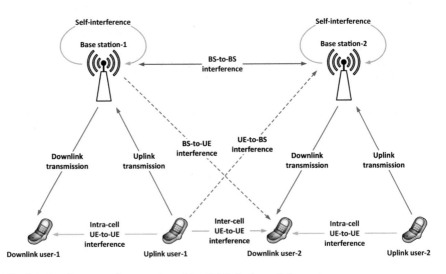

Fig. 5.1 Interference environment in multi-cell full-duplex cellular systems

As cellular networks evolve to become more heterogeneous and to operate independently with less cooperation and less synchronized with each other, intra-cell and/or inter-cell interference become more severe, which are major limiting factors to the spectral efficiency gain from the FD operation. The existence of additional interference terms in FD cellular networks raises the question of whether there is any actual spectral efficiency gain achieved from the FD operation. The authors in [7] have investigated this issue of whether to deploy FD nodes in macro-cells or micro-cells under the additional interference sources generated by FD communication. It has been shown that unless proper countermeasures for interference management are taken, the FD communication is infeasible to operate on the macro-cell networks. On the other hand, when the micro BSs are mounted by a certain distance, the FD communication enhances the spectral efficiency of micro-cell networks. In particular, the actual net gain enabled by the FD operation in cellular networks will strongly depend on interference mitigation/management, network deployments, the density of UEs, and propagation effects in mobile channels. To that end, in this chapter, we will discuss efficient interference management and mitigation methods for FD cellular systems to help realizing the full potential of FD systems.

There have been a significant amount of studies on interference cancellation/coordination in traditional HD systems. For example, various solutions to suppress or coordinate the interference between the interfering neighboring cells have been proposed [8]. However, with the new additional FD interference sources generated in FD mode as shown in Fig. 5.1, uplink and downlink channel resources can no longer be allocated separately, and thus a joint allocation is required in order to support a higher number of simultaneous transmission/reception links with different channel characteristics. Therefore, the existing interference cancellation/coordination techniques proposed for the conventional HD systems cannot be applied readily to the FD systems. The main contributions of this chapter are:

- We discuss interference management and mitigation methods for the base station to base station interference (BS-to-BS interference) in FD cellular networks in Sect. 5.3, which comprises:

  - Elevation beam nulling,
  - Power control based.

- We discuss interference management and mitigation methods for the user-equipment to user-equipment interference (UE-to-UE interference) in FD cellular networks in Sect. 5.4, which comprise:

  - Scheduling-based,
  - Medium access control (MAC) techniques,
  - Interference alignment,
  - Beamforming,
  - Partitioning.

- We discuss interference management and mitigation methods in 3rd Generation Partnership Project (3GPP) from a standardization perspective in Sect. 5.5

- We discuss key interference mitigation technologies, which can be used to suppress the huge surge of interference in FD communications in Sect. 5.6, which comprise:

  - Massive MIMO,
  - Millimeter-Wave (mmWave).

5.2 Interference Management

In FD systems, the uplink and downlink transmissions/receptions may occur simultaneously at the same time and frequency resources in all cells, resulting in a far more complicated interference environment than that of HD systems. The main challenge in a single-cell FD network, besides the self-interference, is the intra-cell UE-to-UE interference (also called as CCI), which is created from the uplink transmission to the downlink reception as shown in Fig. 5.1. Moving from a single-cell to a realistic multi-cell scenario introduces a huge surge of interferences including both intra-cell UE-to-UE interference and inter-cell interference (both inter-cell UE-to-UE interference and BS-to-BS interference between all BS nodes and UE nodes as shown in Fig. 5.1). In particular, the overall interferences experienced by a downlink user and an uplink user of a cell operating in an FD mode are given, respectively, as:

$$I_{DL} = I_{\text{intra-cell-UE-to-UE}} + I_{\text{inter-cell-UE-to-UE}} + I_{\text{inter-cell-BS-to-UE}}, \qquad (5.1)$$

$$I_{UL} = I_{\text{inter-cell-UE-to-BS}} + I_{\text{inter-cell-BS-to-BS}}, \qquad (5.2)$$

where the first term in (5.1) denotes the intra-cell UE-to-UE interference from the uplink users of the same cell, the second term denotes the inter-cell UE-to-UE interference from the uplink users of the neighboring cells, and the third term denotes the conventional downlink interference (e.g., BS-to-UE) from the neighboring BSs. Moreover, in (5.2), the first term denotes the conventional uplink interference (e.g., UE-to-BS) from the neighboring uplink users and the second term denotes the BS-to-BS interference from the BSs of the neighboring cells.

As mentioned above and shown in Fig. 5.1, besides the inter-cell BS-to-UE interference and UE-to-BS interference that already exists in HD networks, an FD network experiences an additional inter-cell BS-to-BS interference, and intra- and inter-cell UE-to-UE interference, which we will review below[1]:

---

[1]Designing and implementing an FD terminal requires a significant change in the hardware resulting in a higher cost and power consumption. Therefore, it is more practical and less costly to upgrade the infrastructure elements such as BSs to operate in FD mode, while the mobile users (e.g., downlink and uplink users) can still operate in HD mode [3], which we have assumed in Fig. 5.1.

- **Base station to base station interference (BS-to-BS interference)**: In FD cellular systems, a downlink transmission of a BS may greatly interfere with an uplink reception of a neighboring cell. As the BSs are deployed at higher elevations with few obstacles/obstructions in between, the channel between the BSs is close to line-of-sight (LoS) resulting in much smaller path-loss between the BSs compared to the path-loss between the BSs and the UEs. Moreover, the transmission power of a BS is much larger than that of a UE which increases the interference further when combined with the small path-loss between the BSs. Therefore, if the BS-to-BS interference is not managed/mitigated properly, it can lead to a poor performance by dominating the weak intended uplink transmission from the uplink UEs. On the other hand, as the BSs are not mobile, the BS-to-BS interference channel is static, which allows static interference mitigation techniques (e.g., elevation beam nulling [9], semi-static uplink/downlink power control [10]) to be applied for the BS-to-BS interference mitigation.
- **User-equipment to user-equipment interference (UE-to-UE interference)**: In FD cellular systems, an uplink transmission from an uplink user may generate an interference on a nearby downlink user, called as UE-to-UE interference. The UE-to-UE interference can exist between two UEs in the same cell (e.g., intra-cell UE-to-UE interference) and/or in different cells (e.g., inter-cell UE-to-UE interference). In contrast to static BS-to-BS interference channel, UE-to-UE interference channel is more dynamic as the downlink/uplink users are mobile. Therefore, the UE-to-UE interference mitigation can be realized through smart scheduling which avoids scheduling nearby uplink and downlink users (e.g., creating high UE-to-UE interference) to transmit and receive on the same time/frequency resources so that the potential capacity gain from the FD operation can be achieved. For example, the authors in [11, 12] have proposed uplink user and downlink user pairing based on a *distance* metric, while the authors in [9, 13] have incorporated power control mechanism in their proposed scheduling mechanism.

Interference management techniques are necessary to tackle various types of interference sources and to make sure that FD networks achieve better performance than the HD networks, e.g., in terms of throughput. Therefore, it is critical to carefully design protocols and algorithms to deal with the newly introduced interference in FD networks. In the next two sections, we will discuss practical and efficient interference mitigation methods for the BS-to-BS interference and the UE-to-UE interference in FD cellular systems which limit the exploitation of FD gains in cellular systems.

5.3   BS-to-BS Interference Mitigation

The BS-to-BS interference can be much stronger (e.g., 40 dB [10]) than the conventional uplink interference in HD systems, which requires practical and efficient BS-to-BS interference mitigation techniques to make sure that the uplink reception is not degraded by the BS-to-BS interference, and thus the benefits of FD operation can be realized in FD cellular networks.

In this section, we will discuss two methods to suppress the BS-to-BS interference: (1) Elevation beam nulling and (2) Power control based.

5.3.1   Elevation Beam Nulling

The BSs deployed today can employ elevation antenna down-tilting to optimize the signal strength within their coverage area and create different beam patterns via multiple antennas. The authors in [9] have revisited the same concept of elevation antenna down-tilting to mitigate the BS-to-BS interference (especially the LoS component of the interference channel) in FD cellular networks. Particularly, as the altitudes of BSs are fixed and the incoming direction of the BS-to-BS interference is relatively stable, the authors have proposed to use elevation beam nulling to suppress the BS-to-BS interference, where the BS applies 3D beamforming/beam nulling techniques to null at an arbitrary azimuth angle and elevation angle. It is shown that when every BS has a similar altitude, the BS-to-BS interference can be greatly suppressed by forming nulls at the vicinity of 90° in elevation. For BSs with different altitudes, a wider null width is required to suppress the BS-to-BS interference.

By using the elevation nulling approach, around 70 dB to 80 dB BS-to-BS interference suppression can be achieved for small cell and large cell deployments, respectively [10], which is weaker than the conventional uplink interference level, and thus the benefits of FD operation can be realized.

One drawback of the elevation nulling approach is that it may cause degradation on received power levels of the UEs on the cell edge. Therefore, a cell size reduction may be required while applying elevation nulling to guarantee a fair performance for a UE on the cell edge. Moreover, in a multi-path environment, there may be beams reflected from the main lobe causing strong BS-to-BS interference, which may prevent from achieving the full nulling gain [10]. Therefore, additional interference mitigation methods may be needed to combat the residual BS-to-BS interference.

5.3.2   Power Control Based

The BS-to-BS interference can be further suppressed by power control (e.g., uplink and/or downlink) mechanisms. For example, when an uplink user increases its transmission power, it can dominate the BS-to-BS interference (e.g., increased uplink signal-to-interference plus noise ratio (SINR)), but at the same time it can also create additional increased interference on a nearby downlink user (e.g., decreased downlink SINR) in addition to increasing its power consumption. In another example, when a BS decreases its downlink transmission power to create less BS-to-BS interference, this may result in a reduced coverage and deteriorated performance for the downlink users.

Although the authors in [14] have proposed a power control algorithm to suppress the BS-to-BS interference, they have not considered the impact of their power control mechanism on the downlink users as they have not taken the UE-to-UE interference into account in their proposed power control algorithm. In [10], an open-loop uplink power control mechanism has been proposed for FD cellular systems which considers both the BS-to-BS interference and the UE-to-UE interference in the power control design to minimize the uplink SINR degradation. In particular, the authors propose to boost the target received power level used in Long-Term Evolution (LTE) systems (e.g., P_0) based on the distribution of the BS-to-BS interference over conventional uplink UE-to-BS interference, and conventional downlink BS-to-UE interference over UE-to-UE interference. For example, as each BS may experience different levels of BS-to-BS interference, as the BS-to-BS interference increases (decreases), a higher (lower) target uplink received power (i.e., P0) may be selected by each cell independently, which is dynamic.

5.4 UE-to-UE Interference Mitigation

In this subsection, we discuss the efficient and practical methods to mitigate the UE-to-UE interference in FD cellular networks.[2] Handling the UE-to-UE interference is much more challenging than handling the BS-to-BS interference as it is between distributed users, who cannot share data information without sacrificing bandwidth resources. As the level of UE-to-UE interference depends on the UE locations and their transmission powers, coordination mechanisms are needed to mitigate the negative effect of the interference on the spectral efficiency of the system [13].

Below, we will discuss various methods, such as scheduling, interference alignment, beamforming, partitioning, etc., to mitigate/suppress the UE-to-UE interference.

---

[2]Note that some of the techniques discussed in this section can also be applied to mitigate the BS-to-BS interference.

5.4.1 Scheduling-Based

Scheduling has been a commonly used approach in multi-user communication systems to optimize system performance such as link data rate, network capacity, and coverage. In conventional orthogonal frequency division multiple access (OFDMA)-based cellular systems operating in HD mode, the scheduling decision at a BS is made independently for uplink and downlink transmissions. Moreover, users served by the same cell are allocated orthogonal resource blocks (RBs) and/or time slots so that there exists no interference between them.

As mentioned in Sect. 5.2, as the uplink and downlink transmissions are coupled and simultaneous in FD systems, a scheduling-based interference management is required. In particular, a scheduling problem in an FD system boils down to pair one uplink UE and one downlink UE properly from all active UEs and to select a frequency channel (e.g., RB, subcarrier, sub-band) to determine which UEs should be scheduled for simultaneous uplink and downlink transmissions over the same frequency channels. The existing scheduling methods used in HD cellular networks, such as proportional fairness (PF) and round-robin, can also be used in the FD cellular networks directly. For example, the scheduler in the FD network can employ the well-known PF procedure for an independent selection of the downlink and uplink UEs and pair them for simultaneous transmission and reception. However, the ignorance of the UE-to-UE interference in such a method could degrade the performance. For example, when an uplink user creating strong UE-to-UE interference on a downlink user is selected to be served on the same RB for a simultaneous transmission, the downlink SINR of the downlink user is highly degraded. Therefore, interference awareness in scheduling should be an important feature for the UE pairing process.

Within each selected pair, the transmission of the uplink user will cause CCI to the downlink user, and this interference varies largely with the mutual distance between the uplink and downlink users of each pair and transmission powers, etc. The uplink and downlink user pairing, subcarrier, and power allocation among different pairs of users need to be properly managed to achieve the optimal performance in the network. In particular, frequency channel (e.g., RB, subcarrier) allocation involves allocating the different subsets of subcarriers (or RBs or bands) to different users by taking into account the self-interference at the BS and the CCI between the uplink and downlink users within each user pair. This is significantly different from the traditional subcarrier allocation problem and presents further research challenges in resource allocation. For example, a scheduler selecting the best pair of uplink and downlink users based on a joint scheduling metric optimizing both uplink and downlink transmissions is required. When the joint scheduling metric which takes the UE-to-UE interference into account is optimized, the UE-to-UE interference is directly managed by selecting the best pair of users creating the least UE-to-UE interference on the frequency channel (e.g., subcarrier, RB). However, due to the combinatorial nature of pairing multiple uplink users, downlink users, and subcarriers, and also the complexity of optimal power allocation to each subcarrier-transceiver pair, resource allocation in such an FD OFDMA network can be very challenging [15].

The joint scheduling metric can be PF scheduler, which is generally used in current cellular systems, or a maximum-sum-throughput scheduler, or a maximum-minimum-throughput scheduler. Since an FD BS schedules both uplink and downlink transmissions, the objective function (e.g., a function of achievable downlink and uplink data rates) should consider the performance of both uplink and downlink. When the scheduler selects a pair of uplink and downlink users on an RB, the scheduler searches across all candidate uplink and downlink user pairs on the RB and selects the pair achieving the highest joint scheduling metric (or with the lowest interference level for transmission and reception on the same scheduling RB). The selection procedure can be an exhaustive search or an optimization algorithm (e.g., greedy search, dynamic programming, Hungarian algorithm, etc.).

The authors in [10] show that when the inter-cell interference dominates the intra-cell interference, a scheduler that independently schedules the uplink and downlink transmission directions can provide the FD gain. On the other hand, when the intra-cell interference dominates the inter-cell interference, a scheduler that jointly schedules the uplink and downlink transmission directions by avoiding a pair of downlink and uplink users with a high UE-to-UE interference being paired together can improve the downlink SINR.

The management of the UE-to-UE interference caused by the intra-cell and inter-cell users may require extra signaling, feedback, and coordination between BSs. For example, to deal with both intra- and inter-cell interference, each cell first selects or pairs the appropriate users to maximize its spectral efficiency, then the scheduler coordinates with other cells to allocate the power levels of selected users such that the aggregate network spectral efficiency is maximized. This may result in a high computational complexity and signaling overhead diminishing the gains obtained by the FD operation. Therefore, distributed scheduling algorithms are required to mitigate the UE-to-UE interference [10].

5.4.1.1 Related Work

In [16], the authors study the joint subcarrier and power allocations to maximize the sum rate in FD OFDMA networks. In [17], a suboptimal user pairing and resource allocation algorithm that maximizes the sum achievable rate while guaranteeing a quality-of-service (QoS) of each user have been proposed. An application of an opportunistic interference suppression technique at the user side to both mitigate the UE-to-UE interference and maximize the system spectral efficiency has been proposed in [18]. A simple opportunistic joint uplink and downlink scheduling algorithm has been proposed in [19], which suppresses, asymptotically, the UE-to-UE interference in a single-cell MIMO FD network as if there were no UE-to-UE interference. In [20], the problem of pairing uplink and downlink users with the joint power allocation along with the consideration of user fairness has been studied. Moreover, a resource allocation method using matching theory to optimally allocate the subcarriers among uplink/downlink user pairs for a single-cell FD OFDMA network was proposed in [21].

In these works, the authors have only considered FD communication in a single-cell scenario and ignored the inter-cell interference in their analysis. However, in the multi-cell scenario, it is not clear how much throughput gain can be achieved by the FD communication as the number of interference links increases.

Recently, several works have investigated the usage of FD communication in a multi-cell scenario. For example, to mitigate the UE-to-UE interference, a scheduling approach maximizing the achievable sum-rate in FD multi-cellular networks has been proposed in [13, 22]. A suboptimal joint user selection and power allocation strategy has been proposed in [13, 22], which can yield around 65% throughput gain over the traditional HD system. A radio resource allocation algorithm assigning uplink and downlink transmissions jointly has been proposed in [23], by taking both intra-cell and inter-cell interference into consideration. In [24], the authors have obtained closed-form throughput expressions for the FD multi-cell networks with the help of stochastic geometry tools.

Finally, in [25], a UE-to-UE interference coordination approach based on geographical context information (e.g., provided by radio maps and user positions) has been provided, which exploits the signal attenuation from obstacles between UEs so that UE-to-UE interference is minimized by assigning simultaneous uplink and downlink transmissions to UEs in areas that are separated by large obstacles.

5.4.1.2   Space-Time Power Scheduling

Another promising idea for interference management is space-time-frequency power scheduling, also known as space-time (or space-temporal) power scheduling [26–28]. Despite its previous applications in ad hoc networks, the space-time power scheduling idea is particularly useful for small cells, where the channel state information (CSI) of various channels can be jointly exploited. The key idea behind space-time power scheduling is to best distribute the interferences within the (complete) space-time physical world that we have. In other words, channels with weak interferences should be exploited, channels with strong interferences should be avoided, and all interferences should be optimally distributed via a well-designed algorithm for optimal space-time power scheduling. The idea of space-time power scheduling has also recently been applied to FD systems [29–31]. Further research of space-time power scheduling for FD cellular networks should be of a promising direction.

5.4.2   Medium Access Control (MAC) Techniques

Apart from the aforementioned physical-layer solutions, FD research opportunities have also been explored in the context of efficient MAC protocols for addressing the challenges of long end-to-end delays of network congestion and the hidden terminal problems. As the FD networks generate new additional inter-node collisions, conventional MAC protocols such as the carrier-sense multiple access with collision avoidance (CSMA/CA)-based HD MAC protocols cannot be readily applied to reduce these additional inter-node collisions.

To that end, the authors in [32] propose an optimal location-aware node selection algorithm minimizing the intra-node interference while still maintaining the signal strength of the intended link. Moreover, transmitting RTS/CTS (Request to Send/Clear to Send) signaling or busy tones before establishing the FD connection has been proposed to reduce the inter-node collisions due to the hidden nodes.

The authors in [33] propose a simple method that avoids the UE-to-UE interference by selecting nodes that are completely hidden from each other. Another method in [34] optimizes user pairing by considering the UE-to-UE interference based on the information about the UE-to-UE interference and traffic demands reported from all pairs of users. Goyal et al. in [35] develop a centralized MAC protocol by taking the UE-to-UE interference between two users into account. In [36], the UE-to-UE interference is managed by a random access MAC protocol using distributed power control.

It is more practical to develop an adaptable MAC protocol based on the channel/network conditions (e.g., inter-cell, intra-cell interference) to switch to the desired (FD or HD) transmission mode opportunistically. To that end, Chen et al. [37] have proposed a distributed FD MAC protocol for a BS that can adaptively switch between FD and HD modes based on the channel conditions so that the negative impact of the (intra- and inter-cell) interference can be removed. In particular, when intra-cell interference is low (high), the BS selects FD (HD) mode for transmission [11, 22].

Another MAC technique that can be applied for interference mitigation is clustering in small cell networks. An FD operation is more suitable for small cells due to low transmission power at the BSs resulting in a relatively easy self-interference cancellation. However, as the small cells are densely deployed, the uplink to downlink and downlink to uplink inter-cell interference become even more critical. Therefore, efficient clustering methods/techniques that allow nearby small cells to coordinate their FD transmissions need be designed. In an example, a cluster of small cells may coordinate such that each small cell within the cluster operates in the FD mode in a sequential manner, while the other small cells operate in the HD mode until their turns come up. This would alleviate the additional interference sources generated in FD mode [32].

Another MAC technique that can be applied for interference mitigation can be cooperative FD device-to-device (D2D) transmission. While D2D links, which enable direct communication between nearby users, are utilized for HD data transmissions, they may also be utilized for interference suppression in an FD mode. In an example, a D2D transmitter can overhear interference signals on an FD receiver and can feed back/forward the interference information to its intended receiver along with the data packets for an interference suppression [32].

5.4.3 Interference Alignment

Another technique that is used for mitigating the intra- and inter-cell interference is interference alignment (IA) [38, 39]. Initially proposed by the seminal works in [40, 41], IA is a coding technique that efficiently deals with interference and is known to achieve the optimal degrees-of-freedom (DoF) for various interference networks [42]. Especially, it is shown that IA can be successfully applied to mitigate interference in various cellular networks [42].

The authors in [38, 39] employ signal space IA schemes optimized for FD networks to manage inter-user interference and fully utilize the wireless spectrum with FD operation. In particular, in [38, 39], transmit beamformers, on the uplink channel, are designed for aligning the UE-to-BS interference, and intra- and inter-cell UE-to-UE interference. On the other hand, on the downlink channel, each BS designs the transmit beamformers to align the downlink multi-user interference, BS-to-UE interference, and BS-to-BS interference. It is shown that the achievable sum-DoF obtained with the IA technique in an FD multi-cell system increases with the number of cells, users, and BS antennas. In particular, the key idea of the proposed schemes is to carefully allocate the uplink and downlink information streams using IA and beamforming techniques. The uplink data is sent to the BS using IA such that the inter-user interference is confined within a tolerated number of signal dimensions, while the BS transmits in the remaining signal dimensions via IA beamforming or zero-forcing beamforming for the downlink transmission. This would result in nulling out the downlink and uplink intra-cell and inter-cell interferences.

To suppress the UE-to-UE interference in FD communication systems, the authors in [43] propose an interference management strategy enabling the network to suppress the UE-to-UE interference while obtaining achievable rate gains. In their following work, the authors propose an IA scheme to suppress the UE-to-UE interference for MIMO FD communication systems in order to achieve rate gains over conventional cellular systems in terms of DoF [44]. As the downlink mobile user observes a MAC of two users (e.g., BS and the uplink user) in FD systems, the authors in [45] propose to apply successive interference suppression (SIC) so that the downlink user has an opportunity to remove the UE-to-UE interference according to the transmission rates and its received powers. In their following work, the authors in [46] reuse the interference suppression method applied for the X-interference channels for a superposition coding based UE-to-UE interference suppression.

Although IA does not reduce the spectral efficiency, it requires accurate CSI between communicating and interfering users. Such CSI comes at the cost of high signaling overhead. Since interference alignment operates on a number of time slots that depends on the CSI, it also adds a non-deterministic transmission delay.

5.4.4   Beamforming

One of the techniques used for mitigating the intra- and inter-cell UE-to-UE interference and/or BS-to-BS interference is beamforming design. In FD systems, the uplink and downlink transmission optimization problems can no longer be separated due to the joint dependency of throughput functions on the uplink and downlink transmissions. Therefore, the uplink and downlink transmissions in FD networks have adverse effects on each other's performance. For example, if the downlink transmit power increases to combat the UE-to-UE interference at the downlink users, the self-interference also increases at the BS, and thus the uplink sum rate decreases. On the other hand, if the uplink users increase their transmit powers to dominate the self-interference at the BS, the UE-to-UE interference increases at the downlink users. Thus, in an FD multi-user network, transmission strategies at the BS and the uplink users are coupled, and thus have to be designed to address both UE-to-UE interference and self-interference simultaneously, which poses a jointly coupled optimization problem.

As mentioned above, it is attractive and useful to analyze the mutual effect of uplink and downlink transmissions in FD networks by designing precoding matrices for both uplink and downlink transmissions to manage the newly generated interference sources. In a beamforming design, an interference management can be performed without explicitly avoiding the UE-to-UE interference in the sense that the BSs and the users use their multiple antennas for interference mitigation. The scheduler collects all the required CSI and then maximizes the various optimization metrics (e.g., sum power, harmonic sum, sum rate, etc.) of the network by employing optimal transmit and receiver beamformers [47–57] for mitigating the self-interference and the UE-to-UE interference in the FD multi-user systems.

Finding a feasible point for the various optimization metrics is already challenging due to the nonconvexity and disconnectivity of the feasible set. The authors in [58] exploit the well-known duality between broadcast channel (BC) and MAC and reformulate the sum-rate maximization problem in the BC as an equivalent optimization problem for MAC, resulting in a concave function.

Channel Estimation   The mitigation of the BS-to-BS interference and the UE-to-UE interference requires accurate knowledge of the channel(s) between the BSs and the UEs (between uplink and downlink users), respectively, so that a joint power-rate optimization [9, 13] or a joint transmit/receive beamforming [12] can be performed by the scheduler. Estimation of an interference channel requires pilot (e.g., preamble) transmissions and/or feedback of the estimated interference measurement.

The levels of interference awareness for scheduling and/or beamforming may vary. For example, tracking the UE-to-UE interference channel which is dynamic and fast varying results in a knowledge of a short-term interference giving the best performance, but at the same time resulting in the highest overhead and signaling, especially as the number of users increases. On the other hand, tracking long-term statistics of the UE-to-UE interference channel (e.g., the path-loss, large-scale fading terms) can be obtained easily from the relative positions of the UEs, converting the scheduling problem into a distance-aware scheduling problem [59, 60]. Tracking long-term statistics of the BS-to-BS interference channel with low overhead and less signaling is also realistic due to the static channel between the BSs.

For the joint scheduling and/or beamforming, the scheduler has to obtain an accurate downlink data rate estimate which can be acquired via feedback of rank indicator (RI), precoding matrix indicator (PMI), and channel quality indicator (CQI). As the UE-to-UE interference has less impact on the transmit direction and rank for beamforming at the BS, the same RI and PMI feedback mechanisms in legacy systems (e.g., LTE) can be reused for FD networks. However, the CQI report reflects the received downlink SINR, which is highly based on and sensitive to the UE-to-UE interference. Therefore, for every downlink and uplink user pair, the scheduler must acquire an accurate CQI reflecting the UE-to-UE interference [10].

In LTE, the BSs require sounding reference signals (SRS) regularly to estimate the uplink channel quality at different frequency bands for uplink scheduling. The authors in [22] propose to reuse the uplink SRS at the downlink UEs for the UE-to-UE interference channel estimation between themselves and nearby UEs [61]. All UEs within a cell are informed about the subframes that will be used for SRS. The biggest challenge in reusing the SRS in FD systems as a neighbor discovery is to distinguish between different (both in the same and in the different cell) UEs during SRS transmission. One way to help the UEs distinguish different UEs is to assign different SRS combination sets to neighboring cells while assigning orthogonal combinations to UEs within the same cell which are scheduled to transmit simultaneously [22]. In addition, this allocation of SRS combinations can be passed to UEs through the downlink shared channel [61].

There are other ways to design neighbor discovery, such as the ones used in D2D communications [62, 63]. In an example, each UE can estimate the (interference) channels from strong UE interferer, and in response to completing the estimation, the UE can forward this estimation to its serving BS at the beginning of each scheduling slot. When the serving BS receives this information from many UEs, it can make its scheduling decision at each scheduling time slot.

5.4.5 Partitioning

To reduce the effect of the UE-to-UE interference and maximize the spectral efficiency, an intelligent scheduler is required to pair the uplink and downlink users as well as to compute the optimal transmit powers of the BS and users. A natural and efficient UE-to-UE interference mitigation approach is to separate the uplink and downlink users as far as possible. To that end, the authors in [64] propose the use of cell sectorization for the separation. As shown in Fig. 5.2, the uplink and downlink users are scheduled in two opposite (e.g., 120°) sectors of time-frequency resources to avoid scheduling an uplink user and a downlink user on a same time-frequency resource, reducing, if not eliminating, the UE-to-UE interference.

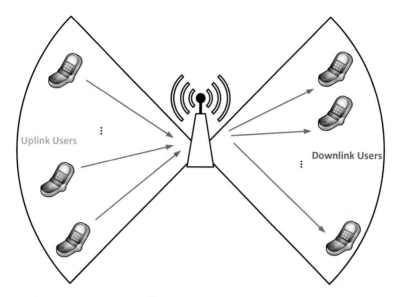

Fig. 5.2 Cell sectorization to reduce the UE-to-UE interference. ©2018 IEEE. Reprinted, with permission, from [64]

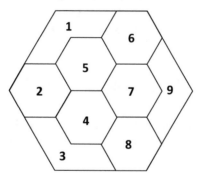

Fig. 5.3 Cell partitioning to reduce the UE-to-UE interference

Once user scheduling in opposite sectors is performed, the optimal powers can be computed for the BS and uplink users.

Similarly, Shao et al. [60] propose partitioning method that the whole cell is divided into several partitions where the same frequency resource is assigned/allocated properly to the two users who are far enough from each other. For example, as shown in Fig. 5.3, by dividing the frequency resources into nine groups, group 1 can be assigned to use frequency-1 on the uplink and frequency-2 on the downlink, while frequency-2 can be reused in the uplink for the group 9 while frequency-1 can be reused in the downlink in the group 4. This would alleviate the intra-cell UE-to-UE interference between users in the FD cellular system as the co-channel reverse links are apart enough from each other. However, this method may not work when the cell gets smaller.

The existing inter-cell interference coordination (ICIC) methods are used to suppress the inter-cell interference in HD systems to improve the performance of cell-edge users in LTE systems [65] via static, semi-static, or dynamic resource coordination, and cannot be directly used in FD cellular systems to suppress the additional interference sources. Therefore, in [65], the authors have proposed new ICIC designs to mitigate the inter-cell UE-to-UE interference in FD cellular systems. In particular, for the inter-cell UE-to-UE interference suppression, the total available bandwidth is divided into multiple (e.g., three) orthogonal sub-bands (zones) under different interference levels in each zone. For example, the first zone can be an uplink-only zone, in which only uplink transmission can be performed, which is beneficial for weak uplink UEs; the second zone can be downlink-only zone, in which only downlink transmission can be performed, which is beneficial for downlink users exposed to high interference levels from uplink users; and the third zone is downlink-centric and uplink-centric zones where major downlink-victims are restricted to transmit only in downlink-centric sub-bands and the uplink-aggressors can only transmit in orthogonal uplink-centric sub-bands. In addition, the proposed ICIC algorithm allows adjustable BS transmission power in different zones. In an example, the BS can set its power to a higher level in downlink-centric zone, and to a lower level in uplink-centric zone to further improve the throughput of downlink-victims and uplink-aggressors.

5.5 Interference Mitigation in the 3GPP

In this section, we discuss technical solutions for an LTE-compatible FD cellular network, featuring intra- and inter-cell interference mitigation techniques.

5.5.1 Frame Structure

Frame structure is the basic uplink/downlink operation framework in wireless communication systems specifying the timing and location of the transmission of the control signaling and uplink/downlink data. Due to the additional interference sources in an FD cellular system, the existing frame structure needs to be tailored for interference mitigation capabilities in an FD communication.

In the LTE/LTE-Advanced (LTE-A) systems, reference signals (RSs) such as cell-specific reference signals (CRSs), SRS, and downlink channel-state-information RS (CSI-RS) are designed for channel estimation [66]. However, since the current frame structures for LTE and LTE-A were originally designed for the HD modes such as frequency division duplexing (FDD) and time division duplexing (TDD), the resource elements (REs) carrying the downlink RSs (e.g.,

CRSs, CSI-RSs) are interfered by the REs in the uplink channel co-located with the downlink RSs, and thus self-interference channel estimation and/or BS-to-BS interference based on the downlink RSs is degraded.

To deal with this problem, the authors in [67] propose an uplink nulling technique that prevents a certain portion of REs allocated to, for example, the downlink RSs (e.g., CRSs), the physical broadcast channel (PBCH), the primary synchronization signal (PSS), and/or the secondary synchronization signal (SSS), being used by the REs in the uplink in order not to degrade the self-interference and BS-to-BS interference channel estimation. Since some of the uplink resources are not used, this will result in a throughput loss and resource waste. However, the loss will not be significant due to the traffic asymmetry between the downlink and uplink [67]. After the self-interference channel and/or BS-to-BS interference channel is estimated, the BS can perform the self-interference cancellation. Moreover, the BS-to-BS interference channel information can be feedback to the adjacent BSs, which pre-nulls the BS-to-BS interference via transmit beamforming. The uplink nulling technique requires minimal frame structure changes to the existing LTE and LTE-A standards. Moreover, accurate self-interference channel and/or BS-to-BS interference channel estimations require frame synchronization between the downlink and uplink, which can be achieved by using the existing uplink timing alignment mechanism in LTE/LTE-A systems which limits the uplink–downlink misalignment within the cyclic prefix [67].

A more general downlink and uplink frame structure utilizing both an uplink nulling and downlink nulling techniques is illustrated in Fig. 5.4, where the regions are defined as [60]:

- Downlink common control region (uplink blank region): The downlink cell common control information, e.g., for initial access, such as PSS, SSS, and PBCH, physical downlink control channel (PDCCH), and RSs are transmitted in this region. Therefore, the uplink transmission needs to be nulled to avoid any interference that can be caused by an uplink UE.
- Downlink UE specific control region (uplink opportunistic data region): The UE specific control information such as downlink/uplink data scheduling is transmitted in this region. The wireless device may perform an opportunistic uplink transmission if the level of interference can be tolerated or is below a threshold.
- Uplink control region (downlink opportunistic data region): In this region, the UE may transmit a physical uplink control channel (PUCCH) or a physical uplink shared channel (PUSCH). The BS may schedule a downlink transmission opportunistically as long as the UE-to-UE interference can be managed by an intelligent scheduler.
- Uplink special control region (downlink blank region): The UE transmits important signals/channels such as physical random access channel (PRACH), SRS, demodulation RS (DM-RS) in this region. Therefore, the downlink transmission needs to be nulled to avoid any interference that can be caused by a BS. This nulling can also be used to estimate the UE-to-BS interference channel.

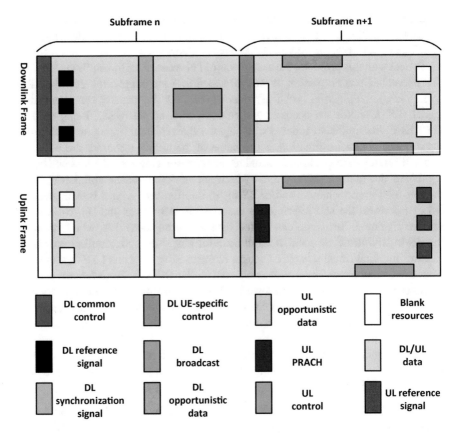

Fig. 5.4 Frame structure in full-duplex systems. ©2014 IEEE. Reprinted, with permission, from [59]

- Downlink data region (uplink data region): In this region, the BS can serve both uplink and downlink users at the same time and frequency resources.

With the new frame structure, the UE-to-UE interference channel can be mitigated via (1) an interference channel measurement, and (2) large-scale fading (or relative locations) [60]. The steps of the first technique, which is based on the interference channel measurement, are described below:

1. The BS broadcasts the number of uplink UEs (e.g., N1) transmitting orthogonal RS sequences to distinguish between different UEs.
2. Uplink UEs transmit the uplink RSs and the BS transmits the downlink RSs in the same time slot, but on orthogonal frequency resources such that downlink users can measure uplink RSs and downlink RSs simultaneously to calculate their corresponding SINRs.
3. Based on the measurements in Step 2, a downlink UE feeds back the lowest N2 (N2 < N1) interference power measurement and the corresponding uplink RS

index broadcasted in the Step 1. In some cases, uplink RSs may not be received at a downlink UE due to high path-loss, which will have an interference power of zero (e.g., 0).

4. The BS performs a joint uplink/downlink scheduling and power allocation based on the downlink UEs' feedback in Step 3.

The second technique (e.g., based on large-scale fading) has a lower overhead. The steps of the second technique can be described as below:

1. The BS first estimates the distance from an uplink UE to itself (Distance-UL) and from itself to a downlink UE (Distance-DL) based on existing techniques.
2. Based on the estimations in Step 1, the BS computes the minimum distance between an uplink UE and a downlink UE by computing an absolute value of (Distance-UL minus Distance-DL), which is associated with the highest UE-to-UE interference level.
3. Based on the computations in Step 2, the BS schedules the UE pairs that can transmit simultaneously on the same time/frequency resources, where the UEs with the minimum distance are not paired.

The drawback of the second technique is that slow-fading channel characteristics are not taken into account in the scheduling decision at the BS, which lessens the scheduling flexibility.

5.5.2 Flexible Duplexing

The fifth-generation (5G) is expected to support various services, such as voice over IP (VoIP), online video, virtual reality, social networking, real-time video sharing, etc. As each of these services requires different uplink and downlink traffic demand, which cannot be supported with the existing semi-static uplink/downlink configuration in LTE, during Release-14 new radio (NR) study item, flexible duplexing (e.g., dynamic TDD) has been studied, and a flexible frame structure is agreed to be supported in Release-15, which enables both semi-static and dynamic TDD operation (e.g., dynamic uplink and downlink assignments), latter of which allows dynamic change of uplink/downlink transmission direction per slot [68–70]. With the flexible duplexing, the ratio between uplink and downlink traffic can change over time (e.g., over each slot, subframe) dynamically which allows support of different services. For example, a symmetric downlink/uplink partitioning in a TDD mode of LTE may not be effective to address an ultra-reliable and low latency communication (uRLLC) service when the partitioning is either downlink heavy or uplink heavy.

As each cell may have a different instantaneous traffic condition, the transmission direction of a cell can be dynamically changed in flexible duplexing, resulting in cross-link interference (CLI) when the transmission directions of neighboring cells in the network are misaligned (e.g., transmitting in opposite links/directions),

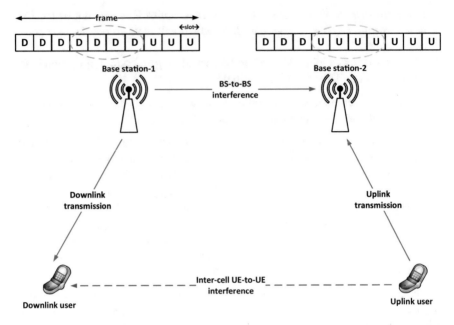

Fig. 5.5 Flexible duplexing

which may severely degrade uplink and/or downlink performance. For example, in Fig. 5.5, from fourth slot to seventh slot of cell-1 and cell-2 may be downlink and uplink, respectively, which are not aligned. While base station of cell-1 is scheduling downlink, the base station of cell-2 is scheduling uplink resulting in UE-to-UE interference from cell-2 to cell-1 and BS-to-BS interference from cell-1 to cell-2, which is similar to interference terms in FD cellular networks discussed for Fig. 5.1.

Interference management is crucial for flexible duplexing to provide performance improvement over semi-static TDD operation, and thus various CLI mitigation schemes, such as interference sensing, scheduling and link adaptation, the advanced receiver (e.g., minimum mean square error interference rejection combining (MMSE-IRC), maximum likelihood (ML), and iterative ML, etc.), coordinated beamforming, and power control have been discussed in [68, 70]. In the next subsections, we will discuss interference sensing/measurement and link adaptation as the other interference mitigation techniques have been already discussed in the previous sections.

5.5.2.1 Interference Sensing

As sensing/listen-before-talk (LBT)-like based mechanisms have proven to be effective distributed protocols for minimizing channel access collision for spectrum sharing scenarios, e.g., Wi-Fi, licensed assisted access (LAA), etc., CLI can be mitigated by channel sensing (such as energy detection and/or signal detection) performed by the transmitting nodes (e.g., by the base station for the BS-to-BS inter-ference and/or by the UE for the UE-to-UE interference) before transmission [70].

The channel sensing may be performed based on instantaneous measurement, which is more accurate for a dynamic environment and efficient than statistical measurement, which can recognize the temporal channel state. In an instantaneous measurement performed based on the energy detection, the transmitting node can perform the sensing of the CLI in a pre-defined time slot dynamically. The length of the pre-defined slot may depend on the QoS and/or the priority of the service, where the service with a higher priority has a shorter time slot. In an instantaneous measurement performed based on the signal detection, the transmitting node can perform the sensing of the CLI based on pre-defined RSs (e.g., SRS, CSI-RS, DM-RS, etc.) carrying some information (e.g., cell index, transmission direction (uplink or downlink), etc.) transmitted by the other BSs and/or UEs. Based on the sensing results, power control, scheduling coordination, and beamforming may be adjusted to reduce the CLI.

5.5.2.2   Link Adaptation

Link adaptation is another method to mitigate CLI, which can be based on the sensing results discussed in Sect. 5.5.2.1. For example, if the transmitting node detects a strong CLI, the transmitting node can drop the transmission of the data on the scheduled slot. In another example, if the transmitting node detects a strong CLI, the scheduling of the data can be adjusted to mitigate CLI by reducing the transmit power, changing the carrier or the transmitting beam, adjusting the modulation and coding scheme (MCS) or the transport block size (TBS), etc. [70]. For example, based on the sensing result, the BS can transmit another downlink control information (DCI) to the UE adjusting the initial scheduling parameters (e.g., MCS, TBS, beam, etc.) to mitigate the CLI.

5.6   Key Technologies for Interference Mitigation

5.6.1   Massive MIMO

The huge surge of interference in FD communications can be further suppressed by using the massive MIMO technology as the increased number of antennas help to form highly directional beams [71] and help to exploit the very large DoF [72].

When the number of simultaneous (e.g., at the same time/frequency resource) active uplink and downlink users increases, the level of the UE-to-UE interference from uplink users to downlink users increases accordingly. The authors in [56] have shown that multi-user precoding is an efficient approach for the UE-to-UE interference control. However, as the number of users is envisioned to be much larger in the massive MIMO systems, the acquisition and the estimation of the UE-to-UE interference channels via the cooperation among the users becomes challenging. Moreover, as the numbers of BS antennas and users increase, the high computational complexity in the computation of multi-user precoding hinders the application of FD technology in massive MIMO systems. Therefore, new interference mitigation methods taking the high number of BS antennas and users into account must be developed for FD massive MIMO systems [64].

For example, the authors in [64] have studied a joint user scheduling and power control algorithm to suppress the UE-to-UE interference that maximizes the achievable spectral efficiency. The key point of the proposed algorithm is that it requires only the knowledge of the slow time-varying channel which makes it practical in the massive MIMO systems as it reduces the overhead and computational complexity. In particular, the proposed algorithm exploits the fact that the instantaneous uplink/downlink SINR converges to its deterministic equivalent approximation, which is a function of slow time-varying parameters, e.g., the large-scale fading of channels [73].

In [74], the authors study the region of DoF under the assumption that BSs have full coordination, which transforms the multi-cell problem into a tractable single-cell problem. This transformation would enable the utilization of the IA method to obtain the highest possible DoF. Although the proposed algorithm provides insights and theoretical maximum limits of DoF, as it assumes full coordination among the BSs, it is impractical and challenging to develop in practice due to high overhead and high signaling. To that end, the authors in [72] consider the scenario when there is no BS coordination among the BSs, which does not allow the transformation from the multi-cell problem to a single-cell problem and study the use of a large number of antennas for intra- and inter-cell interference management in FD cellular networks. It has been shown that as the number of BS antennas increases, the transmit power of BSs and UEs can be reduced proportionally to maintain a fixed asymptotic rate (e.g., QoS). Moreover, the additional interference terms (e.g., residual self-interference, intra-cell and inter-cell interference) in the multi-cell multi-user MIMO FD cellular systems disappear as the number of BS antennas becomes infinitely large. Finally, the authors show that the FD system asymptotically achieves two times spectral efficiency gain over the HD system when the CSI knowledge is perfect or when the users also operate in FD mode under the imperfect CSI knowledge assumption.

The centralized scheme may be useful for sparse scenario with few users. However, it incurs a significant amount of overhead as the number of users grows, with the inter-node interference being the dominant bottleneck. Therefore, a distributed power control method for the UE-to-UE interference management in a single FD massive MIMO cell has been proposed in [75].

5.6.2   Millimeter Wave

MmWave is an emerging technology that improves the spectral efficiency of wireless networks by exploiting an enormous amount of spectrum. In mmWave bands, the signals suffer from huge propagation loss and are sensitive to blockage, which requires highly directional beams in achieving high SINR (or high antenna gain) at the user. In the omni-directional case, the signal is transmitted in all directions and, as a result, interferes with all other terminals in the network. Therefore, focusing the signal to a certain direction reduces the number of terminals that are affected by the interfering signal, i.e., the terminals that lie in the transmitted direction. Furthermore, compared to the omni-directional case, the directed transmitted signal can achieve a longer distance with the same power and can also reach the receiver with a higher power. As the beamwidth decreases, the gain of the signal increases and the possibility of interfering with other terminals decreases.

Hence, combining the mmWave technology with FD technology reduces the self-interference significantly because a highly directional beam from a transmitter antenna has a weak LOS component suffering from higher attenuation towards its own receiver antennas. This would result in the uplink users transmitting with a reduced power leading to a reduction of the intra-cell UE-to-UE interference [76], which is dominant when compared with inter-cell UE-to-UE interference [22]. Furthermore, FD small cell-based networks can benefit from the employment of mmWave communications, since the distances that signals have to travel as well as the transmissions powers reduce, resulting in a reduced interference compared to legacy systems in microwave bands [77].

5.7   Conclusion

This chapter discusses several key design issues in FD cellular networks (e.g., both in single-cell and multi-cell scenarios) and provides a brief overview of the state-of-the-art developments in FD cellular communications. Firstly, the newly introduced interference terms (e.g., intra- and inter-cell interference) were analyzed, including BS-to-BS interference and UE-to-UE interference. Secondly, various promising interference mitigation schemes for the intra- and inter-cell UE-to-UE interference and BS-to-BS interference have been discussed, with particular attention given to elevation beam nulling and joint uplink/downlink power control for the BS-to-BS interference; and scheduling, MAC protocols, interference alignment and beamforming for the UE-to-UE interference. With the growing research interests and efforts in the upcoming years in FD technology, FD technology is very likely to become one of the key technologies studied/specified in 5G standardization. Therefore, we have discussed technical solutions for an LTE-compatible FD cellular network from a 3GPP standardization perspective. Although FD cellular networks have a complicated interference environment compared to existing HD cellular networks, the techniques discussed in this chapter can help to resolve the complicated interference environment to fully unlock the potential of FD communication, which can increase the achievable rates, reduce latency, and enable flexible scheduling.

References

1. IMT-2020 (5G) Promotion Group. (2015, Feb.). 5G vision and requirements. White paper. [Online]. Available: http://www.imt-2020.cn/zh/documents/download/11
2. D. Kim, H. Lee, and D. Hong, "A survey of in-band full-duplex transmission: From the perspective of PHY and MAC layers," *IEEE Commun. Surveys & Tutorials*, vol. 17, no. 4, pp. 2017–2046, Fourthquarter 2015.
3. A. Sabharwal, P. Schniter, D. Guo, D. Bliss, S. Rangarajan, and R. Wichman, "In-band full-duplex wireless: Challenges and opportunities," *IEEE J. Sel. Areas Commun.*, vol. 32, no. 9, pp. 1637–1652, Sept 2014.
4. Y. Hua, P. Liang, Y. Ma, A. C. Cirik and Q. Gao, "A method for broadband full-duplex MIMO radio," *IEEE Signal Process. Lett.*, vol. 19, no. 12, pp. 793–796, Dec 2012.
5. Y.-S. Choi and H. Shirani-Mehr, "Simultaneous transmission and reception: algorithm, design and system level performance," *IEEE Trans. Wireless Commun.*, vol. 12, no. 12, pp. 5992–6014, Dec 2013.
6. Y. Hua, Y. Ma, A. Gholian, Y. Li, A. Cirik, P. Liang, "Radio self-interference cancellation by transmit beamforming, all-analog cancellation and blind digital tuning," *Signal Processing*, vol. 108, pp. 322–340, 2015.
7. R. Mungara, I. Thibault and A. Lazano, "Full-duplex MIMO in cellular networks: System-level performance," *IEEE Trans. Wireless Commun.*, vol. 16, no. 5, pp. 3124–3137, May 2017.
8. A. Hamza, S. Khalifa, H. Hamza, and K. Elsayed, "A survey on inter-cell interference coordination techniques in OFDMA-based cellular networks," *IEEE Communications Surveys Tutorials*, vol. 15, no. 4, pp. 1642–1670, March 2013.
9. Y.-S. Choi and H. Shirani-Mehr, "Simultaneous transmission and reception: Algorithm, design and system level performance," *IEEE Trans. Wireless Commun.*, vol. 12, no. 12, pp. 5992–6010, Dec. 2013.
10. S. Yeh, J. Bai, P. Wang, F. Xue, Y.-S. Choi, S. Talwar, S. Chiu, and V. Kristem, Full-Duplex System Design for 5G Access. 5G Networks: Fundamental Requirements, Enabling Technologies, and Operations Management, *IEEE Wiley Series*, Oct. 2018.
11. R. Li, Y. Chen, G. Y. Li and G. Liu, "Full-duplex cellular networks," *IEEE Commun. Magazine*, vol. 55, no. 4, pp. 184–191, April 2017.
12. S. Han, C. Yang, and P. Chen, "Full duplex-assisted intercell interference cancellation in heterogeneous networks," *IEEE Trans. Commun.*, vol. 63, no. 12, pp. 5218–5234, Dec. 2015.
13. S. Goyal, P. Liu, S. Panwar, R. A. DiFazio, R. Yang, and E. Bala, "Full duplex cellular systems: Will doubling interference prevent doubling capacity?" *IEEE Commun. Magazine*, vol. 53, no. 5, pp. 121–127, May 2015.
14. H. Shirani-Mehr, Y.-S. Choi, R. Yang, and A. Papathanassiou, "Method and apparatus for power control in full-duplex wireless systems with simultaneous transmission reception," U.S. Patent 8861443B2, Oct. 14, 2014.
15. L. Song, Y. Li, and Z. Han, "Resource allocation in full-duplex communications for future wireless networks," *IEEE Wireless Commun. Mag.*, vol. 22, no. 4, pp. 88–96, Aug. 2015.
16. C. Nam, C. Joo, and B. S., "Joint subcarrier assignment and power allocation in full-duplex OFDMA networks," *IEEE Trans. Wireless Commun.*, vol. 14, no. 6, pp. 3108–3119, June 2015.
17. D. Wen and G. Yu, "Time-division cellular networks with full-duplex base stations," *IEEE Commun. Letters*, vol 20, no 2, pp. 392–395, Feb. 2016.

18. G. Yu, D. Wen, and F. Qu, "Joint user scheduling and channel allocation for cellular networks with full duplex base stations," *IET Commun.*, vol. 10, no. 5, pp. 479–486, Mar. 2016.
19. C. Karakus and S. N. Diggavi, "Opportunistic scheduling for full-duplex uplink-downlink networks," *IEEE Int. Symp. Inf. Theory (ISIT)*, pp. 1019–1023, Jun. 2015.
20. J. M. B. da Silva, G. Fodor, and C. Fischione, "Spectral efficient and fair user pairing for full-duplex communication in cellular networks," *IEEE Trans. Wireless Commun.*, vol. 15, no. 11, pp. 7578–7593, Nov. 2016.
21. B. Di, S. Bayat, L. Song, and Y. Li, "Radio resource allocation for full-duplex OFDMA networks using matching theory," *IEEE Conf. Computer Commun. Workshops (INFOCOM WKSHPS)*, pp. 197–198, 2014.
22. S. Goyal, P. Liu, and S. S. Panwar, "User selection and power allocation in full-duplex multicell networks," *IEEE Trans. Veh. Technol.*, vol. 66, no. 3, pp. 2408–2422, Mar. 2017.
23. J. Yun, "Intra and inter-cell resource management in full-duplex heterogeneous cellular networks," *IEEE Trans. Mobile Computing*, vol. 15, no. 2, pp. 392–405, Feb. 2016.
24. S. Wang, V. Venkateswaran, and X. Zhang, "Exploring full-duplex gains in multi-cell wireless networks: A spatial stochastic framework," *IEEE Conf. Computer Commun. (INFOCOM)*, pp. 855–86, Apr. 2015.
25. M. Duarte, A. Feki, and S. Valentin, "Inter-user interference coordination in full-duplex systems based on geographical context information," *IEEE Int. Conf. Commun. (ICC)*, pp. 1–7, May 2016.
26. Y. Rong and Y. Hua, "Space-time power scheduling of MIMO links - Fairness and QoS considerations," *IEEE J. Sel. Topics in Signal Process. Special Issue on MIMO-Optimized Transmission Systems for Delivering Data and Rich Content*, vol. 2, no. 2, pp. 171–180, April 2008.
27. Y. Rong, Y. Hua, A. Swami, and A. L. Swindlehurst, "Space-time power schedule for distributed MIMO links without instantaneous channel state information at the transmitting nodes," *IEEE Trans. Signal Process.*, vol. 56, no. 2, pp. 686–701, Feb 2008.
28. Y. Rong and Y. Hua, "Optimal power schedule for distributed MIMO links," *IEEE Trans. Wireless Commun.*, vol. 7. no. 8, pp. 2896–2900, August 2008.
29. B. Day, A. Margetts, D. Bliss, and P. Schniter, "Full-duplex bidirectional MIMO: Achievable rates under limited dynamic range," *IEEE Trans. Signal Process.*, vol. 60, pp. 3702–3713, July 2012.
30. B. Day, A. Margetts, D. Bliss, and P. Schniter, "Full-duplex MIMO relaying: Achievable rates under limited dynamic range," *IEEE J. Sel. Areas in Commun.*, vol. 30, pp. 1541–1553, September 2012.
31. A. C. Cirik, Y. Rong, and Y. Hua, "Achievable rates of full-duplex MIMO radios in fast fading channels with imperfect channel estimation," *IEEE Trans. Signal Process.*, vol. 62, no. 15, pp. 3874–3886, Aug. 2014.
32. K. M. Thilina, H. Tabassum, E. Hossain, and D. I. Kim, "Medium access control design for full-duplex wireless systems: Challenges and approaches," *IEEE Communications Magazine*, vol. 53, no. 5, pp. 112–120, May 2015.
33. A. Sahai, G. Patel, and A. Sabharwal, "Pushing the limits of full-duplex: Design and real-time implementation," arXiv preprint arXiv:1107.0607, 2011.
34. J. Y. Kim, O. Mashayekhi, H. Qu, M. Kazandjieva, and P. Levis, "Janus: A novel mac protocol for full duplex radio," *University of Stanford - Computer Science Department, Tech. Rep. CSTR*, vol. 2, no. 7, p. 23, 2013.
35. S. Goyal, P. Liu, O. Gurbuz, E. Erkip, and S. Panwar, "A distributed MAC protocol for full duplex radio," *Asilomar Conf. Signals, Systems and Computers*, pp. 788–792, Nov. 2013.
36. W. Choi, H. Lim, and A. Sabharwal, "Power-controlled medium access control protocol for full-duplex WiFi networks," *IEEE Trans. Wireless Commun.*, vol. 14, no. 7, pp. 3601–3613, July 2015.
37. S. Y. Chen, T. F. Huang, K. C. J. Lin, Y. W. P. Hong, and A. Sabharwal, "Probabilistic based adaptive full-duplex and half-duplex medium access control," *IEEE Global Commun. Conf. (GLOBECOM)*, pp. 1–6, Dec 2015.

38. S. H. Chae, S.-W. Jeon and S. H. Lim, "Fundamental limits of spectrum sharing full-duplex multicell networks," *IEEE J. Select. Areas Commun.*, vol. 34, no. 11, pp. 3048–3061, Nov. 2016.

39. S. H. Chae, S. H. Lim, and S. W. Jeon, "Degrees of freedom of full-duplex multiantenna cellular networks," *IEEE Trans. Wireless Commun.*, vol. 17, no. 2, pp. 982–995, Feb. 2017.

40. S. A. Jafar and S. Shamai (Shitz), "Degrees of freedom region for the MIMO X channel," *IEEE Trans. Inf. Theory*, vol. 54, no. 1, pp. 151–170, Jan. 2008.

41. V. R. Cadambe and S. A. Jafar, "Interference alignment and degrees of freedom for the K-user interference channel," *IEEE Trans. Inf. Theory*, vol. 54, no. 8, pp. 3425–3441, Aug. 2008.

42. C. Suh and D. Tse, "Downlink interference alignment," *IEEE Trans. Commun.*, vol. 59, no. 9, pp. 2616–2626, Sep. 2011.

43. A. Sahai, S. Diggavi, and A. Sabharwal, "On uplink/downlink full-duplex networks," *Asilomar Conf. Signals, Systems and Computers*, pp. 14–18, Nov. 2013.

44. A. Sahai, S. Diggavi, and A. Sabharwal, "On degrees-of-freedom of full-duplex uplink/downlink channel," *IEEE Information Theory Workshop (ITW)*, pp. 1–5, Sept 2013.

45. W. Bi, X. Su, L. Xiao, and S. Zhou, "On rate region analysis of full-duplex cellular system with inter-user interference cancellation," *IEEE Int. Conf. Communication Workshop (ICCW)*, pp. 1166–1171, June 2015.

46. W. Bi, X. Su, L. Xiao, and S. Zhou, "Superposition coding based inter-user interference cancellation in full duplex cellular system," *IEEE Wireless Communications and Networking Conference (WCNC)*, pp. 1–6, April 2016.

47. P. Aquilina, A. C. Cirik and T. Ratnarajah, "Weighted sum rate maximization in full-duplex multi-user multi-cell MIMO network," *IEEE Trans. Commun.*, vol. 65, no. 4, pp. 1590–1608, Apr. 2017.

48. A. C. Cirik, "Fairness considerations for full duplex multi-user MIMO systems," *IEEE Wireless Commun. Lett.*, vol. 4, no. 4, pp. 361–364, Aug. 2015.

49. A. C. Cirik, S. Biswas, S. Vuppala, and T. Ratnarajah, "Beamforming design for full-duplex MIMO interference channels: QoS and energy-efficiency considerations,"*IEEE Trans. Commun.*, vol. 64, no. 11, pp. 4635–4651, Nov. 2016.

50. A. C. Cirik, M. J. Rahman and L. Lampe, "Robust fairness transceiver design for a full-duplex MIMO multi-cell system," *IEEE Trans. Communications*, vol. 66, no. 3, pp. 1027–1041, March 2018.

51. M. J. Rahman, A. C. Cirik and L. Lampe, "Power-efficient transceiver design for full-duplex MIMO multi-cell systems with CSI uncertainty," *IEEE Access*, vol. 5, pp. 22689–22703, 2017.

52. Y. Sun, D. W. K. Ng, J. Zhu, and R. Schober, "Multi-objective optimization for robust power efficient and secure full-duplex wireless communication systems," *IEEE Trans. Wireless Commun.*, vol. 15, no. 8, pp. 5511–5526, Aug. 2016.

53. Y. Jiang, F. C. M. Lau, I. W. H. Ho, H. Chen, and Y. Huang, "Max min weighted downlink SINR with uplink SINR constraints for full-duplex MIMO systems," *IEEE Trans. Signal Process.*, no. 12, pp. 3277–3292, Jun. 2017.

54. D. Nguyen, L.-N. Tran, P. Pirinen, and M. Latva-aho, "On the spectral efficiency of full-duplex small cell wireless systems," *IEEE Trans. Wireless Commun.*, vol. 13, no. 9 pp. 4896–4910, Sep. 2014.

55. –, "Precoding for full duplex multiuser MIMO systems: spectral and energy efficiency maximization," *IEEE Trans. Signal Process.*, vol. 61, no. 16, pp. 4038–4050, Aug. 2013.

56. S. Huberman and T. Le-Ngoc, "Full-duplex MIMO precoding for sum-rate maximization with sequential convex programming," *IEEE Trans. Veh. Technol.*, vol. 64, no. 11, pp. 5103–5112, Nov. 2015.

57. –, "MIMO full-duplex precoding: a joint beamforming and self-interference cancellation structure," *IEEE Trans. Wireless Commun.*, vol. 14, no. 4, pp. 2205–2217, Apr. 2015.

58. J. Kim, W. Choi, and H. Park, "Beamforming for full-duplex multiuser MIMO systems," *IEEE Trans. Veh. Technol.*, vol. 66, no. 3, pp. 2423–2432, Mar. 2017.

59. S. Han, I. Chih-Lin, Z. Xu, C. Pan, and Z. Pan, "Full duplex: Coming into reality in 2020?," *IEEE Global Commun. Conf. (GLOBECOM)*, pp. 4776–4781, Dec. 2014.

60. S. Shao, D. Liu, K. Deng, Z. Pan, and Y. Tang, "Analysis of carrier utilization in full-duplex cellular networks by dividing the co-channel interference region," *IEEE Commun. Lett.*, vol. 18, no. 6, pp. 1043–1046, June 2014.
61. H. Tang, Z. Ding, and B. Levy, "Enabling D2D communications through neighbor discovery in LTE cellular networks," *IEEE Trans. Signal Processing*, vol. 62, no. 19, pp. 5157–5170, Oct 2014.
62. K. Lee, W. Kang, and H.-J. Choi, "A practical channel estimation and feedback method for device-to-device communication in 3GPP LTE system," *8th Int. Conf. Ubiquitous Information Management and Communication (ICUIMC)*, ACM, Jan. 2014.
63. F. Baccelli, N. Khude, R. Laroia, J. Li, T. Richardson, S. Shakkottai, S. Tavildar, and X. Wu, "On the design of device-to-device autonomous discovery," *Fourth Int. Conf. Communication Systems and Networks (COMSNETS)*, pp. 1–9, Jan 2012.
64. X. Xia, K. Xu, Y. Wang and Y. Xu, "A 5G-Enabling technology: Benefits, feasibility, and limitations of in-band full-duplex mMIMO," *IEEE Vehicular Technology Magazine*, vol. 13, no. 3, pp. 81–90, Sept. 2018.
65. J. Bai, S. Yeh and Y. Choi, "Interference mitigation and traffic adaptation in full-duplex small cell networks," *IEEE Int. Symp. Personal, Indoor, and Mobile Radio Commun. (PIMRC)*, pp. 1–6, Oct. 2017.
66. S. Sesia, I. Toufik, and M. Baker. *LTE: The UMTS Long Term Evolution From Theory to Practice*. John Wiley and Sons, Inc., 2009
67. G. Noh, H. Wang, C. Shin, S. Kim, Y. Jeon, H. Shin, J. Kim, and I. Kim, "Enabling technologies toward fully LTE-compatible full-duplex radio," *IEEE Commun. Mag.*, vol. 55, no. 3, pp. 188–195, March 2017.
68. 3GPP, "TR38.802: Study on new radio access technology Physical layer aspects," TR38.802 V14.2.0, Release-14, September 2017, http://www.3gpp.org/DynaReport/38-series.htm.
69. RP-182864, "Revised WID on Cross Link Interference (CLI) handling and Remote Interference Management (RIM) for NR," LG Electronics, 3GPP TSG RAN Meeting 82, Dec., 2018.
70. R1-1700270, "Discussion on dynamic TDD and cross-link interference mitigation schemes," ZTE, 3GPP TSG RAN WG1 AH-NR 1 Meeting, Jan. 2016.
71. Y. Li, P. Fan, A. Leukhin and L. Liu, "On the spectral and energy efficiency of full-duplex small-cell wireless systems with massive MIMO," *IEEE Trans. Veh. Technol.*, vol. 66, no. 3, pp. 2339–2353, Mar. 2017.
72. J. Bai and A. Sabharwal, "Asymptotic analysis of MIMO multi-cell full-duplex networks," *IEEE Trans. Wireless Commun.*, vol. 16, no. 4, pp. 2168–2180, Apr. 2017
73. F. Rusek, D. Persson, B. K. Lau, E. Larsson, T. Marzetta, O. Edfors, and F. Tufvesson, "Scaling up MIMO: Opportunities and challenges with very large arrays," *IEEE Signal Process. Mag.*, vol. 30, no. 1, pp. 40–60, Jan. 2013.
74. M. Amir Khojastepour, K. Sundaresan, S. Rangarajan, and M. Farajzadeh-Tehrani, "Scaling wireless full-duplex in multi-cell networks,"*IEEE Conf. Computer Communications (INFOCOM)*, pp. 1751–1759, April 2015.
75. W. Ouyang, J. Bai, and A. Sabharwal, "Leveraging one-hop information in massive MIMO full-duplex wireless systems," *IEEE/ACM Trans. Networking*, vol. 25, no. 3, pp. 1528–1539, June 2017.
76. V. V. Mai, J. Kim, S. W. Jeon, S. W. Choi, B. Seo, and W. Y. Shin, "Degrees of freedom of millimeter wave full-duplex systems with partial CSIT," *IEEE Communications Letters*, vol. 20, no. 5, pp. 1042–1045, May 2016.
77. A. Yadav, G. I. Tsiropoulos, and O. A. Dobre, "Full-duplex communications: performance in ultradense mm-wave small-cell wireless networks," *IEEE Veh. Technol. Mag.*, vol. 13, no. 2, pp. 40–47, Jun. 2018.

Chapter 6
Robust Interference Management and Network Design for Heterogeneous Full-Duplex Communication Networks

Nurul Huda Mahmood, Gilberto Berardinelli, Hirley Alves, Dong Min Kim, and Marta Gatnau Sarret

Abstract The challenging and diverse design requirements of 5G New Radio open the door for investigating novel solutions targeting enhanced mobile broadband and ultra-reliable low latency communication service classes. This chapter discusses challenges and opportunities of in-band full-duplex communication as a promising solution for next generation radio technologies. Realistic network assumptions are considered. The main challenges relate to self-interference cancellation and the increase of inter/intra-cell interference resulting from simultaneous transmission and reception, as well as the asymmetric nature of network traffic, thus limiting the potential gains in terms of throughput and latency reduction. Particular focus is given to system aspects including the effects on TCP/UDP protocols and symmetric/asymmetric traffic types. We discuss efficient techniques to overcome the challenges of full-duplex communication and propose interference management alternatives such as the virtual full-duplex concept. Up to 70% success probability gains over conventional full-duplex communication can be observed with the proposed virtual full-duplex technique. In addition, we foresee significant benefits and relevance of full-duplex technology in applications other than throughput enhancement and latency reduction, such as relaying, self-backhauling, autonomous device-to-device discovery, and physical layer security.

N. H. Mahmood (✉) · H. Alves
Centre for Wireless Communication, University of Oulu, Oulu, Finland
e-mail: nurulhuda.mahmood@oulu.fi; hirley.alves@oulu.fi

G. Berardinelli
Wireless Communications Network Section, Aalborg University, Aalborg, Denmark
e-mail: gb@es.aau.dk

D. M. Kim
Department of IoT, Soonchunhyang University, Asan-si, South Korea
e-mail: dmk@sch.ac.kr

M. G. Sarret
Lleida.net, Lleida, Spain
e-mail: mgatnau@lleida.net

© Springer Nature Singapore Pte Ltd. 2020
H. Alves et al. (eds.), *Full-Duplex Communications for Future Wireless Networks*,
https://doi.org/10.1007/978-981-15-2969-6_6

6.1 Introduction

The fifth generation (5G) wireless system, also known as 5G New Radio (NR), introduced New Radio Release-15 introduces a new service class, namely ultra-reliable low latency communication (URLLC), alongside enhanced mobile broadband (eMBB) services. A wide range of design requirements in terms of the data rate, latency, reliability, and energy efficiency are targeted by 5G new radio (NR). Novel technology components are needed to meet such challenging requirements. For example, full-duplex (FD) communication, i.e., simultaneous transmission and reception over the same frequency band, opens the door to significant improvements in terms of throughput, reliability, and latency as well as secrecy rate [1, 2], provided the challenges associated with FD transmissions are well managed. It has been demonstrated that FD capabilities result in significant aggregate throughput gains [3] and latency reductions [4] over baseline half-duplex (HD) scenario in ideal conditions.

There are two major challenges in realizing the potential of FD communication in practice. The first relates to FD node design itself, while the other involves interference and traffic management. FD communication generates strong loopback interference from the transmitter to the receiver-end of the same radio device, which can be in the order of 100–120 dB higher than the noise floor. Though recent advances in self-interference cancellation (SIC) techniques, featuring a combination of active (analog and digital) and passive cancellation allow suppressing the loopback interference to tolerable limits, self-interference cannot be completely eradicated [1]. From the network perspective, FD communication doubles the number of concurrent transmissions, resulting in an increase in the overall network interference. Furthermore, the opportunities arising from simultaneous uplink and downlink transmission can only be exploited when packets are readily available for transmission in both directions. Hence, the envisioned performance gains cannot be fully realized without addressing such practical aspects of FD communication.

This chapter focuses on the network aspects of FD communication. We first evaluate the potential performance gains of FD communication through system level simulations under a practical network scenario. The dense small cell scenario is specifically assumed since densification is identified as one of the main tools in meeting the ambitious 5G design targets [5]. Instead of the generic full buffer traffic model, realistic traffic protocols like the transmission control protocol (TCP) and the user datagram protocol (UDP) with variable traffic profiles, and physical/medium access control (PHY/MAC) layer techniques like hybrid automatic repeat request (HARQ) retransmissions and link adaptation are considered. The considered degree of realism in our assumptions allows us to get insights into the potential performance gains of FD communication in dense small cell networks under realistic network conditions.

We then present techniques to improve the performance of FD communication through efficient intra- and inter-cell interference management and FD-aware network design. In particular, we propose the virtual FD technique, which is a way to mimic FD transmission using HD transceivers, thereby avoiding the need for SIC and the detrimental impact of residual self-interference power. Our results demonstrate that simultaneous scheduling of uplink and downlink users in two neighboring cells through cooperation between the access points (APs) can outperform FD transmission and results in improved reception. More specifically, up to 70% success probability gains over conventional FD communication can be observed with the proposed interference management technique.

In addition to its role in throughput enhancement and latency reduction, FD communication—by its nature—lends itself to applications in other domains, such as spectrum sensing, physical layer security, and device-to-device (D2D) communication. As a final contribution, we discuss how FD communication can provide gains in relaying, self-backhauling, autonomous D2D discovery, and physical layer security.

6.2   Full-Duplex Network Design: Challenges and Performance Trends

A major appeal of FD technology is the possibility of boosting the network throughput. This is particularly attractive when operating in the centimeter-wave spectrum region, given the scarcity of available frequency resources. A further less intuitive benefit is latency reduction for services running over unpaired bands, where time division duplexing (TDD) is the traditional operational mode. The possibility of simultaneously transmitting and receiving removes delays associated with uplink/downlink ratios of the TDD frames. This has significant potential in scenarios where a transmitter needs to sense the channel, for example, in carrier sense multiple access (CSMA), D2D discovery applications, and cognitive radios.

6.2.1   Challenges in Full-Duplex Network Design

The second major challenge of FD communication is in the networking aspect. Simultaneous transmission and reception enabled by FD leads to an increase in the interference footprint with respect to traditional duplexing. While traditional network deployments are affected by interference from a single direction at a given time/frequency slot, base stations (BS) and devices can experience interference from both uplink and downlink transmissions in neighbor cells with FD communication, as illustrated in Fig. 6.1. This is particularly significant in dense deployments (e.g., dense small cells), given the short intersite distance and the similar power level in uplink and downlink directions. In the case of large cells with significant uplink/downlink power asymmetry, the downlink interference is predominant and performance is deemed to be similar to that of half-duplex (HD) with uncoordinated uplink/downlink switching point between neighbor cells [6].

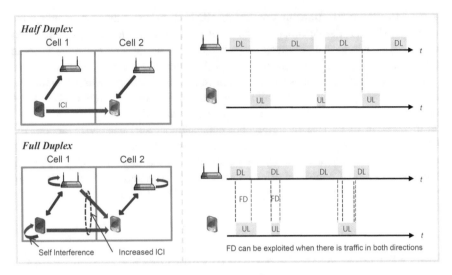

Fig. 6.1 Full-duplex communication results in increased inter-cell interference, incurs residual self-interference, and is fully exploited with symmetric full buffer traffic in both directions

Two different FD modes are usually considered in the literature, namely base station full-duplex (BS FD) and in-band bidirectional full-duplex (IB-FD). In the former case, only the BS is FD capable, while the terminals operate in traditional HD mode. In such a case, FD communication benefits from the ability to schedule uplink and downlink users simultaneously, at the expense of additional intra-cell interference at the downlink receiver. On the other hand, IB-FD assumes that both BS and user terminals are FD capable. The number of concurrent transmissions for both modes is doubled in comparison with HD operation, resulting in a substantial increase in the overall inter-cell interference (ICI). In particular, mutual coupling between cells specific to the network deployment obviously has a major impact on the FD potential due to such increased ICI.

The traffic flow density in the uplink and the downlink directions also impacts the expected performance enhancement of FD communication. More precisely, in order to benefit from the opportunity to schedule uplink and downlink traffic simultaneously, there should be available traffic in both directions, i.e., uplink and downlink traffic profile should be symmetric. However, network traffic is generally skewed in favor of either transmit directions (e.g., content uploading, video streaming). Even when the residual self-interference power and the increased intra- and inter-cell interference with FD communication are efficiently managed, the full potential of FD communication can only be availed in scenarios with symmetric traffic profiles. Figure 6.1 pictorially depicts that asymmetric traffic lowers the probability of exploiting FD potential by sending UL and DL data simultaneously.

6.2.2 Performance Evaluation of Full-Duplex Communication Under Realistic Network Assumptions

We have highlighted how increased ICI may present a significant challenge for harvesting the potential gain of FD. In this section we evaluate the performance of FD transmission in a dense network of small cells characterized by tight interference coupling. The objective is to observe FD performance under realistic dense small cell scenario. The presented results are extracted from an event-driven system level simulator. We consider a single floor scenario with 20 small cells of $10\,m \times 10\,m$ dimension arranged in a 10×2 formulation, as specified by 3GPP [7]. Each cell contains a number of user terminals and a single access point placed randomly, both transmitting at a $10\,dBm$ power. The same frame structure is assumed for both link directions. Rayleigh fading model is assumed, while the path loss is given by the Winner II indoor office model.

As a further tier of robustness to the interference, we also consider an HARQ mechanism to be in place. It has been identified in [4] that FD has an attractive interaction with high layer protocols such as TCP, given the nature of its congestion control mechanism. In the initial stage of a service, the TCP congestion window which defines the amount of data to be sent through the channel grows exponentially with the reception of TCP acknowledgements (ACK). When a certain threshold is reached, a contention avoidance phase starts, where the congestion window grows linearly. The congestion window might grow faster and reach the congestion avoidance phase sooner with FD transmission, allowing a larger amount of data transmission within a single transmission slot. The TCP implementation is New Reno, and includes the recovery and congestion control mechanisms. However, handshake procedures are not considered since they are not relevant for our studies.

In order to support FD communication, a radio resource management (RRM) module design that dynamically schedules the transmission between FD and HD mode at each TTI is considered. The scheduling decision at the AP and the UE is taken independently based on the information received from the PHY, MAC, and radio link control (RLC) layers. Thereafter, links with scheduled traffic in both directions are allocated to FD mode. We refer to [4] for further details on the simulation setup.

6.2.2.1 Analysis of the Traffic Constraint Limitation in Isolated Cell

In order to first characterize the benefits of FD communication in interference free scenarios, let us consider for the moment an isolated cell in the aforementioned setup. Figure 6.2 presents the system performance (in terms of average cell session throughput and average packet delay) with UDP and TCP protocols, assuming a IB-FD setup with symmetric and asymmetric traffic profiles, respectively. Low, medium, and high loads correspondingly refer to resource usage of 25%, 50%, and 75%.

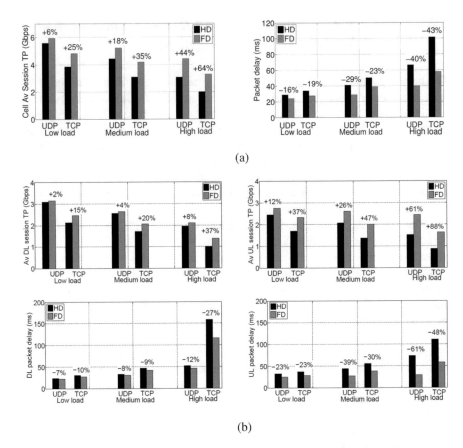

Fig. 6.2 Throughput gain and delay reduction of IB-FD over HD with TCP and UDP traffic in a single cell scenario. (**a**) Symmetric traffic profile. (**b**) Asymmetric traffic profile (6:1). *Source*: Ref. [4]

For the symmetric scenario, a larger gain with FD communication is observed with TCP traffic. The reasons are twofold: firstly, the TCP congestion window grows faster with FD traffic, allowing larger amount of data transfer compared to HD communication. Secondly, the accumulation of traffic in the buffer in the slow start phase of the TCP congestion window build-up allows for better exploitation of the FD capability compared to UDP traffic, in low traffic load scenarios where the FD gain is 25% when TCP is used with respect to the 6% gain of UDP. FD further allows faster transmission of ACKs, which also contributes to the performance gains.

A traffic ratio of 6:1 in favor of downlink traffic is assumed for the asymmetric scenario. The gains are expectedly different for the different transmission directions. In general FD communication benefits the lightly loaded direction more. It is important to note that even though we consider an isolated cell (hence no ICI) and ideal SIC, the performance gains with FD communication are below the theoretical levels reported in [3] as a result of considering practical TCP/UDP traffic protocols.

6.2.2.2   FD Performance Under the Impact of Increased Interference and Traffic Constraints

We next investigate the impact of increased ICI. The results are presented in Table 6.1. A 5 dB wall loss is assumed. Only the medium load scenario and symmetric traffic is considered for the sake of brevity. The throughput performance (in Mbps) of HD, IB-FD, and BS FD with symmetric TCP and UDP traffic in a multi-cell scenario are shown.

The average uplink and downlink performances are similar under the IB-FD case. In the case of UDP traffic, around 40% and 20% outage and median throughput gains are, respectively, observed; whereas TCP traffic demonstrates a huge outage throughput loss of over 70% and median throughput loss of over 60%. Due to the TCP slow start, FD is exploited much more often compared to UDP (in this case 81% TCP transmission were FD compared to 15% with UDP), resulting in increased ICI, which is further exacerbated in the dense small cell scenario. In addition, HARQ retransmissions are triggered more often with FD in TCP, further congesting the network.

In the BS FD mode, similar trends are observed with both TCP and UDP traffic. The downlink direction is affected by the additional interference resulting from the concurrent uplink transmission, while the uplink direction slightly benefits from the increased transmission opportunity accorded by FD communication. Similar performance trends are observed for latency as well. IB-FD results in lower latency with UDP protocol, whereas the delay increases dramatically with TCP. On the other hand, BS FD shows nearly the same results for UDP and TCP. This is because, exploitation in FD mode is limited due to the traffic constraints, which means that the gain that FD can bring on speeding up the congestion window in TCP is barely shown and therefore both schemes perform nearly the same.

It is thus observed that the potential throughput and latency gains with FD communication are rather limited when practical considerations such as the impact of increased ICI and asymmetric traffic profile are accounted for.

Table 6.1 Throughput performance (in Mbps) of HD, IB-FD, and BS FD with symmetric TCP and UDP traffic in a multi-cell scenario with 5 dB wall loss

	Downlink TP					Uplink TP				
	HD	IB-FD	Gain	BS FD	Gain	HD	IB-FD	Gain	BS FD	Gain
	UDP traffic protocol									
Outage	1.1	1.6	45%	0.7	−36%	1.1	1.5	36%	0.7	−36%
Median	2.5	2.9	16%	2.4	−4%	2.5	3.0	20%	2.6	4%
Peak	4.6	5.1	11%	4.3	−6%	4.5	5.1	13%	4.8	7%
	TCP traffic protocol									
Outage	0.4	0.1	−75%	0.3	−25%	0.4	0.1	−75%	0.4	0%
Median	1.3	0.5	−62%	1.2	−8%	1.2	0.4	−66%	1.3	8%
Peak	2.3	1.8	−22%	2.2	−4%	2.2	1.8	−18%	2.2	0%

6.3   Interference Management and Network Design for FD Communication

The previous section highlighted the role of increased intra- and inter-cell interference and traffic asymmetry in limiting the potential of FD communication. Techniques to overcome these challenges and improve the performance of FD communication through efficient intra- and inter-cell interference management and designing the network to have more symmetric traffic conditions are discussed in this section.

6.3.1   Interference Management Techniques in Full-Duplex Communication

The obvious benefit of FD communication are gains in terms of the throughput and latency. In the context of a cellular network where users are served by macro base stations with higher transmit powers, the throughput gains are higher in downlink [8]. However, the inter-mode interference limits the throughput gains that can be obtained in uplink, which can in fact be negative. In order to achieve its full potential, the FD node should be RRM-aware, and therefore, intelligently pair and schedule downlink and uplink users with proper transmission powers so as to reduce the intra- and inter-cell interference [9].

In [6], a smart scheduling between FD and HD mode termed as α-duplex is proposed to limit the adverse effect on the uplink interference. The authors propose to partially overlap between the uplink and the downlink frequency bands, so as to have FD operation in only the overlapped bands. The amount of the overlap is controlled via the design parameter α to balance the trade-off between the uplink and downlink performances.

For small cell networks with similar uplink and downlink transmit powers, the increased ICI equally impacts the performance gains in both directions. Limiting the interference coupling among cells is the best way to deal with the increased interference in this case. This can be done through physical isolation among cells, for example, installing different cells in different rooms/apartments for indoor networks, or by using directional antennas. With BS FD operation, where the different uplink and downlink users transmit simultaneously, taking the resulting uplink to downlink interference into account during scheduling decision can address the interference management problem to a large extent [10].

6.3.2   Inducing Traffic Symmetry Through Network Design

It has been shown in [11] that asymmetry in traffic profile between the uplink and downlink transmission directions impacts the achieved performance gains with FD.

While this is not an impediment resulting from FD transmission itself, this fact has to be carefully considered in order to fully utilize FD networks. In this respect, FD communication has the highest potential in scenarios with symmetric traffic, such as relaying and backhaul networks and D2D communication.

FD relaying is perhaps the most successful application of FD communication, and one reason is due to the inherent availability of symmetric traffic. In a cellular deployment, for instance, a FD small cell can act as a relay for macro-cell edge users. In cellular networks, the backhaul link connects the small cell and macro BS to the core network. A practical low complexity backhaul solution is to leverage the radio access network spectrum simultaneously for the access link as well as the backhaul link, with advantages in terms of easy installation, reduced costs, and flexibility. Due to the limited spectrum availability, FD transmission is recently being considered as a viable option for self-backhauling since it accommodates efficient reuse of the scarce spectrum, with limited additional hardware cost [12]. Furthermore, the backhaul link can be designed such that the interference generated to the users operating in the network is limited, which can be achieved by allowing RRM coordination between the small cell and the macro BS [9]. With coordinated RRM between BS and relay, employing low cost FD relays for in-band backhauling is found to provide large gains over an equivalent HD baseline [9].

D2D communication is known to provide benefits such as network offloading, potential coverage extension, and reduction of control overhead and latency. In order to establish direct communication with potential neighbors, a discovery phase is required at each device. A typical autonomous discovery procedure consists of broadcasting of discovery messages (often referred as beacons) by each node. Upon awareness of the neighbors' presence, an eventually dedicated communication phase can start. The discovery procedure needs to be repeated with a certain occurrence in order to track new appearing nodes in the network. FD communication has significant potential in terms of latency reduction in the device discovery phase since every device can receive the beacons from its neighbors while simultaneously transmitting its own beacons [13]. On the other hand, since D2D communication is between two (usually) physically proximate and isolated users, the increase in intra- and inter-cell interference due to simultaneous transmissions, as discussed earlier, is rather limited.

6.4   Virtual Full-Duplex

The main challenge to the realization of FD wireless transceivers is the high imple-mentation complexity that is required to cope with the induced self-interference. This is particularly one of the main motivations behind the BS FD setup described earlier, where only the BS has to cope with the complexity of SIC while the user terminals operate in HD mode.

To resolve the self-interference issue in FD implementation, the use of inter-connected HD macro-cell base stations is recently proposed [14, 15]. A distributed implementation of the FD functionality is realized by simultaneously serving users that have opposite uplink and downlink connections each. The interconnection between coordinating BSs is assumed to have sufficient capacity. The connected BSs form a *virtual* FD BS. Thus, the complexity of SIC and the impact of residual self-interference power [1] are readily avoided. The main benefit of virtual FD comes from shorter transmission distance between UEs and their serving APs.

The virtual FD concept can be extended to the small cell scenario as well. Virtual FD using small cell APs can be realized by several interconnection architectures. As in previous works, two APs can be connected directly. Alternately, two or more APs can be connected by a central unit similar to the cloud-radio access network (C-RAN) architecture. Due to the low mobility of users in small cell scenario, we can assume that the channel state information (CSI) can be perfectly shared among the connected APs.

In Fig. 6.3, signals are depicted as blue arrows, interference as red arrows, and the interconnection between nodes as a bold line. HD-AP2 sends the signal to UE2, while HD-AP1 receives signal from UE1 HD-AP1 receives interference from HD-AP2, while UE2 receives interference from UE1. However, HD-AP2 can use the high-bandwidth interconnection to HD-AP1 to send the same data contained in the packet to UE2, such that HD-AP1 can regenerate the interference signal exploiting the received data and the available CSI, and cancel the interference from HD-AP2 perfectly. On the other hand, the signal from UE1 remains as interference to UE2. It can therefore be concluded that the setup on Fig. 6.3 operates equivalently with the setup containing a single FD AP. By utilizing TDD, the traffic directions of UE1 and UE2 are reversed in the next time slot, rendering it possible two-way communication for both UEs.

In Fig. 6.4, the transmission success probabilities of virtual FD and FD are depicted. The curves are drawn by Monte Carlo simulation to model a 150×150 m two-dimensional area. All small cell APs are randomly deployed and each UE is associated with its nearest AP. It is assumed that the transmission is successful

Fig. 6.3 Virtual full-duplex topology with two interconnected half-duplex access points

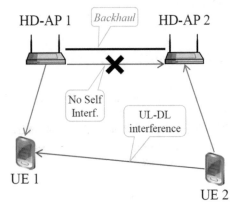

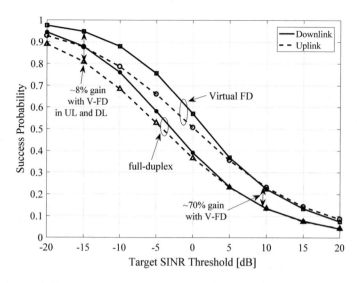

Fig. 6.4 Transmission success probability as a function of target SINR threshold

when the received signal-to-interference-and-noise ratio (SINR) is above the target threshold. The transmission power of UEs and APs are the same (10 dBm). The distance between virtual FD pair is less than or equal to 10 m, a restriction added to model a dense small cell environment. In the FD scenario, the APs are deployed according to a Poisson point process with half the density of virtual FD. This is because two HD-APs in virtual FD form a single virtual AP. In each cell, two UEs (one in UL, the other in DL) are served simultaneously resulting in the same number of UEs for the FD and the virtual FD scenarios.

We observe that the success probability of virtual FD is superior to FD over the entire target SINR range, with gains of up to 70%. This is because UEs are located closer to the serving APs than FD. The above simulation assumes an ideal connection between cooperating APs. Since the two distant APs must exchange information in real time, the performance of the link connecting the two APs is important. In the case of a high density network, wired connection between these base stations may be more challenging. In this case, a wireless connection using millimeter wave (mmW) technology can be considered.

6.5   Other Applications of Full-Duplex

The role of FD communication in enhancing the throughput and/or latency has so far been highlighted in this contribution. In this remaining section, we present other application areas where FD communication has demonstrated the potential to provide significant performance gains.

6.5.1 Physical Layer Security

In the last few years wireless networks have become ubiquitous and an indispensable part of our daily life due to a broad range of applications. As a result, devices are even more vulnerable to security attacks, such as eavesdropping. Due to the broadcast nature of the wireless medium, devices expose their transmission attempts to passive as well as to malicious nodes raising several security concerns.

One alternative to prevent eavesdropping is the use of information-theoretic security, also known as physical layer security, which allows the transmitter to send unbreakable secure messages [2]. FD communication has so far appeared as a way to boost throughput, but in this context the FD transceiver has an additional capability since it can act simultaneously as receiver and as a jammer for unintended eavesdroppers, thus improving the security and reliability of the legitimate communication link, as illustrated in Fig. 6.5. For example, the ergodic secrecy rate with FD communication is found to grow linearly with the logarithm of the direct channel signal power, as opposed to the flattened out secrecy rate with conventional HD communication [2].

Those gains become even more evident as the number of antennas grows or more FD nodes are available to cooperate, thus allowing for the use of relay selection, beamforming, and artificial noise techniques which increases interference at the unintended eavesdroppers while simultaneously improving the performance of the legitimate link. Such gains come at a cost of self-interference and increased interference in multiple cell deployments. Nonetheless recent advances in transceiver design and self-interference mitigation have provided solutions that mitigate the interference to the noise floor, while network coordination potentially reduces the overall interference profile of the network, as discussed in Sect. 6.3.

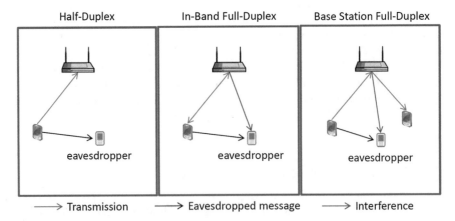

Fig. 6.5 Different placement of Alice, Bob, and Eve under consideration in evaluating the physical layer security potential of full-duplex communication

6.5.2   Cooperative Communication

As discussed above, relaying is one of the most successful applications of FD communication, and one more reason is that relaying allows not only to increase throughput but also solves the issue of multiplexing loss existent in HD cooperative networks. Besides, it also increases link reliability, and it is a potential solution for enhancing cell coverage. One often overlooked point is that FD relaying also reduces the latency since communication is resolved within one time slot compared to its HD counterpart. Therefore, FD relaying is an attractive solution for latency-constrained applications.

However, on that case the FD relay node only acts as a helper forwarding messages from source to destination. Nonetheless, whenever the FD relay has backlog information from one or more users it can forward it together with the message being received, thus exploiting one more capability and providing in-band backhaul.

6.5.3   Wireless Backhaul

Backhaul has become a major concern for future wireless networks, especially for dense networks where conventional solutions, such as fiber or xDSL, are not fully available or their deployment comes with an exacerbate price tag. In this context, FD appears as an alternative potential solution that alleviates such costs of capital expenses and operating expenses for the network operator while providing enhanced throughput and coverage. Besides, another advantage is that the FD node can potentially allocate even traffic over both link directions, thus making better use of the transmission opportunities, improving spectral usage, and reducing the latency. Therefore, FD relaying/backhaul can be an attractive solution for latency-constrained applications. Even though some efforts have been made, there are several research challenges ahead, as in smart resource scheduling and prioritization of traffic contained on distinct network functions, as well as in efficient RRM, especially for dense urban scenarios.

6.5.4   Cognitive Radio

In conventional cognitive radio networks the secondary users should not harm the communication of the primary network. Thus, an often used protocol first listens for primary transmissions, then if not occupied the secondary network talks, at cost of reduced transmission time due to sensing of primary network activities.

However, a transceiver operating in FD fashion can potentially increase the transmission time, while providing accurate sensing for transmission opportunities, due to its capability of simultaneous transmission and reception [16]. Moreover, collisions with primary network are potentially reduced, thus improving throughput of the primary network as well.

All in all, FD communication goes beyond increasing throughput as evinced by applications discussed in this chapter, thus rendering increased spectral efficiency, reducing collisions and enabling efficient jamming according to each application, and reducing latency as well. Several research challenges still remain though, in particular with respect to network coordination so to reduce additional interference from increased FD transmissions. Traffic balancing and smart scheduling are also an open problem since traffic is preferred over one direction in current deployments. Prioritization of the traffic flows are as well an open problem, especially when information flow is latency constrained. Even though research on FD communication has significantly advanced lately, there are still many potential applications and challenges ahead.

6.6   Conclusions and Outlook

Simultaneous transmission and reception enabled by full-duplex communication promises significant throughput gains and latency reduction and is considered as a potential solution for next generation radios. Under ideal assumptions, the gains of FD communication are close to 100%. However, such impressive achievements are significantly degraded in realistic network settings, primarily due to strong self-interference, resultant increase in inter-cell interference, and asymmetric nature of network traffic. For example, our extensive system level simulation analysis in close-to-real-life dense network scenario shows best median gains of 15–20% considering the UDP traffic protocol.

Nonetheless, with efficient techniques to overcome such limitations, and application in appropriate scenarios, FD technology can be exploited to harness significant performance benefits. This chapter has presented a number of interference management techniques for FD communication, the virtual FD concept—which mimics FD communication with HD nodes—being a particular example. We showed that the reliability of virtual FD can be superior to FD over the entire target SINR range with success probability gains of up to 70%. Furthermore, we showed that the potential of FD technology is not limited to harnessing throughput gain and latency reduction. Rather, owing to its nature of simultaneous transmission and reception, it has valuable applications in other areas such as physical layer security, D2D communication, and use cases requiring channel sensing, e.g., cognitive radios.

References

1. D. Korpi, J. Tamminen, M. Turunen, T. Huusari, Y. S. Choi, L. Anttila, S. Talwar, and M. Valkama, "Full-duplex mobile device: pushing the limits," *IEEE Communications Magazine*, vol. 54, no. 9, pp. 80–87, Sep. 2016.
2. N. H. Mahmood, I. S. Ansari, P. Popovski, P. Mogensen, and K. A. Qaraqe, "Physical-layer security with full-duplex transceivers and multiuser receiver at eve," *IEEE Transaction on Communications*, vol. 65, no. 10, pp. 4392–4405, Oct. 2017.
3. S. Goyal, P. Liu, S. Panwar, R. Yang, R. A. DiFazio, and E. Bala, "Full duplex operation for small cells," *CoRR*, vol. abs/1412.8708, 2014. [Online]. Available: http://arxiv.org/pdf/1412.8708.pdf
4. M. G. Sarret, G. Berardinelli, N. H. Mahmood, M. Fleischer, P. E. Mogensen, and H. Heinz, "Analyzing the potential of full duplex in 5G ultra-dense small cell networks," *EURASIP Journal on Wireless Communications and Networking*, vol. 2016, no. 284, Dec. 2016.
5. J. G. Andrews, S. Buzzi, W. Choi, S. V. Hanly, A. Lozano, A. C. K. Soong, and J. C. Zhang, "What will 5G be?" *IEEE Journal on Selected Areas in Communications*, vol. 32, no. 6, pp. 1065–1082, Jun. 2014.
6. A. AlAmmouri, H. ElSawy, O. Amin, and M.-S. Alouini, "In-band α-duplex scheme for cellular networks: A stochastic geometry approach," *IEEE Transactions on Wireless Communications*, vol. 15, no. 10, pp. 6797–6812, Oct. 2016.
7. "Further advancements for E-UTRA physical layer aspects (Release9)," 3rd Generation Partnership Project, Tech. Rep. TR 36.814, V9.0.0, 2010.
8. Z. Tong and M. Haenggi, "Throughput analysis for full-duplex wireless networks with imperfect self-interference cancellation," *IEEE Transactions on Communications*, vol. 63, no. 11, pp. 4490–4500, Nov. 2015.
9. R.-A. Pitaval, O. Tirkkonen, R. Wichman, K. Pajukoski, E. Lähetkangas, and E. Tiirola, "Full-duplex self-backhauling for small-cell 5G networks," *IEEE Wireless Communications*, vol. 22, no. 5, pp. 83–89, Oct. 2015.
10. I. Randrianantenaina, H. Dahrouj, H. Elsawy, and M.-S. Alouini, "Interference management in full-duplex cellular networks with partial spectrum overlap," *IEEE Access*, vol. 5, pp. 7567–7583, 2017.
11. N. H. Mahmood, G. Berardinelli, F. Tavares, and P. Mogensen, "On the potential of full duplex communication in 5G small cell networks," in *Proc. IEEE 81st Vehicular Technology Conference: VTC-Spring*, Glasgow, Scotland, May 2015.
12. H. Tabassum, A. H. Sakr, and E. Hossain, "Analysis of massive MIMO-enabled downlink wireless backhauling for full-duplex small cells," *IEEE Transactions on Communications*, vol. 64, no. 6, pp. 2354–2369, 2016.
13. M. G. Sarret, G. Berardinelli, N. H. Mahmood, B. Soret, and P. Mogensen, "Can full duplex reduce the discovery time in D2D communication?" in *Proc. IEEE 13th International Symposium on Wireless Communication Systems (ISWCS)*, Poznan, Poland, Sep. 2016.
14. H. Thomsen, P. Popovski, E. De Carvalho, N. Pratas, D. Kim, and F. Boccardi, "CoMPflex: CoMP for in-band wireless full duplex," *IEEE Wireless Commun. Lett.*, vol. 5, no. 2, pp. 144–147, Apr. 2016.
15. H. Thomsen, D. M. Kim, P. Popovski, N. K. Pratas, and E. De Carvalho, "Full duplex emulation via spatial separation of half duplex nodes in a planar cellular network," in *IEEE International Workshop on Signal Processing Advances in Wireless Communications (SPAWC), 2016*, July 2016.
16. Y. Liao, T. Wang, L. Song, and Z. Han, "Listen-and-talk: Protocol design and analysis for full-duplex cognitive radio networks," *IEEE Transactions on Vehicular Technology*, vol. 66, no. 1, pp. 656–667, Jan. 2017.

Chapter 7
Full-Duplex Non-Orthogonal Multiple Access Systems

Mohammadali Mohammadi, Batu K. Chalise, Himal A. Suraweera, and Zhiguo Ding

Abstract Full-duplex communications and non-orthogonal multiple access (NOMA) each can deliver high spectral efficiency for modern wireless systems. In this chapter, we investigate the joint performance of full-duplex and NOMA when applied in multiple antenna systems, systems based on the antenna selection and relaying systems based on the cognitive radio principle. First, we explore the current research progress reported in the recent literature. Next, we consider downlink NOMA operation in which a central transmitter communicates with a pair of near and far located users with the help of a multi-antenna equipped full-duplex relay. In particular, optimum and suboptimal beamforming schemes are proposed to counter the self-interference effect and inter-user interference at the strong user. Next, we study the problem of selecting best transmit/receive antennas in full-duplex NOMA systems. To this end, a multi-antenna equipped access point and a full-duplex relay setup is considered in which a base station serves a user located nearby while at the same time, a relay is used to assist communication between the access point and a far user. Finally, we investigate the integration of full-duplex and

---

Portions of this chapter are reprinted from Refs. [22, 23, 43], with permission from IEEE.

M. Mohammadi (✉)
Faculty of Engineering, Shahrekord University, Shahrekord, Iran
e-mail: m.a.mohammadi@sku.ac.ir

B. K. Chalise
Department of Electrical and Computer Engineering, New York Institute of Technology, Old Westbury, NY, USA
e-mail: batu.k.chalise@ieee.org

H. A. Suraweera
Department of Electrical and Electronic Engineering, University of Peradeniya, Peradeniya, Sri Lanka
e-mail: himal@ee.pdn.ac.lk

Z. Ding
School of Electrical and Electronic Engineering, The University of Manchester, Manchester, UK
e-mail: zhiguo.ding@manchester.ac.uk

© Springer Nature Singapore Pte Ltd. 2020
H. Alves et al. (eds.), *Full-Duplex Communications for Future Wireless Networks*,
https://doi.org/10.1007/978-981-15-2969-6_7

NOMA in cognitive relay systems and pose an optimization problem to maximize the near/far user information rate region. The presented results clearly demonstrate that full-duplex operation at terminals with NOMA is capable of ushering in performance gains for the systems studied. We also identify several future research directions that are useful in designing future full-duplex enabled NOMA systems.

7.1 Introduction

The proliferation of wireless technologies, the ever-increasing growth of devices such as smartphones, tablets, and internet-of-things (IoT), and the emergence of audio/video intensive multimedia applications have made a case to improve the spectral efficiency beyond what is used in current wireless systems. Two effective solutions embraced by the industry to satisfy the growing demand for spectral efficiency in wireless communication applications are full-duplex communications [1, 2] and non-orthogonal multiple access (NOMA) [3–5]. NOMA principle uses power-domain multiplexing in general to perform non-orthogonal resource allocation among multiple users at the expense of increased signal processing and receiver complexity [3]. Using full-duplex mode concurrent transmission and reception of radio waves over the same frequency band can be achieved at a price of performance loss due to self-interference (SI) [1, 2, 6].

A full-duplex device is difficult to realize in practice due to the signal leakage phenomenon at the transceiver [1]. Since the early days of full-duplex hardware design and development, passive isolation methods such as addition of radio frequency (RF) absorber material between the transceiver antennas and directional antennas with high gain and narrow beam width properties are used to suppress the effect of SI [6]. However, passive isolation alone cannot lower the SI up to the noise level. On the other hand, recent success in analog/digital signal processing and spatial domain suppression/cancellation solutions through the use of beamforming techniques have championed an increasing attention on deploying full-duplex transceivers in modern wireless systems [6–10]. To this end in the literature, single and multiple antenna full-duplex implementations based on new SI cancellation techniques have been proposed. Single antenna implementations are based on combinations of analog suppression and digital cancellation methods capable of providing high amount of isolation to suppress the SI [11] while multiple antenna implementations rely on spatial suppression via the application of beamforming methods, zero-forcing (ZF), null-space projection, and antenna selection [10, 12, 13].

Non-orthogonal Multiple Access A multiple access scheme allows several mobile users to communicate, for example, a stream of data with a cellular base station (BS) using a shared radio resource. Since the introduction of wireless systems in the twentieth century, several multiple access schemes that operate over shared time, frequency, or space resources have been used [14]. Specifically, orthogonal multiple access (OMA) schemes, such as frequency division multiple access (FDMA), time division multiple access (TDMA), and code division multiple access (CDMA) were exploited in the first generation (1G), second generation (2G), and third generation (3G) cellular standards, respectively. More recently, orthogonal frequency division multiple access (OFDMA) was adopted for fourth generation (4G) cellular systems, namely, Long Term Evolution-Advanced. Among multiple access techniques, TDMA and FDMA are special cases of orthogonal multiple access, where only a single user is serviced within a certain time/frequency resource block. On the other hand, CDMA can serve multiple users over the same time/frequency resource by allocating orthogonal spreading codes for different users so that the multiuser interference can be effectively suppressed for better performance. However, CDMA demands the chip rate to be larger than the data rate which can burden hardware implementation, especially, if the information data rate is in the order of gigabytes. Therefore, OMA techniques alone are not adequate to meet the increasing demands for higher throughput and strict quality-of-service (QoS) requirements of modern wireless systems. In this regard, NOMA techniques capable of allowing multiple users to occupy the same spectrum have been proposed for 5G to improve the access efficiency [4].

NOMA techniques can be grouped into two broader classes as power-domain NOMA and code-domain NOMA [3, 4]. Power-domain NOMA which is applied in this chapter multiplexes multiple users within a given time/frequency resource block with different power levels at the transmitter and applies multiuser detection algorithms such as successive interference cancellation (SIC) at the receivers to detect incoming signals as shown in Fig. 7.1. Accordingly, spectral efficiency can be improved, although a simple receiver cannot be used as compared to conventional OMA. As a diversion to the well-known "water-filling" principle, power-domain NOMA scheme allocates less power to users with good channel conditions (termed as NOMA strong users) and higher power to users that have poor channel conditions

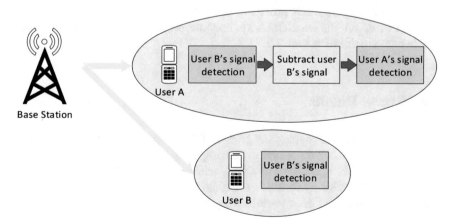

Fig. 7.1 A downlink NOMA communication model with one central base station and User A and User B

(termed weak users), thereby incurring an extra fairness constraint. Since NOMA strong users decode their own signals by removing the signals intended to NOMA weak users using SIC, the high power interference caused by transmissions to the NOMA weak users does not affect the reception quality of the NOMA strong users. On the contrary, NOMA weak users experience poor channel conditions, and thus the interference caused by NOMA strong users' superimposed signals at NOMA weak users is significantly suppressed. As such, the reception ability of the weak users is not affected. Some beneficial characteristics of NOMA can be listed as follows [4]:

- Improved spectral efficiency and throughput at the cell fringe.
- Promotion of massive connectivity.
- Low-latency communication and signaling cost.
- Guaranteeing user fairness in terms of their throughput.
- Relaxed channel feedback.

Besides the aforementioned advantages, NOMA can be integrated into the existing wireless systems with minimal modifications. Moreover, due to its flexibility and compatibility with advances such as the multiple-input multiple-output (MIMO) technology, NOMA has attracted plenty of interest from both academic and industrial communities. In the literature, different aspects of the application of NOMA in uplink/downlink (UL/DL) cellular networks [15], cooperative one-way and two-way relaying networks [16], cognitive and spectrum sharing networks [17], machine-to-machine communications [18], and cloud radio access networks [19] have been investigated.

We have organized the chapter as follows: In Sect. 7.2 we discuss recent results on full-duplex NOMA systems. In Sect. 7.3 the performance of a full-duplex NOMA relay system is analyzed. Antenna selection performance of full-duplex cooperative NOMA systems is investigated in Sect. 7.4. In Sect. 7.5 beamforming design and power allocation for a full-duplex NOMA cognitive radio system is explored. Several interesting research problems and future directions for the design and deployment of full-duplex NOMA systems are presented in Sect. 7.6, while the conclusions are drawn in Sect. 7.7.

7.2 Recent Results

The combination of full-duplex and NOMA shows significant promise to improve the spectrum efficiency compared to contemporary wireless communications that rely on half-duplex operation and OMA schemes. Therefore, a sizable body of research work has already focused on the modeling, analysis, and optimization of full-duplex enabled NOMA communication systems.

7.2.1 Full-Duplex NOMA Topologies

In the present literature, there are two basic full-duplex transmission topologies benefiting from the NOMA concept [5]. The first topology is the full-duplex network setup where a full-duplex BS simultaneously serves multiple UL and DL users in the same spectrum via NOMA. The performance analysis of a full-duplex single-cell NOMA network where UL and DL transmissions occur at the same time has been provided in [20], which demonstrated that the marriage between full-duplex and NOMA can deliver appreciable performance gains over half-duplex and NOMA operation or half-duplex and OMA operation, if the co-channel interference between UL and DL transmissions can be sufficiently suppressed.

The second topology is the cooperative topology in which a full-duplex infrastructure relay [21–23] or a full-duplex NOMA near user [24] is employed to enhance transmissions between a source and NOMA users. In the relay-assisted cooperative NOMA setup, as shown in Fig. 7.2a, an infrastructure relay is used to assist the NOMA far user located away from the source. In the user-assisted cooperative NOMA setup, as shown in Fig. 7.2b, NOMA strong user is available to help the NOMA weak user, since NOMA strong user is in a position to decode the information intended for both users. The analysis in [21] shows that the full-duplex cooperative NOMA system can improve the ergodic sum capacity in comparison to half-duplex cooperative NOMA operation at low-to-moderate signal-to-noise ratio (SNR) values. In [23], for a full-duplex relay system, antenna selection schemes that jointly take QoS priorities of the NOMA near/far users into account have been proposed and analyzed. In addition, diversity analysis reported in [24] for user-assisted cooperative full-duplex NOMA systems reveals that a direct link if available can compensate for the lack of diversity of the NOMA far user.

In [25], a NOMA-based two-way relay network with secrecy considerations has been investigated. In the considered system model, a trusted full-duplex relay is used to facilitate the NOMA signal exchange between two users in the presence of an eavesdropper. Different decoding schemes based on SIC were proposed for the two users, relay, and eavesdroppers and the achievable ergodic secrecy rates under both single/multiple eavesdropper cases were presented in [25]. A full-duplex protocol for a cooperative relay sharing network has been studied in [26], wherein

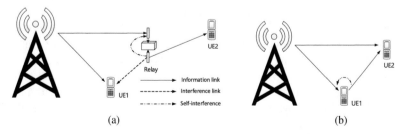

(a)   (b)

Fig. 7.2 Full-duplex cooperative NOMA topologies. (**a**) Relay-assisted. (**b**) User-assisted

two sources can communicate concurrently with their respective destination nodes via a full-duplex relay and utilizing NOMA. Performance of virtual full-duplex relaying in cooperative NOMA systems, where two half-duplex relays mimic the virtual operation of full-duplex relaying, has been investigated in [27, 28]. The authors in [29] proposed an adaptive solution which switches between NOMA, cooperative NOMA, and OMA schemes, respectively, depending on the SI level at the full-duplex user and the link quality in the investigated network. The outage probability and sum rate of a NOMA system with cooperative full-duplex relaying have been derived in [30], where the NOMA near user operates as a full-duplex relay and thereby retransmits to the weak user. In [31], a two-user cooperative NOMA downlink system has been studied, where either full-duplex or half-duplex relaying is employed at the NOMA near user to augment direct communication between a BS in the cell center and a far user in the cell boundary. The proposed scheme when applied in two-user NOMA systems can improve the outage performance at the cell boundary as well as can deliver a significantly high sum throughput.

7.2.2 Resource Allocation and Optimization in Full-Duplex NOMA

In general, due to the scarce nature of radio resources such as bandwidth and power, efficient resource allocation and optimization should be performed in wireless networks. Therefore, resource allocation in full-duplex enabled NOMA systems has been studied in the existing literature [30, 32–37]. Specifically, in [38], optimal power allocation at the BS and relay has been derived for a cooperative NOMA system, to maximize the minimum achievable user rates. Full-duplex operation can be used at a BS to simultaneously service multiple half-duplex UL and DL users. In [32], a resource allocation scheme for such a full-duplex multicarrier NOMA (MC-NOMA) system has been proposed. In [33], the problem of duplex mode selection, user association, and power allocation in a multi-cell NOMA system has been investigated. Specifically, a time-averaged UL and DL rate maximization problem has been formulated, and the advantages of operating in half-duplex or full-duplex, as well as in OMA or NOMA modes, depending on network parameters and conditions, and SI cancellation capabilities have been quantified. In [34], a power minimization problem for a multi-cell NOMA setup was studied. A robust and secure resource allocation algorithm for full-duplex multi-antenna equipped MC-NOMA systems has been designed in [35]. The considered optimization problem in [35] maximizes the weighted system throughput, while imperfectness of the channel state information (CSI) of the eavesdropper channels and user QoS is accounted for the problem. Power allocation and channel assignment for a single-cell full-duplex NOMA system was studied in [36], where NOMA is used in the DL and OMA is utilized in the UL.

In [37], resource allocation schemes for a full-duplex NOMA network have been presented, where a full-duplex access point (AP) transmits information and power wirelessly to multiple full-duplex users in the DL, while at the same time, users transmit information to the AP through relays in the UL.

7.2.3 Applications of Full-Duplex NOMA

Full-duplex enabled NOMA systems have found applications in energy-constrained systems and networks as reported in the literature [38–42], cognitive radio networks [43], HetNets [44], etc. An energy-constrained cooperative NOMA system has been considered in [38], where near user is powered using radio frequency signals to relay the far user's information in full-duplex fashion. A full-duplex cooperative NOMA system where a NOMA near user is exploited as a relay capable of self-energy recycling (the relay is equipped with related hardware to harvest energy from SI) to help a far user is explored in [39]. Specifically, the performance several popular relaying protocols have been studied. A wireless-powered NOMA system has been studied in [40], where a full-duplex transmitter harvests energy from a power beacon and its loop channel, while transmitting information to the NOMA users concurrently. Given the QoS requirements of the NOMA users and the constraint of the full-duplex transmitter's power, the minimal transmit power of the dedicated energy source and the corresponding optimal beamforming vectors have been obtained in [40].

In [41], the outage probability and ergodic rate applicable to an energy harvesting full-duplex relaying system is derived in closed form. The energy efficiency maximization problem considered in [42] jointly optimizes beamforming design and the power splitting ratio subject to a minimum required target rate at the far user.

In [43] assuming a full-duplex relay-assisted NOMA cognitive radio network, an optimization problem to maximize the near user rate has been posed. Moreover, authors have provided a solution by performing joint beamforming and power allocation design at the relay under a far user rate threshold.

Table 7.1 summarizes the above recent results as pertained to different full-duplex NOMA systems.

7.3 Full-Duplex Cooperative NOMA Systems

In NOMA systems, due to the co-existence of near and far users, loss of performance at the far users should be expected [16, 45]. The performance loss to a certain level can be removed through the deployment of solutions such as user cooperation [16] or dedicated relays [46]. In user-assisted NOMA systems, a user with a better channel condition is selected to assist the far user with a poor link as reported in

Table 7.1 Summary of recent results on full-duplex NOMA systems

Literature	Topology	Contribution
[20]	Cellular	Performance analysis
[21]	Cooperative (relay-assisted)	Performance analysis
[22]	Cooperative (relay-assisted)	Performance analysis and beamforming design
[23]	Cooperative (relay-assisted)	Antenna selection schemes and performance analysis
[24]	Cooperative (user-assisted)	Performance analysis
[25]	Cooperative (relay-assisted)	Ergodic secrecy rates analysis under single eavesdropper, multiple non-colluding, and colluding eavesdroppers
[26]	Cooperative (relay-assisted)	Performance analysis
[27, 28]	Cooperative	Virtual full-duplex relaying
[29]	Cooperative (user-assisted)	Performance analysis
[30]	Cooperative (user-assisted)	Performance analysis; Optimal power allocation to minimize outage probability
[31]	Cooperative (user-assisted)	Performance analysis
[32]	Cellular	Optimal joint power and subcarrier allocation to maximize the weighted sum system throughput
[33]	Cellular	Resource allocation
[34]	Cellular	Developing a power efficient two-tier NOMA scheme
[35]	Cellular	Secure resource allocation algorithm design for weighted system throughput maximization
[36]	Cellular	Distributed power control and channel assignment
[37]	Cooperative (relay-assisted)	SWIPT, Maximizing the minimum sum of DL and UL transmit rates, with jointly allocating subcarriers and powers, and relay selection
[38]	Cooperative (user-assisted)	Beamforming design and energy harvesting
[39]	Cooperative (relay-assisted)	Self-energy recycling and outage probability analysis
[40]	Cellular	Beamforming design, energy harvesting, and self-energy recycling
[41]	Cooperative(user-assisted)	Performance analysis, energy harvesting
[42]	Cooperative(user-assisted)	Beamforming design, energy harvesting
[43]	Cooperative (relay-assisted)	Beamforming design and power allocation

Sect. 7.2. In relay-assisted NOMA systems, a fixed infrastructure relay is installed to assist the far user away from the BS and as such, the system designer has the choice of deploying full-duplex operation at the fixed infrastructure relay [21] or at a relaying user [29].

We now describe such a system in which a full-duplex multi-antenna relay assists transmission from an AP to NOMA far users located at the cell rim. We employ stochastic geometry, a popular tool used in the literature to statistically model the locations of the nodes/users and study user selection problem. Further, performance gains of the system are presented to demonstrate the superior performance of the full-duplex mode.

7.3.1 System Model

Figure 7.3 shows a cooperative NOMA system with an AP and two user sets distributed in the cell: near and far users. The AP is located at the middle of the cell, near users $\{U_{1,i}\}$, $i = 1, \ldots, N_{U_1}$ are located within a disc of radius R_1, and the far users $\{U_{2,i}\}$, $i = 1, \ldots, N_{U_2}$ are randomly distributed inside a closed region of inner and outer radii R_2 and R_3. The positions of near and far users form two independent homogeneous Poisson point processes (PPPs) Φ_n and Φ_f, respectively, with the densities λ_n and λ_f. The AP uses the NOMA principle to serve two NOMA users simultaneously in downlink. Specifically, users are selected according to: (1) random near user and random far user (RNRF) selection strategy; (2) nearest near user and nearest far user (NNNF) selection strategy. In the model, a direct communication link between the AP and far user does not exist. Therefore, K full-

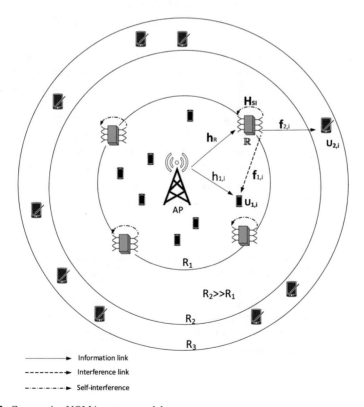

Fig. 7.3 Cooperative NOMA system model

duplex DF relays, denoted as $\{\mathbb{R}_k\}$, $k = 1, \ldots, K$, are used to assist transmission between AP and selected far user. Specifically, relays are symmetrically located at a distance R_1 from middle of the cell in a circular arrangement. Infrastructure-based relays are assumed and each of them has N_T transmit antennas and N_R receive antennas to carry out full-duplex communications. Moreover, in order to mimic realistic propagation conditions encountered in wireless cellular systems, the bounded path loss model $\ell(X, Y) = \frac{1}{1+d_{XY}^\alpha}$ between node X and Y is used where $\alpha \geq 2$ denotes the path loss exponent [45].

Let $x_{k,i}$, $k \in \{1, 2\}$ be the symbol for $U_{k,i}$ and P_S presents the AP transmit power. Following the NOMA principle [3], the AP transmits superposition coded signal $s[n] = \sqrt{P_S a_{1,i}} x_{1,i}[n] + \sqrt{P_S a_{2,i}} x_{2,i}[n]$ to $U_{1,i}$ and the selected relay $\mathbb{R}$, where $a_{k,i}$ denotes the NOMA power allocation coefficient, such that $a_{1,i} + a_{2,i} = 1$ and $a_{1,i} < a_{2,i}$. After receiving the signal at $\mathbb{R}$, it uses the vector $\mathbf{w}_r$ to combine the received signal prior to the estimation of $s[n]$. Then, modeling $x_{2,i}$ as interference, $\mathbb{R}$ decodes the symbol of $U_{2,i}$ [21]. Finally, relay forwards $\mathbf{w}_{t,i} x_{2,i}[n - \delta]$ to $U_{2,i}$, where $\mathbf{w}_{t,i}$ is the relay transmit beamformer and δ denotes the delay caused by full-duplex processing [10]. The received SINR at $\mathbb{R}$ can be characterized as

$$\gamma_R = \frac{P_S a_{2,i} \ell(\mathbb{R}) |\mathbf{w}_r^\dagger \mathbf{h}_R|^2}{P_S a_{1,i} \ell(\mathbb{R}) |\mathbf{w}_r^\dagger \mathbf{h}_R|^2 + P_R |\mathbf{w}_r^\dagger \mathbf{H}_{SI} \mathbf{w}_{t,i}|^2 + \sigma_n^2}, \tag{7.1}$$

where P_R is the transmit power at $\mathbb{R}$ and $\mathbf{h}_R \in \mathscr{C}^{N_R \times 1}$ denotes the channel between the AP and $\mathbb{R}$ whose elements are identically independent distributed (i.i.d.) RV $\mathscr{CM}(0, 1)$. The entries of $N_R \times N_T$ SI channel $\mathbf{H}_{SI}$ are modeled as i.i.d., $\mathscr{CM}(0, \sigma_{SI}^2)$ RVs [10]. σ_n^2 is the variance of the zero-mean additive white Gaussian noise (AWGN) at $\mathbb{R}$. The superscripts $(\cdot)^\dagger$ stands for the conjugate transpose and $\mathscr{CM}(0, \sigma^2)$ denotes a circularly symmetric complex Gaussian RV with variance σ^2.

At the same time, $U_{1,i}$ receives $s[n]$ and $x_{2,i}[n - \delta]$ due to the full-duplex transmission at $\mathbb{R}$. $U_{1,i}$ performs SIC to detect its own signal $x_{1,i}$ as follows. $U_{1,i}$ first decodes the message intended for $U_{2,i}$, and subsequently remove it from the received signal to detect $x_{1,i}$. Hence, the SINR at $U_{1,i}$ to detect the far user signal $x_{2,i}$ is given by

$$\gamma_{1,i}^{x_{2,i}} = \frac{P_S a_{2,i} \ell(U_{1,i}) |h_{1,i}|^2}{P_S a_{1,i} \ell(U_{1,i}) |h_{1,i}|^2 + P_R \ell(\mathbb{R}, U_{1,i}) |\mathbf{f}_{1,i}^T \mathbf{w}_{t,i}|^2 + \sigma_n^2}, \tag{7.2}$$

where the superscript $(\cdot)^T$ stands for the transpose.

Consider that the transmission rate of $U_{2,i}$ is $\mathscr{R}_2$, where $\mathscr{R}_2 = \log_2(1 + \gamma_t)$ and γ_t is the set threshold for the received SINR of $U_{2,i}$ to decode $x_{2,i}$ correctly. Hence, $U_{1,i}$ could decode his own symbol $x_{1,i}$ if $\gamma_{1,i}^{x_{2,i}} > \gamma_t$, i.e., the SIC has been carried out correctly. In this case, the received SINR at $U_{1,i}$ to detect $x_{1,i}$ can be written as

$$\gamma_{1,i}^{x_{1,i}} = \frac{P_S a_{1,i} \ell(U_{1,i}) |h_{1,i}|^2}{P_R \ell(\mathbb{R}, U_{1,i}) |\mathbf{f}_{1,i}^T \mathbf{w}_{t,i}|^2 + \sigma_n^2}, \tag{7.3}$$

where $h_{1,i}$ is the channel coefficient between the AP and $U_{1,i}$ and $\mathbf{f}_{1,i} \in \mathscr{C}^{N_T \times 1}$ denotes the channel vector between $\mathbb{R}$ and $U_{1,i}$. Moreover, $\ell(\mathbb{R}, U_{1,i}) = 1/(1 + d_{\mathbb{R}U_{1,i}}^{\alpha})$ with $d_{\mathbb{R}U_{1,i}} = \sqrt{R_1^2 + d_{U_{1,i}}^2 - 2R_1 d_{U_{1,i}} \cos(\theta_r - \theta_i)}$, where θ_r is the angle between the x-axis and line from cell center to $\mathbb{R}$ and θ_i is the angle of the $U_{1,i}$ from the reference x-axis, $-\pi \le \theta_r - \theta_i \le \pi$.

The received SNR at $U_{2,i}$ can be characterized as

$$\gamma_{2,i}^{x_{2,i}} = \frac{P_R \ell(\mathbb{R}, U_{2,i}) |\mathbf{f}_{2,i}^T \mathbf{w}_{t,i}|^2}{\sigma_n^2}, \tag{7.4}$$

with $\ell(\mathbb{R}, U_{2,i}) = 1/(1 + d_{\mathbb{R}U_{2,i}}^{\alpha})$, where $d_{\mathbb{R}U_{2,i}} = \sqrt{R_1^2 + d_{U_{2,i}}^2 - 2R_1 d_{U_{2,i}} \cos(\theta_r - \acute{\theta}_i)}$, $\acute{\theta}_i$ denoting the angle of $U_{2,i}$ from reference x-axis, $\mathbf{f}_{2,i} \in \mathscr{C}^{N_T \times 1}$ denoting the channel between $\mathbb{R}$ and $U_{2,i}$.

In the NOMA systems, users are ordered according of their channel conditions or their QoS priorities [4]. To this end, we assume that users do not have stringent QoS constraints and thus are served opportunistically via the proposed RNRF and NNNF strategies. More specifically, the RNRF strategy selects the near user $U_{1,i}$ and far user $U_{2,i}$ from the two sets of user randomly. On the other hand, with the NNNF strategy, the user located nearest to the AP from those located inside the disc of radii R_1, denoted as $U_{1,i}^{\star}$, and the nearest user to AP from those located inside the ring, denoted as $U_{2,i}^{\star}$, are paired. It is worth to stress that the RNRF and NNNF strategies exhibit different system performance-complexity tradeoffs in terms of the reliability, user fairness, and implementation complexity. For example, RNRF does not require the users' CSI, which in turn reduces the system overhead. As a result, users experience low path loss and hence NNNF can deliver the best performance at the expense of potential issues in user fairness.

The relay with the shortest Euclidean distance from the tagged far user is selected to assist transmission from the AP to the far user. Relay selection criterion can be mathematically defined as

$$\min\{\|\mathbb{R}_k, U_{2,i}\|, k \in \{1, \ldots, K\}\}. \tag{7.5}$$

The relay selection criterion in (7.5) can improve the performance of the far users, especially, in scenarios where the far users are near to the cell boundary of the cellular network. Further, selected relay should not cause significant out-of-cell interference to the users in neighboring cells in practice.

7.3.2 Beamforming Design

MIMO technology has gained momentum now for a considerable period to boost the performance and reliability of contemporary wireless systems. Use of multiple antennas at relay also enables beamforming design, which, for example, allows the power to be directed towards a user and accordingly enhances the received SINR. In the case of full-duplex NOMA communication, beamforming design is complicated [22]. Challenges for system design arise in particular, since beamforming affects spatial SI cancellation while it also influences the signal quality towards the users. According to the SINR expressions in (7.1)–(7.4), it becomes clear that different choices for the transmit and receive beamformer at $\mathbb{R}$ allow to mimic different performance gains.

We now present optimum and sub-optimum beamforming designs. In the optimum beamforming scheme, receive/transmit beamformers at $\mathbb{R}$ are jointly optimized, while the application of several linear receiver/transmitter combiners is investigated at $\mathbb{R}$.

7.3.2.1 Optimum Beamforming Design

In what follows we focus on optimum beamforming design at $\mathbb{R}$ and consider the problem of maximizing the near user SINR, while satisfying the SINR constraint γ_t for CU_2, with the following mathematical formulation:

$$\max_{\mathbf{w}_{t,i}, \mathbf{w}_r} \quad \min\left(\gamma_{1,i}^{x_{2,i}}, \gamma_{1,i}^{x_{1,i}}\right) \tag{7.6a}$$

$$\text{s.t.} \quad \min\left(\gamma_R, \gamma_{2,i}^{x_{2,i}}\right) \geq \gamma_t, \tag{7.6b}$$

$$||\mathbf{w}_{t,i}|| = ||\mathbf{w}_r|| = 1. \tag{7.6c}$$

By inspecting the constraint (7.6b), we observe that, for a given $\mathbf{w}_{t,i}$, the optimum $\mathbf{w}_r$ should be designed such that the γ_R is maximized, i.e.,

$$\max_{||\mathbf{w}_r||=1} \frac{\mathbf{w}_r^\dagger \mathbf{h}_R \mathbf{h}_R^\dagger \mathbf{w}_r}{\mathbf{w}_r^\dagger \mathbf{C} \mathbf{w}_r}, \tag{7.7}$$

where $\mathbf{C} \triangleq P_S a_{1,i} \ell(\mathbb{R}) \mathbf{h}_R \mathbf{h}_R^\dagger + P_R \mathbf{H}_{SI} \mathbf{w}_{t,i} \mathbf{w}_{t,i}^\dagger \mathbf{H}_{SI}^\dagger + \sigma_n^2 \mathbf{I}$. The solution of the generalized Rayleigh ratio problem (7.7) can be obtained as

$$\mathbf{w}_r = \frac{\mathbf{C}^{-1} \mathbf{h}_R}{||\mathbf{C}^{-1} \mathbf{h}_R||}. \tag{7.8}$$

Accordingly, by substituting $\mathbf{w}_r$ into (7.6), after some manipulation, the problem can be transferred into the standard semi-definite relaxation (SDR) problem which can be efficiently solved. The procedure to obtain the optimum $\mathbf{w}_t$ was given in [22].

7.3.2.2 Sub-optimum Beamforming Design

The optimum beamforming design requires semi-definite programming (SDP) that has a high computational complexity. Therefore, in this subsection, three low-complexity sub-optimum beamforming designs, namely transmit ZF (TZF), receive ZF (RZF), and maximum ratio combining (MRC)/maximal ratio transmission (MRT) [12, 13, 47] are studied and outage performance analysis of the corresponding systems is presented.

TZF Scheme The intention behind the TZF scheme is to use the relay's transmit antenna to cancel the SI, while the far user's SNR is maximized [12]. To make it feasible, the relay should be equipped with $N_T > 1$ transmit antennas. Furthermore, at the relay receiver MRC is deployed, i.e., $\mathbf{w}_r^{\mathsf{MRC}} = \frac{\mathbf{h}_R}{\|\mathbf{h}_R\|}$. As such, the optimal transmit beamformer $\mathbf{w}_{t,i}$ can be found by solving the following problem:

$$\max_{\|\mathbf{w}_{t,i}\|=1} \quad |\mathbf{f}_{2,i}^T \mathbf{w}_{t,i}|^2,$$

$$\text{s.t.} \quad \mathbf{h}_R^\dagger \mathbf{H}_{\mathsf{SI}} \mathbf{w}_{t,i} = 0. \tag{7.9}$$

According to [12], the optimal transmit vector $\mathbf{w}_{t,i}$ in (7.9) can be obtained in closed form as $\mathbf{w}_{t,i}^{\mathsf{ZF}} = \frac{\mathbf{A}\mathbf{f}_{2,i}^*}{\|\mathbf{A}\mathbf{f}_{2,i}^*\|}$, where $\mathbf{A} = \mathbf{I}_{N_T} - \frac{\mathbf{H}_{\mathsf{SI}}^\dagger \mathbf{h}_R \mathbf{h}_R^\dagger \mathbf{H}_{\mathsf{SI}}}{\|\mathbf{h}_R^\dagger \mathbf{H}_{\mathsf{SI}}\|^2}$.

RZF Scheme In contrast to the former design, the SI cancellation can be performed at the relay receiver. To make it feasible, relay should have $N_R > 1$ receive antennas. Moreover, relay applies MRT at the transmitter side as $\mathbf{w}_{t,i}^{\mathsf{MRT}} = \frac{\mathbf{f}_{2,i}^*}{\|\mathbf{f}_{2,i}\|}$. The optimal receive beamformer $\mathbf{w}_r$ can be found by solving [12]

$$\max_{\|\mathbf{w}_r\|=1} \quad |\mathbf{w}_r^\dagger \mathbf{h}_R|^2,$$

$$\text{s.t.} \quad \mathbf{w}_r^\dagger \mathbf{H}_{\mathsf{SI}} \mathbf{f}_{2,i}^* = 0. \tag{7.10}$$

The solution of (7.10), $\mathbf{w}_r^{\mathsf{ZF}}$, can be expressed as $\mathbf{w}_r^{\mathsf{ZF}} = \frac{\mathbf{B}\mathbf{h}_R}{\|\mathbf{B}\mathbf{h}_R\|}$, where $\mathbf{B} = \mathbf{I}_{N_R} - \frac{\mathbf{H}_{\mathsf{SI}}\mathbf{f}_{2,i}^*\mathbf{f}_{2,i}^T\mathbf{H}_{\mathsf{SI}}^\dagger}{\|\mathbf{H}_{\mathsf{SI}}\mathbf{f}_{2,i}^*\|^2}$, and the superscript $(\cdot)^*$ stands for the conjugate.

MRC/MRT Scheme With MRC/MRT scheme, $\mathbf{w}_r$ is set to match the $\mathbf{h}_R$ channel, i.e., $\mathbf{w}_r^{\mathsf{MRC}} = \frac{\mathbf{h}_R}{\|\mathbf{h}_R\|}$ and $\mathbf{w}_{t,i}$ is set to match the $\mathbf{f}_{2,i}$ channel, i.e., $\mathbf{w}_{t,i}^{\mathsf{MRT}} = \frac{\mathbf{f}_{2,i}^*}{\|\mathbf{f}_{2,i}\|}$. Since utilizing the MRC/MRT scheme in half-duplex relaying systems has drawn a wide attention due to achieving an optimal performance, it serves as an interesting benchmark scheme for the full-duplex relay-assisted NOMA systems.

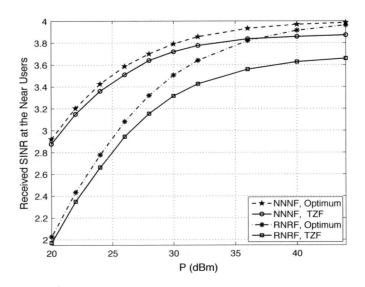

Fig. 7.4 The received SINR at CU_1 versus P for optimum and sub-optimum beamforming designs ($N_T = 4$ and $N_R = 2$)

Figure 7.4 shows the average received SINR at the $U_{1,i}$ for the optimum and TZF beamforming designs and for $a_1 = 0.1, a_2 = 0.9, \alpha = 3$, and $\mathcal{R}_1 = \mathcal{R}_2 = 1$ bps/Hz, where $\mathcal{R}_1$ and $\mathcal{R}_2$ are the transmission rates at CU_1 and CU_2, respectively. Moreover the total power P is allocated between the AP and relay as $P_S = P_R = P/2$. As expected, the optimum scheme yields a comparable performance gain compared to the TZF design. Especially, as P increases, the performance gain with respect to the TZF scheme becomes larger. Further, it can be observed that in case of TZF, there exists a difference in the corresponding SINR values of RNRF and NNNF strategies, whereas with the optimum scheme, the achieved SINR by the RNRF strategy converges to that of the NNNF strategy, when the transmit power is high. Therefore, in the high SNR region, RNRF strategy with optimum scheme yields a good performance/implementation complexity trade-off in high SNR regime. The above result is encouraging for practical implementation.

7.3.3 Performance Analysis

We now characterize the outage probability of the RNRF and NNNF strategies. The RNRF strategy pairs a near user $U_{1,i}$ and a far user $U_{2,i}$ from two sets of users in a random fashion. On the other hand, the users' CSI is exploited in NNNF user selection scheme to pair the near and far users from the two sets of users, which have shortest distance to the AP.

7.3.3.1 Outage Probability of the Near Users

The near user is in outage when it fails to decode $x_{2,i}$ or when it decodes $x_{2,i}$ correctly, but fails to decode $x_{1,i}$. Let $\tau_1 = 2^{\mathscr{R}_1} - 1$ and $\tau_2 = 2^{\mathscr{R}_2} - 1$. The outage probability of $U_{1,i}$ for the RNRF strategy and with sub-optimum beamforming designs is given by [22]

$$\mathsf{P}^i_{out,1} = 1 - \frac{1}{\pi R_1^2} \int_0^{R_1} \int_{-\pi}^{\pi} \frac{e^{-\mu(1+r^\alpha)}}{1 + \frac{q_r \rho_r \mu}{1+\left(R_1^2 + r^2 - 2rR_1 \cos(\theta_r - \theta_i)\right)^{\frac{\alpha}{2}}}(1+r^\alpha)} r \, d\theta_i \, dr,$$

$$(7.11)$$

if $\tau_2 \leq \frac{a_{2,i}}{a_{1,i}}$, otherwise $\mathsf{P}^{\mathsf{TZF}}_{out,1} = 1$, where $i \in \{\mathsf{TZF}, \mathsf{RZF}, \mathsf{MRC}\}$, $\mu = \max\left(\frac{1}{\zeta}, \frac{\tau_1}{\rho_s a_{1,i}}\right)$ with $\zeta = \frac{\rho_s a_{2,i} - \rho_s a_{1,i} \tau_2}{\tau_2}$, $\rho_s = \frac{P_S}{\sigma_n^2}$, and $\rho_r = \frac{P_R}{\sigma_n^2}$.

With the sub-optimum beamforming designs, the outage probability of $U^\star_{1,i}$ for the NNNF strategy is [22]

$$\mathsf{P}^i_{out,1^\star} = 1 - \frac{\upsilon_n}{2\pi} \int_0^{R_1} \int_{-\pi}^{\pi} \frac{e^{-\mu(1+r^\alpha)}}{1 + \frac{q_r \rho_r \mu}{1+\left(R_1^2 + r^2 - 2rR_1 \cos(\theta_r - \theta_i)\right)^{\frac{\alpha}{2}}}(1+r^\alpha)} r e^{-\pi\lambda_n r^2} \, d\theta_i \, dr,$$

$$(7.12)$$

where $i \in \{\mathsf{TZF}, \mathsf{RZF}, \mathsf{MRC}\}$ and $\upsilon_n = \frac{2\pi\lambda_n}{1 - e^{-\pi\lambda_n R_1^2}}$.

As a special case for $\alpha = 2$, the outage probability of $U^\star_{1,i}$ with the NNFF user selection can be derived as

$$\mathsf{P}^{i,\mathsf{P}}_{out,1^\star} = \begin{cases} 1 - \frac{\upsilon_n e^{-\mu}}{2(\mu + \pi\lambda_n)}\left(1 - e^{-R_1^2(\mu + \pi\lambda_n)}\right) & \tau_2 \leq \frac{a_{2,i}}{a_{1,i}}, \\ 1 & \tau_2 > \frac{a_{2,i}}{a_{1,i}}. \end{cases}$$

$$(7.13)$$

From (7.13), as $\lambda_n \to \infty$, we have $\mathsf{P}^{i,\mathsf{P}}_{out,1^\star} \sim 1 - e^{-\mu}$ which indicates that in dense networks, the near user outage decreases exponentially with P_S.

Figure 7.5 compares the outage performance of CU_1 with RNRF and NNNF strategies and for different user densities. We observed that for different near user densities and over the whole transmit power region, the NNNF strategy yields a better outage probability than the RNRF strategy. As shown in Fig. 7.5 the outage probability of the near users with the NNNF strategy improves as λ_n increases, while the outage probability of the near user with RNRF strategy does not depend on λ_n. The number of available near users increases and thus the number of choices for the NNNF strategy increases, which explains the previous observation.

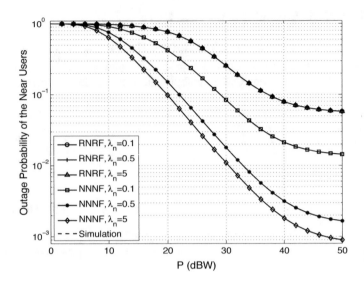

Fig. 7.5 Outage probability of $U_{1,i}$ with different near user densities ($R_1 = 4$ m)

7.3.3.2 Outage Probability of the Far Users

The far user experiences outage events if: (1) $\mathbb{R}$ is unable to successfully decode $x_{2,i}$, or (2) $\mathbb{R}$ can decode $x_{2,i}$ but $U_{2,i}$ cannot decode $x_{2,i}$ correctly. Therefore, the far user outage probability is [21]

$$P_{out,2} = \Pr\left(\gamma_R < \tau_2\right) + \Pr\left(\gamma_R > \tau_2\right) \Pr\left(\gamma_{2,i}^{x_{2,i}} < \tau_2\right), \quad (7.14)$$

where $\Pr(\cdot)$ denotes the probability. Accordingly, outage probability of $U_{2,i}$ with the TZF and RZF schemes and for the RNRF strategy can be expressed as [22]

$$P_{out,2}^{TZF} = 1 - \frac{\pi}{M(R_3 + R_2)\Gamma(N_R)} \Gamma\left(N_R, \frac{(1 + R_1^\alpha)}{\zeta}\right) \sum_{k=0}^{N_T-2} \frac{1}{k!} \left(\frac{\tau_2}{\rho_r}\right)^k$$

$$\times \sum_{m=1}^{M} z_m \sqrt{1 - \phi_m^2} \left(1 + z_m^\alpha\right)^k e^{-\left(\frac{\tau_2}{\rho_r}\right)\left(1 + z_m^\alpha\right)}, \quad (7.15)$$

and

$$P_{out,2}^{RZF} = 1 - \frac{\pi}{M(R_3 + R_2)\Gamma(N_R - 1)} \Gamma\left(N_R - 1, \frac{(1 + R_1^\alpha)}{\zeta}\right)$$

$$\times \sum_{k=0}^{N_T-1} \frac{1}{k!} \left(\frac{\tau_2}{\rho_r}\right)^k \sum_{m=1}^{M} z_m \sqrt{1 - \phi_m^2} \left(1 + z_m^\alpha\right)^k e^{-\left(\frac{\tau_2}{\rho_r}\right)\left(1 + z_m^\alpha\right)}. \quad (7.16)$$

respectively, where $z_m = \frac{R_3 - R_2}{2}(\phi_m + 1) + R_2$, $\phi_m = \cos\left(\frac{2m-1}{2M}\pi\right)$, M is the parameter of the Gaussian–Chebyshev quadrature method used to derive (7.16), and $\Gamma(a, x) = \int_x^\infty e^{-t} t^{\alpha-1} dt$ [48, Eq. (8.350)].

Based on (7.15) and (7.16), it can be inferred that in some antenna configurations, TZF and RZF schemes yield an identical outage performance for $U_{2,i}$. For example, if we pair N_T and N_R as (N_T, N_R), TZF (N_T, N_R) has an identical outage performance with RZF $(N_T - 1, N_R + 1)$. In addition, outage probability analysis reveals that the performance of the far users with TZF and RZF scheme improves for large values of transmit power P_S and P_R. This is because both TZF and RZF schemes totally cancel the SI, which in turn increases the second-hop SNR of the far users for large values of the relay transmit power, P_R.

The outage probability of $U_{2,i}^\star$ with the TZF and RZF schemes and with the NNNF strategy is given by [22]

$$
P_{\text{out},2^\star}^{\text{TZF}} \approx 1 - \frac{\upsilon_f \pi (R_3 - R_2) e^{\pi \lambda_f R_2^2}}{2M \Gamma(N_R)} \Gamma\left(N_R, \frac{(1 + R_1^\alpha)}{\zeta}\right) \sum_{k=0}^{N_T-2} \frac{1}{k!} \left(\frac{\tau_2}{\rho_r}\right)^k
$$
$$
\times \sum_{m=1}^{M} z_m \sqrt{1 - \phi_m^2} (1 + z_m^\alpha)^k e^{-\left(\frac{\tau_2}{\rho_r} + \frac{\tau_2}{\rho_r} z_m^\alpha + \pi \lambda_f z_m^2\right)}, \tag{7.17}
$$

and

$$
P_{\text{out},2^\star}^{\text{RZF}} \approx 1 - \frac{\upsilon_f \pi (R_3 - R_2) e^{\pi \lambda_f R_2^2}}{2M \Gamma(N_R - 1)} \Gamma\left(N_R - 1, \frac{(1 + R_1^\alpha)}{\zeta}\right) \sum_{k=0}^{N_T-1} \frac{1}{k!} \left(\frac{\tau_2}{\rho_r}\right)^k
$$
$$
\times \sum_{m=1}^{M} z_m \sqrt{1 - \phi_m^2} (1 + z_m^\alpha)^k e^{-\left(\frac{\tau_2}{\rho_r} + \frac{\tau_2}{\rho_r} z_m^\alpha + \pi \lambda_f z_m^2\right)}. \tag{7.18}
$$

where $\upsilon_f = \frac{2\pi \lambda_f}{1 - e^{-\pi \lambda_f (R_3^2 - R_2^2)}}$.

Figure 7.6, shows the outage performance of full-duplex and half-duplex relaying versus σ_{SI}^2 for RNRF user selection. The outage probability gain, defined as $G_j(M_T, M_R) = P_{\text{out},2}^{\text{HD}} / P_{\text{out},2}^j, j \in \{\text{TZF, RZF, MRC}\}$, versus the SI strength, σ_{SI}^2, is plotted. As expected when the strength of SI is low the full-duplex significantly outperforms the half-duplex approach. The relative gain is observed to be mildly decreasing when the SI strength is decreased to the extent that in the low SI strength regime ($\sigma_{\text{SI}}^2 < -60$ dBm), the MRC/MRT scheme outperforms the ZF-based schemes, e.g., $G_{\text{TZF}}(3, 2) = 3.6$ as compared to $G_{\text{MRC}}(3, 3) = 28$ at $\sigma_{\text{SI}}^2 = -70$ dBm. In this regime, MRC/MRT(3, 2) has the largest gain.

We will next explore the problem of antenna selection in full-duplex cooperative NOMA systems. Antenna selection in full-duplex systems is a complicated and a hard problem as compared to half-duplex systems widely studied in the current literature. This is because the backward and forward channels at the full-duplex

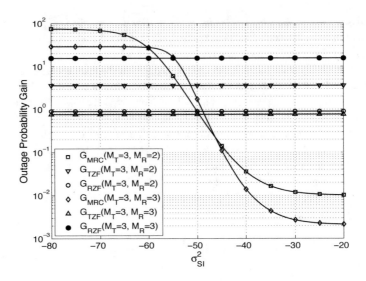

Fig. 7.6 Outage probability gain of the far users for the RNRF ($a_1 = 0.1, a_2 = 0.9$)

nodes are coupled through the SI link [10]. Therefore, antenna selection at the transmit or receive side cannot be performed independently of each other as is the case of half-duplex systems. The coupling introduces new mathematical challenges for performance analysis since the involved random variables of various links now become correlated.

7.4 Full-Duplex Cooperative NOMA Systems with Antenna Selection

MIMO systems need multiple RF chains, consisting of mixers, amplifiers, and analog to digital convertors, which typically incur significant system implementation costs. Moreover, MIMO systems suffer from high signal processing complexity. As an alternative, antenna selection techniques have been proposed as a low-complexity viable solution in the existing literature to maintain system performance at a certain required level. The application of the antenna selection in NOMA system has attracted increasing attention in recent years [23, 49–51]. In the following, antenna selection performance of a full-duplex cooperative NOMA system is studied, where antenna selection should look at the impact of the SI channel when selecting favorable channels toward the NOMA users.

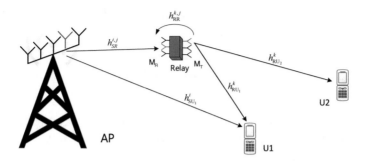

Fig. 7.7 Antenna selection in full-duplex cooperative NOMA system

7.4.1 System Model

Consider a cooperative dual-user NOMA system where an AP communicates directly with near user, CU_1, while a relay, denoted as $\mathbb{R}$, assists the AP in order to deliver its message to far user CU_2, as shown in Fig. 7.7. The AP has M_T antennas, while CU_1 and CU_2 each has a single antenna. $\mathbb{R}$ operates in full-duplex mode and has N_R receive antennas and N_T transmit antennas. Assume that the AP and $\mathbb{R}$ perform single antenna selection in order to further alleviate the cost and implementation complexity [49]. More specifically, the AP selects one (i-th) out of its total set of transmit antennas. Moreover, $\mathbb{R}$ selects one receive antenna and one transmit antennas.

All links are assumed to undergo Rayleigh fading. The channel between i-th transmit and the j-th receive antenna from terminal A to terminal B is denoted by $h_{AB}^{i,j} \sim \mathscr{CN}(0, \sigma_{AB}^2)$ where $A \in \{S, \mathbb{R}\}$ and $B \in \{\mathbb{R}, U1, U2\}$. The SI channel between the j-th receive and the i-th transmit antenna of the full-duplex relay is denoted as $h_{SI}^{k,j}$.

AP transmits a superimposed signal containing symbols of CU_1 and CU_2 via $\mathbb{R}$. The SINR at $\mathbb{R}$ is

$$\gamma_R = \frac{a_2 \gamma_{SR}^{i,j}}{a_1 \gamma_{SR}^{i,j} + \gamma_{SI}^{k,j} + 1}, \tag{7.19}$$

where $\gamma_{SR}^{i,j} = \rho_S |h_{SR}^{i,j}|^2$ and $\gamma_{SI}^{k,j} = \rho_R |h_{SI}^{k,j}|^2$ with $\rho_S = \frac{P_S}{\sigma_n^2}$ and $\rho_R = \frac{P_R}{\sigma_n^2}$, P_S and P_R are the AP and relay transmit power, respectively, and a_i denotes NOMA power allocation coefficient.

$\mathbb{R}$ decodes the received signal and forwards the symbol intended for CU_2. Since the transmit symbol at $\mathbb{R}$ is prior known to CU_1, it is rational to assume that it can be removed at CU_1 [21]. CU_1 cannot perfectly remove $\mathbb{R}$'s signal due to imperfect cancellation and as such the interference channel between $\mathbb{R}$ and CU_1 is modeled as $h_{RU1}^k \sim \mathscr{CN}(0, k_1 \sigma_{RU1}^2)$ where k_1 captures the strength of inter-user interference [21]. Hence, the SINR of CU_2 seen at CU_1 becomes

$$\gamma_{12} = \frac{a_2 \gamma_{SU1}^i}{a_1 \gamma_{SU1}^i + \gamma_{RU1}^k + 1}, \tag{7.20}$$

where $\gamma_{SU1}^i = \rho_S |h_{SU1}^i|^2$ and $\gamma_{RU1}^k = \rho_R |h_{RU1}^k|^2$.

If $\gamma_{12} > \gamma_t$, where $\gamma_t = 2^{\mathcal{R}_2} - 1$, the SIC at CU_1 is successful and CU_1 can completely cancel CU_2's signal. The SINR at CU_1 can be written as

$$\gamma_1 = \frac{a_1 \gamma_{SU1}^i}{\gamma_{RU1}^k + 1}. \tag{7.21}$$

The relay transmits decoded signal towards CU_2, and hence the SNR at CU_2 can be expressed as

$$\gamma_{RU2}^k = \frac{P_R}{\sigma_n^2} |h_{RU2}^k|^2. \tag{7.22}$$

The end-to-end SINR at CU_2 can be expressed as

$$\gamma_2 = \min \left(\frac{a_2 \gamma_{SU1}^i}{a_1 \gamma_{SU1}^i + \gamma_{RU1}^k + 1}, \frac{a_2 \gamma_{SR}^{i,j}}{a_1 \gamma_{SR}^{i,j} + \gamma_{SI}^{k,j} + 1}, \gamma_{RU2}^k \right). \tag{7.23}$$

7.4.2 Antenna Selection Schemes

Two antenna selection schemes are now studied where joint selection of a single antenna at the AP and a single antenna at $\mathbb{R}$ is performed according to the end-to-end SINRs at the near user, CU_1, and the far user, CU_2.

Max-CU$_1$ AS Scheme The max-CU$_1$ AS scheme selects antennas first to maximize the end-to-end SINR at CU_1 (7.21) and then tries to maximize the end-to-end SINR at CU_2 (7.23) with the remaining antenna selection choices. Therefore, we select antennas according to

$$\{i^*, k^*\} = \underset{1 \leq i \leq M_T, 1 \leq k \leq N_T}{\arg \max} \frac{a_1 \gamma_{SU1}^i}{\gamma_{RU1}^k + 1}$$

$$j^* = \underset{1 \leq j \leq N_R}{\arg \max} \frac{a_2 \gamma_{SR}^{i^*,j}}{a_1 \gamma_{SR}^{i^*,j} + \gamma_{SI}^{k^*,j} + 1}. \tag{7.24}$$

Max-CU$_2$ AS Scheme In contrast to the max-CU$_1$ AS scheme, the priority of the max-CU$_2$ AS scheme is to maximize the end-to-end SINR at CU$_2$ according to

$$\{i^*, j^*, k^*\}$$

$$= \underset{\substack{1 \leq i \leq M_T, 1 \leq j \leq N_R, \\ 1 \leq k \leq N_T}}{\arg \max} \min \left(\frac{a_2 \gamma_{\mathsf{SU1}}^i}{a_1 \gamma_{\mathsf{SU1}}^i + \gamma_{\mathsf{RU1}}^k + 1}, \frac{a_2 \gamma_{\mathsf{SR}}^{i,j}}{a_1 \gamma_{\mathsf{SR}}^{i,j} + \gamma_{\mathsf{SI}}^{k,j} + 1}, \gamma_{\mathsf{RU2}}^k \right).$$

$$(7.25)$$

By invoking (7.25), it is evident that all degrees-of-freedom (in terms of antenna selection) are used to maximize the end-to-end SINR at CU$_2$ and the end-to-end SINR at CU$_1$ cannot be maximized through antenna selection. Therefore, max-CU$_2$ AS scheme cannot improve the performance of the near user and acts as a random antenna selection for near user.

7.4.3 Performance Analysis

The comparative performance of the antenna selection schemes is now investigated.

7.4.3.1 Ergodic Sum Rate

Let $\mathscr{R}_{\mathsf{CU}_1}^{\mathsf{AS}} = \mathbb{E}\left\{\log_2(1 + \gamma_{1,\mathsf{AS}})\right\}$ and $\mathscr{R}_{\mathsf{CU}_2}^{\mathsf{AS}} = \mathbb{E}\left\{\log_2(1 + \gamma_{2,\mathsf{AS}})\right\}$. Here $\gamma_{1,\mathsf{AS}}$ and $\gamma_{2,\mathsf{AS}}$ denote the respective end-to-end SINRs at the CU$_1$ and CU$_2$, corresponding to the antenna selection scheme, $\mathsf{AS} \in \{\mathsf{S1, S2}\}$, where $\mathsf{S1}$ refers to the max-CU$_1$ AS scheme and $\mathsf{S2}$ refers to the max-CU$_2$ AS scheme. The ergodic sum rate with antenna selection is given by

$$\mathscr{R}_{sum}^{\mathsf{AS}} = \mathscr{R}_{\mathsf{CU}_1}^{\mathsf{AS}} + \mathscr{R}_{\mathsf{CU}_2}^{\mathsf{AS}}. \qquad (7.26)$$

The ergodic achievable rates of CU$_1$ and CU$_2$ with the max-CU$_1$ AS scheme, can be derived as [23]

$$\mathscr{R}_{\mathsf{CU}_1}^{\mathsf{S1}} = \frac{M_T}{\ln 2} \sum_{p=0}^{M_T-1} \frac{(-1)^p \binom{M_T-1}{p}}{(p+1)\left(\frac{(p+1)\bar{\gamma}_{\mathsf{RU1}}}{N_T a_1 \bar{\gamma}_{\mathsf{SU1}}} - 1\right)}$$

$$\times \left(e^{\frac{1}{\bar{\gamma}_{\mathsf{RU1}}}} E_i\left(\frac{-1}{\bar{\gamma}_{\mathsf{RU1}}}\right) - e^{\frac{(p+1)}{a_1 \bar{\gamma}_{\mathsf{SU1}}}} E_i\left(\frac{-(p+1)}{a_1 \bar{\gamma}_{\mathsf{SU1}}}\right) \right), \qquad (7.27)$$

and

$$\mathscr{R}_{CU_2}^{S1} = \frac{N_R M_T}{\ln 2} \int_0^\infty \frac{e^{-\frac{x}{\bar{\gamma}_{RU2}}}}{1+x} \sum_{p=0}^{M_T-1} \frac{(-1)^p \binom{M_T-1}{p} e^{-\frac{(p+1)x}{\bar{\gamma}_{SU1}(a_2-a_1x)}}}{(p+1)\left(1 + \frac{\bar{\gamma}_{RU1}}{N_T \bar{\gamma}_{SU1}} \frac{(p+1)x}{(a_2-a_1x)}\right)}$$

$$\times \sum_{q=0}^{N_R-1} \frac{(-1)^q \binom{N_R-1}{q} e^{-\frac{(q+1)x}{\bar{\gamma}_{SR}(a_2-a_1x)}}}{(q+1)\left(1 + \frac{\bar{\gamma}_{SI}}{\bar{\gamma}_{SR}} \frac{(q+1)x}{(a_2-a_1x)}\right)} dx, \tag{7.28}$$

respectively, where $\bar{\gamma}_{SR} = \rho_S \sigma_{SR}^2$, $\bar{\gamma}_{SU1} = \rho_S \sigma_{SU1}^2$, $\bar{\gamma}_{RU1} = \rho_R k_1 \sigma_{RU1}^2$, $\bar{\gamma}_{RU2} = \rho_R \sigma_{RU2}^2$, $\bar{\gamma}_{SI} = \rho_R \sigma_{SI}^2$, and $E_i(x) = \int_{-\infty}^x \frac{e^t}{t} dt$ [48, Eq. (8.211.1)].

Furthermore, the ergodic achievable rates of CU_1 and CU_2 with the max-CU_2 AS scheme, are [23]

$$\mathscr{R}_{CU_1}^{S2} = \frac{1}{\ln 2} \frac{a_1 \bar{\gamma}_{SU1}}{(\bar{\gamma}_{RU1} - a_1 \bar{\gamma}_{SU1})} \left(e^{\frac{1}{\bar{\gamma}_{RU1}}} E_i\left(\frac{-1}{\bar{\gamma}_{RU1}}\right) - e^{\frac{1}{a_1 \bar{\gamma}_{SU1}}} E_i\left(\frac{-1}{a_1 \bar{\gamma}_{SU1}}\right)\right), \tag{7.29}$$

and

$$\mathscr{R}_{CU_2}^{S2} = \frac{N_T M_T}{\ln 2} \int_0^\infty \frac{e^{-\frac{x}{\bar{\gamma}_{SU1}(a_2-a_1x)}}}{\left(1 + \frac{\bar{\gamma}_{RU1}}{\bar{\gamma}_{SU1}} \frac{x}{(a_2-a_1x)}\right)(1+x)}$$

$$\times \sum_{p=0}^{M_T-1} \frac{(-1)^p \binom{M_T-1}{p} e^{-\frac{(p+1)x}{\bar{\gamma}_{SR}(a_2-a_1x)}}}{(p+1)\left(1 + \frac{\bar{\gamma}_{SI}}{N_R \bar{\gamma}_{SR}} \frac{(p+1)x}{(a_2-a_1x)}\right)}$$

$$\times \sum_{q=0}^{N_T-1} \frac{(-1)^q \binom{N_T-1}{q} e^{-\frac{(q+1)x}{\bar{\gamma}_{RU2}}}}{(q+1)} dx. \tag{7.30}$$

Figure 7.8 illustrates the ergodic sum rates with different antenna selection schemes. The curves for three baseline cases are also plotted: *(1) Optimum AS scheme:* that requires an exhaustive search to determine the antenna subset that maximizes the ergodic sum rate, *(2) Optimum AS (CU_2) scheme:* that aims to optimally maximize the end-to-end SINR at CU_2 which performs an exhaustive search of all possible antenna combinations to find the optimum subset, and *(3) Random AS scheme:* that randomly selects a single antenna from AP and relay input/output. As expected the antenna selection schemes outperform that of the OMA full-duplex relay system. It is observed that max-CU_2 AS scheme generally has poor performance in all SNR region, whereas max-CU_1 AS scheme has a performance close to the optimum AS scheme in this region. Moreover, the optimum

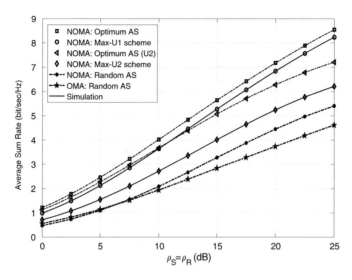

Fig. 7.8 Average sum rate of antenna selection schemes ($M_T = N_R = N_T = 5$, $\sigma_{SI}^2 = 0.3$)

AS (CU$_2$) scheme outperforms the max-CU$_1$ AS scheme in low-to-medium SNR region, however, in the high SNR region, the performance is reversed.

7.4.3.2 Outage Probability

The CU$_1$ is in outage when it fails to decode the CU$_2$'s signal, or when it decodes the CU$_2$'s signal, but is unable to successfully decode its own signal. The outage probability of CU$_1$ with the max-CU$_1$ and max-CU$_2$ AS schemes can be respectively expressed as [23]

$$P_{out,1}^{S1} = 1 - M_T \sum_{p=0}^{M_T-1} \frac{(-1)^p \binom{M_T-1}{p} e^{-\frac{(p+1)\zeta}{\bar{\gamma}_{SU1}}}}{(p+1)\left(1 + \frac{\bar{\gamma}_{RU1}}{\bar{\gamma}_{SU1}} \frac{(p+1)\zeta}{N_T}\right)}, \tag{7.31}$$

and

$$P_{out,1}^{S2} = 1 - \frac{e^{-\frac{\zeta}{\bar{\gamma}_{SU1}}}}{1 + \frac{\bar{\gamma}_{RU1}}{\bar{\gamma}_{SU1}}\zeta}, \tag{7.32}$$

where $\zeta = \max\left(\frac{\theta_2}{a_2 - a_1\theta_2}, \frac{\theta_1}{a_1}\right)$, $\theta_1 = 2^{\mathscr{R}_1} - 1$ and $\theta_2 = 2^{\mathscr{R}_2} - 1$ with $\mathscr{R}_1$ and $\mathscr{R}_2$ being the transmission rates at CU$_1$ and CU$_2$, respectively.

Moreover, the CU$_2$ is in outage if $\mathbb{R}$ fails to decode CU$_2$' signal, or $\mathbb{R}$ can decode CU$_2$'s signal, but CU$_2$ is unable to decode its signal. The outage probability of CU$_2$

with the max-CU$_1$ and max-CU$_2$ AS schemes is [23]

$$P_{out,2}^{S1} = 1 - N_R e^{-\frac{\theta_2}{\gamma_{RU2}}} \sum_{q=0}^{N_R-1} \frac{(-1)^q \binom{N_R-1}{q} e^{-\frac{(q+1)\theta_2}{\bar{\gamma}_{SR}(a_2-a_1\theta_2)}}}{(q+1)\left(1+\frac{\bar{\gamma}_{SI}}{\bar{\gamma}_{SR}} \frac{(q+1)\theta_2}{a_2-a_1\theta_2}\right)}, \tag{7.33}$$

and

$$P_{out,2}^{S2} = 1 - N_T M_T \sum_{q=0}^{N_T-1} \frac{(-1)^q \binom{N_T-1}{q} e^{-\frac{(q+1)\theta_2}{\gamma_{RU2}}}}{(q+1)}$$

$$\sum_{p=0}^{M_T-1} \frac{(-1)^p \binom{M_T-1}{p} e^{-\frac{(p+1)\theta_2}{\bar{\gamma}_{SR}(a_2-a_1\theta_2)}}}{(p+1)\left(1+\frac{\bar{\gamma}_{SI}}{N_R\bar{\gamma}_{SR}} \frac{(p+1)\theta_2}{(a_2-a_1\theta_2)}\right)}. \tag{7.34}$$

Figure 7.9 shows the impact of SI strength on the performance of the antenna selection schemes. For comparison, we have also plotted the outage performance of the random antenna selection. Clearly the outage probability of both users with max-CU$_1$ AS and max-CU$_2$ AS scheme degrades when the SI strength increases. Moreover, the max-CU$_1$ AS scheme exhibits the best outage performance for CU$_1$, while the max-CU$_2$ AS scheme cannot improve the outage performance of the CU$_1$. This is intuitive since max-CU$_2$ AS uses all degrees-of-freedom to maximize the CU$_2$ SINR. Finally, all antenna selection schemes attain a zero diversity gain order

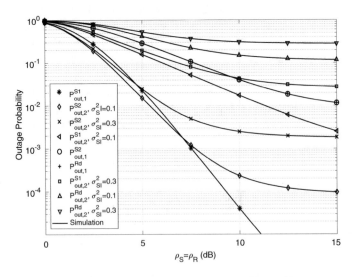

Fig. 7.9 Outage probability versus transmit power ($M_T = 3$, $N_R = N_T = 5$, $\mathscr{R}_1 = \mathscr{R}_2 = 0.5$ bps/Hz)

and exhibit an asymptotic error floor due to SI at $\mathbb{R}$ and inter-user interference at CU_1. However, the error floor can be reduced through the antenna selection schemes.

Alternative antenna selection schemes, such as one which maximizes the far/near user performance while ensuring a pre-defined QoS level at near/far user, are also possible to design. In this case the set of transmit or/and receive antennas which guarantee the pre-defined performance must be first selected, among which the best transmit/receive antenna is selected.

In what follows next we will consider another kind of full-duplex cooperative NOMA system, namely, a full-duplex NOMA cognitive relaying system. Cognitive radio paradigm has attracted wide research attention now for a considerable period as an effective solution to the so-called spectrum scarcity problem in wireless communications. To this end, some studies such as [52] have shown that the introduction of full-duplex operation into cognitive radio systems can improve the spectral efficiency. Furthermore, by integrating NOMA with cognitive radio, additional gains in terms of the spectral efficiency can be reaped. Hence full-duplex, NOMA, and cognitive radio in combination are worthwhile to study in detail so that their suitability to implement future wireless systems can be established.

7.5   Cognitive NOMA Systems with Full-Duplex Relaying

NOMA can be viewed as a form of the spectrum sharing concept, where a user with a better channel operates in the spectrum resided by a user with a weak channel [53]. The NOMA strong user and NOMA weak user can be conceptualized as a cognitive user (CU) and primary user (PU) in a cognitive radio network. Therefore, the transmit power of the NOMA strong user is limited by the NOMA weak user's SINR. Following this point of view, a variation of NOMA coined as cognitive radio inspired NOMA (CR-NOMA) has been studied in [53], where both the PU and CU can be served simultaneously in the same spectrum. Therefore, compared to the conventional cognitive radio systems, a higher spectral efficiency can be realized.

7.5.1   System Model

Deploying NOMA and full-duplex communication in spectrum sharing networks is a promising alternative to further improve the spectral efficiency [43]. Figure 7.10 shows a full-duplex cognitive relay network, where the PUs and CUs operate on the same frequency band. The secondary network consists of a cognitive AP, a DF full-duplex relay, $\mathbb{R}$, and two users, CU_1 and CU_2. A direct communication link between the AP and the near user, CU_1, can be established. However, the AP communicates with the far user, CU_2, through a full-duplex relay. The AP, CU_1, and CU_2 are single antenna terminals, while the relay is equipped with N_T transmit antennas and with N_R receive antennas.

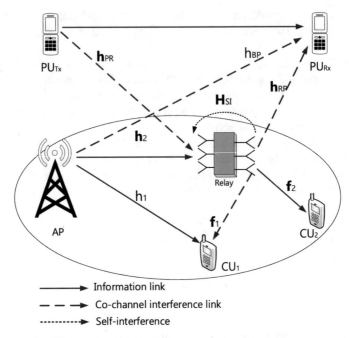

Fig. 7.10 Cognitive NOMA system with full-duplex relaying. The AP of the secondary network serves CU_1 (near user) and CU_2 (far user) through a multi-antenna full-duplex relay while a primary radio transmitter-receiver pair operates in close proximity

AP transmits superposition coded signal $s[n] = \sqrt{P_S a_1} x_1[n] + \sqrt{P_S a_2} x_2[n]$, to the near user CU_1 and the relay $\mathbb{R}$, where P_S is the AP transmit power, a_k is the NOMA power allocation coefficient, and $x_k, k \in \{1, 2\}$ is the symbol intended to NOMA users. CU_1 uses SIC to cancel the interference from CU_2 and decode its own signal at rate $\mathcal{R}_1 = \log(1 + \gamma_t)$, where γ_t is the set threshold of the received SINR of CU_2. However, due to the full-duplex operation of $\mathbb{R}$, a copy of the CU_2's signal is received at CU_1. Therefore, the effective SINR of CU_2 observed at CU_1 is

$$\gamma_{1,2} = \frac{P_S a_2 |h_1|^2}{P_S a_1 |h_1|^2 + P_R |\mathbf{f}_1^T \mathbf{w}_t|^2 + \sigma_n^2}, \tag{7.35}$$

where P_R is the relay transmission power, h_1 is the channel between the AP and CU_1, $\mathbf{f}_1 \in \mathscr{C}^{N_T \times 1}$ is the channel between the full-duplex relay and CU_1, $\mathbf{w}_t \in \mathscr{C}^{N_T \times 1}$ denotes the transmit beamformer at the full-duplex relay, and σ_n^2 is the variance of AWGN at CU_1. If CU_1 cancels the CU_2's signal, i.e., $\gamma_{1,2} \geq \gamma_t$, the SINR at CU_1 can be written as

$$\gamma_1 = \frac{P_S a_1 |h_1|^2}{P_R |\mathbf{f}_1^T \mathbf{w}_t|^2 + \sigma_n^2}. \tag{7.36}$$

Since the full-duplex relay assists transmissions from AP to CU_2, it treats the symbol of CU_1 as interference. Therefore, the SINR at the full-duplex relay can be expressed as

$$\gamma_R = \frac{P_S a_2 |\mathbf{w}_r^\dagger \mathbf{h}_2|^2}{P_S a_1 |\mathbf{w}_r^\dagger \mathbf{h}_2|^2 + P_R |\mathbf{w}_r^\dagger \mathbf{H}_{SI} \mathbf{w}_t|^2 + P_U |\mathbf{w}_r^\dagger \mathbf{h}_{PR}|^2 + \sigma_n^2}, \tag{7.37}$$

where $\mathbf{w}_r \in \mathscr{C}^{N_R \times 1}$ is the relay receive beamformer, $\mathbf{h}_2 \in \mathscr{C}^{N_R \times 1}$ is the channel vector between the AP and relay, P_U is the primary transmitter power, and $\mathbf{h}_{PR} \in \mathscr{C}^{N_R \times 1}$ is the channel vector between the primary transmitter and relay. Moreover, the elements of the $N_R \times N_T$ SI channel $\mathbf{H}_{SI}$ are modeled as i.i.d. $\mathscr{CN}(0, \sigma_{SI}^2)$ [10].

The SNR at CU_2 is given by

$$\gamma_{R,2} = \frac{P_R}{\sigma_n^2} |\mathbf{f}_2^T \mathbf{w}_t|^2, \tag{7.38}$$

where $\mathbf{f}_2 \in \mathscr{C}^{N_T \times 1}$ is the channel between the relay and CU_2.

Since the AP and full-duplex relay simultaneously operates on the same spectrum band as the primary network, the generated interference on the primary receiver must remain below the interference threshold I_{th} in order to maintain the primary receiver's QoS. Therefore, AP and relay transmit power should be capped as in [54]

$$P_S |h_{BP}|^2 + P_R |\mathbf{h}_{RP}^T \mathbf{w}_t|^2 \leq I_{th}, \tag{7.39}$$

where h_{BP} is the channel between the AP and primary receiver and $\mathbf{h}_{RP} \in \mathscr{C}^{N_T \times 1}$ is the channel between the $\mathbb{R}$ and primary receiver.

The channel coefficients h_{BP} and h_1 corresponding to the AP-primary receiver link and AP-CU_1 link are modeled as i.i.d. complex Gaussian RVs with zero mean and variance λ_{BP} and λ_{h_1}, respectively. The elements of $\mathbf{h}_{RP}$, $\mathbf{h}_2$, $\mathbf{f}_1$, $\mathbf{f}_2$, and $\mathbf{h}_{RP}$ are i.i.d. zero-mean complex Gaussian RVs with variance λ_{RP}, λ_{h_2}, λ_{f_1}, λ_{f_2}, and λ_{PR}, respectively.

7.5.2 Beamforming Design and Power Allocation

In this subsection, joint receive and transmit beamforming design at the full-duplex relay and AP/relay transmit power allocation is pursued to improve the system performance.

7.5.2.1 Joint Beamforming Design and Power Allocation

Consider the problem of joint transmit and receive beamformer design and power allocation such that the achievable rate of CU_1 is maximized, while guaranteeing a required minimum rate, $\bar{r}$, at CU_2. The optimization problem can be posed as

$$\max_{\mathbf{w}_t,\mathbf{w}_r,P_S,P_R} C_1(\mathbf{w}_t, P_S, P_R), \tag{7.40a}$$

$$\text{s.t.} \quad C_2(\mathbf{w}_t, \mathbf{w}_r, P_S, P_R) \geq \bar{r}, \tag{7.40b}$$

$$P_S|h_{BP}|^2 + P_R|\mathbf{h}_{\mathsf{RP}}^T\mathbf{w}_t|^2 \leq I_{\text{th}}, \tag{7.40c}$$

$$\|\mathbf{w}_r\| = \|\mathbf{w}_t\| = 1, \tag{7.40d}$$

$$P_S, P_R \geq 0, \tag{7.40e}$$

where

$$C_1(\mathbf{w}_t, P_S, P_R) = \log_2\left(1 + \gamma_1(\mathbf{w}_t, P_S, P_R)\right),$$

$$C_2(\mathbf{w}_t, \mathbf{w}_r, P_S, P_R) = \log_2\Big(1 + \min\big(\gamma_{1,2}(\mathbf{w}_t, P_S, P_R),$$

$$\gamma_R(\mathbf{w}_t, \mathbf{w}_r, P_S, P_R), \gamma_{R,2}(\mathbf{w}_t, P_R)\big)\Big). \tag{7.41}$$

and constraint (7.40c) ensures the interference to the primary receiver does not exceed the I_{th}. The joint optimization problem in (7.40) is non-convex. However, it can be recast as a rank-constrained semi-definite relaxation (SDR) problem [43]. A detailed algorithm to compute transmit/receive beamformers and power allocation coefficients is given in [43].

7.5.2.2 Power Allocation for Fixed Beamforming Design

Optimal power allocation with fixed $\mathbf{w}_t$ and $\mathbf{w}_r$ is an interesting research problem in the considered network. The rationale of deploying a fixed design for $\mathbf{w}_t$ and $\mathbf{w}_r$ is that these beamformers reduce the implementation complexity. For instance, MRT/MRC beamforming is suitable for full-duplex systems as it does not need the CSI of SI link. Furthermore, MRC/MRT scheme is widely utilized in half-duplex relaying systems to achieve an optimal performance, and thus it is also worthwhile to present the performance in the full-duplex case as a benchmark.

For a given $\mathbf{w}_t$ and $\mathbf{w}_r$, the optimization problem (7.40) can be re-expressed as

$$\max_{P_S, P_R} \quad C_1(\mathbf{w}_t, P_S, P_R),$$

$$\text{s.t.} \quad C_2(\mathbf{w}_t, \mathbf{w}_r, P_S, P_R) \geq \bar{r},$$

$$P_S |h_{BP}|^2 + P_R |\mathbf{h}_{RP}^T \mathbf{w}_t|^2 \leq I_{\text{th}},$$

$$P_S, P_R \geq 0. \tag{7.42}$$

The optimum solutions of (7.42) are given by [43]

$$P_{R,\text{OPT}} = \frac{\sigma_n^2 \bar{r}}{|\mathbf{f}_2^T \mathbf{w}_t|^2}, \tag{7.43a}$$

$$P_{S,\text{OPT}} = \frac{I_{\text{th}} - \frac{\sigma_n^2 \bar{r}}{|\mathbf{f}_2^T \mathbf{w}_t|^2} |\mathbf{h}_{RP}^T \mathbf{w}_t|^2}{|h_{BP}|^2}. \tag{7.43b}$$

7.5.2.3 ZF-Based Fixed Beamforming Schemes

The joint optimization design has high computational complexity due to the SDR. To lower the computational complexity, ZF-based beamforming designs can be deployed at the $\mathbb{R}$, where the multiple receive/transmit antennas at the $\mathbb{R}$ are utilized to completely cancel the SI [12].

TZF Scheme Transmit antennas of the relay are exploited to cancel the SI. To guarantee feasibility of the TZF scheme, we need to deploy at least two relay transmit antennas, i.e., the condition $N_T > 1$ should be satisfied. In addition, at the relay input MRC is applied , i.e., $\mathbf{w}_r^{\text{MRC}} = \frac{\mathbf{h}_2}{\|\mathbf{h}_2\|}$. Accordingly, the optimal transmit beamformer $\mathbf{w}_t$ can be obtained by solving

$$\max_{\|\mathbf{w}_t\|=1} |\mathbf{f}_2^T \mathbf{w}_t|^2, \quad \text{s.t.} \quad \mathbf{h}_2^\dagger \mathbf{H}_{SI} \mathbf{w}_t = 0. \tag{7.44}$$

The optimal transmit vector $\mathbf{w}_t$ is obtained in [47] as

$$\mathbf{w}_t^{\text{ZF}} = \frac{\mathbf{B}\mathbf{f}_2^*}{\|\mathbf{B}\mathbf{f}_2^*\|}, \tag{7.45}$$

where $\mathbf{B} = \mathbf{I}_{N_T} - \frac{\mathbf{H}_{SI}^\dagger \mathbf{h}_2 \mathbf{h}_2^\dagger \mathbf{H}_{SI}}{\|\mathbf{h}_2^\dagger \mathbf{H}_{SI}\|^2}$.

RZF Scheme With this scheme, $\mathbf{w}_t$ can be fixed following MRT as $\mathbf{w}_t^{\mathrm{MRT}} = \frac{\mathbf{f}_2^*}{\|\mathbf{f}_2\|}$ and $\mathbf{w}_r$ is designed to satisfy $\mathbf{w}_r^\dagger \mathbf{H}_{\mathrm{SI}} \mathbf{w}_t = 0$. To guarantee feasibility of the RZF scheme, we should have the condition that $N_R > 1$. The optimal receive beamformer $\mathbf{w}_r$ is the solution of

$$\max_{\|\mathbf{w}_r\|=1} |\mathbf{w}_r^\dagger \mathbf{h}_2|^2, \qquad \text{s.t.} \qquad \mathbf{w}_r^\dagger \mathbf{H}_{\mathrm{SI}} \mathbf{f}_2^* = 0. \tag{7.46}$$

Let us denote $\mathbf{C} = \mathbf{I}_{N_R} - \frac{\mathbf{H}_{\mathrm{SI}} \mathbf{f}_2^* \mathbf{f}_2^T \mathbf{H}_{\mathrm{SI}}^\dagger}{\|\mathbf{H}_{\mathrm{SI}} \mathbf{f}_2^*\|^2}$. The solution of (7.46) can be obtained as

$$\mathbf{w}_r^{\mathrm{ZF}} = \frac{\mathbf{C}\mathbf{h}_2}{\|\mathbf{C}\mathbf{h}_2\|}. \tag{7.47}$$

Both TZF and RZF schemes neglect the near user performance which is degraded due to the interference caused by the full-duplex relay. This motivates designing an alternative beamforming scheme, namely the RTZF scheme [43].

RTZF Scheme The RTZF scheme aims to design $\mathbf{w}_t$ and $\mathbf{w}_r$ such that the interference at CU_1 and SI at the $\mathbb{R}$ are totally cancelled. The optimal $\mathbf{w}_t$ is found by solving

$$\max_{\|\mathbf{w}_t\|=1} |\mathbf{f}_2^T \mathbf{w}_t|^2, \qquad \text{s.t.} \qquad \mathbf{f}_1^T \mathbf{w}_t = 0. \tag{7.48}$$

The weight vector $\mathbf{w}_t^{\mathrm{RTZF}}$ can be attained as $\mathbf{w}_t^{\mathrm{RTZF}} = \frac{\mathcal{E}^\perp \mathbf{f}_2^*}{\|\mathcal{E}^\perp \mathbf{f}_2^*\|}$, where $\mathcal{E}^\perp = \left(\mathbf{I}_{N_T} - \mathbf{f}_1(\mathbf{f}_1^T \mathbf{f}_1)^{-1}\mathbf{f}_1^T\right)$ is the projection idempotent matrix.

Furthermore, the optimal receive beamformer $\mathbf{w}_r$ is found by solving the problem

$$\max_{\|\mathbf{w}_r\|=1} |\mathbf{w}_r^\dagger \mathbf{h}_2|^2, \qquad \text{s.t.} \qquad \mathbf{w}_r^\dagger \mathbf{H}_{\mathrm{SI}} \mathbf{w}_t^{\mathrm{RTZF}} = 0, \tag{7.49}$$

and $\mathbf{w}_r^{\mathrm{RTZF}}$ is given by $\mathbf{w}_r^{\mathrm{RTZF}} = \frac{\Pi^\perp \mathbf{h}_2}{\|\Pi^\perp \mathbf{h}_2\|}$, where $\Pi^\perp = \mathbf{I}_{N_R} - \frac{\mathbf{H}_{\mathrm{SI}} \mathbf{w}_t^{\mathrm{RTZF}} (\mathbf{w}_t^{\mathrm{RTZF}})^\dagger \mathbf{H}_{\mathrm{SI}}^\dagger}{\|\mathbf{H}_{\mathrm{SI}} \mathbf{w}_t^{\mathrm{RTZF}}\|^2}$ is the projection idempotent matrix.

Figure 7.11 shows the far user rate versus the near user rate with the optimum scheme and for different values of N_R and N_T at $\mathbb{R}$. For comparison, the rate region of the RZF scheme is also included. The optimum design substantially outperforms the sub-optimum TZF scheme. Moreover, it is evident that by increasing the number of receive and transmit antennas at the full-duplex relay, the achievable rates of CU_1 and CU_2 are significantly improved. The main reason is that by increasing N_T, new choices of fading paths become available toward the far user. Consequently, the received SNR at the CU_2 is increased.

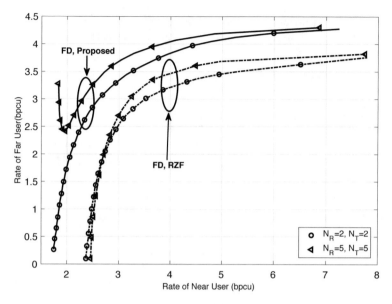

Fig. 7.11 Rate-region of the optimum and RZF schemes ($\rho_{th} = 15$ dB, $\sigma_{SI}^2 = 0.03$)

7.5.3 Performance Analysis

Under uniform power assignment between the AP and relay, the outage performance of the fixed beamforming design, TZF, RZF, and RTZF schemes is investigated. Transmit power of the AP and relay in this case are given by [54]

$$P_S^{EPA} = \frac{I_{th}}{2|h_{BP}|^2}, \qquad P_R^{EPA} = \frac{I_{th}}{2|\mathbf{h}_{RP}^T\mathbf{w}_t|^2}, \tag{7.50}$$

which satisfies the interference constraint (7.39).

7.5.3.1 Outage Probability at the Near User

The CU_1 experiences outage events if CU_1 fails to detect signal intended to far user, i.e., x_2, and hence cannot cancel it, or when CU_1 can detect x_2 but fails to decode signal intended to near user, i.e., x_1. The outage probability of CU_1 can be written as [43]

$$\begin{aligned}
P_{out,1} &= 1 - \Pr\left(\bar{\mathscr{O}}_{CU_1}^1 \bigcap \bar{\mathscr{O}}_{CU_1}^2\right) \\
&= 1 - \Pr\left(\gamma_{1,2}^i \geqslant \theta_2, \gamma_1^i \geqslant \theta_1\right), \tag{7.51}
\end{aligned}$$

where $i \in \{\text{TZF, RZF, RTZF}\}$, $\theta_2 = 2^{\mathscr{R}_2} - 1$, and $\theta_1 = 2^{\mathscr{R}_1} - 1$.

It can be readily checked that the statistics of $\gamma_{1,2}^{\text{RZF}}$ (γ_1^{RZF}) and $\gamma_{1,2}^{\text{TZF}}$ (γ_1^{TZF}) are identical. Thus, RZF and RZF provide the same outage performance at CU_1. Let $\mu = \max\left(\frac{\theta_2}{b_0 - b_1\theta_2}, \frac{\theta_1}{b_1}\right)$ with $b_0 = a_2\frac{\lambda_{h_1}}{\lambda_{BP}}, b_1 = a_1\frac{\lambda_{h_1}}{\lambda_{BP}}$. The outage probability of the near user with the TZF and RZF schemes can be derived as [43]

$$P_{\text{out},1}^{\text{TZF}} = P_{\text{out},1}^{\text{RZF}} = 1 - \frac{1}{(1 - \mu b_2) + 2\mu/\rho_{\text{th}}}$$
$$\times \left(1 + \frac{\mu b_2}{(1 - \mu b_2) + 2\mu/\rho_{\text{th}}} \ln\left(\frac{\mu b_2}{1 + 2\mu/\rho_{\text{th}}}\right)\right), \qquad (7.52)$$

if $0 < \theta_2 \leq \frac{a_2}{a_1}$, otherwise $P_{\text{out},1}^{\text{TZF}} = P_{\text{out},1}^{\text{RZF}} = 1$, where $b_2 = \frac{\lambda_{f_1}}{\lambda_{RP}}$, and $\rho_{\text{th}} = \frac{I_{\text{th}}}{\sigma_n^2}$. Moreover, with the RTZF scheme, outage probability can be derived as

$$P_{\text{out},1}^{\text{RTZF}} = 1 - \frac{\rho_{\text{th}}}{\rho_{\text{th}} + 2\zeta}. \qquad (7.53)$$

7.5.3.2 Outage Probability at the Far User

The CU_2 is in outage when $\mathbb{R}$ fails to decode x_2 or when $\mathbb{R}$ can decode x_2 but CU_2 is unable to successfully decode x_2 correctly. Hence, the outage probability at CU_2 is given by

$$P_{\text{out},2} = \text{Pr}\left(\min\left\{\gamma_R^i, \gamma_{R,2}^i\right\} < \theta_2\right), \qquad (7.54)$$

where $i \in \{\text{TZF}, \text{RZF}, \text{RTZF}\}$.

The exact outage probability of CU_2 for the TZF, RZF, and RTZF schemes can be expressed as [43]

$$P_{\text{out},2}^i$$
$$= \begin{cases} \left(\frac{c_3}{\theta_2} + 1\right)^{-N_i} + \left(1 - \left(\frac{c_3}{\theta_2} + 1\right)^{-N_i}\right)\int_0^\infty \left(\frac{\rho_{\text{th}}(c_0 - \theta_2 c_1)}{\theta_2(c_2\rho_{\text{th}}x + 2)} + 1\right)^{-M_i} e^{-x} dx, & \theta_2 < \frac{a_2}{a_1}, \\ 1, & \theta_2 > \frac{a_2}{a_1}, \end{cases}$$
$$(7.55)$$

where $i \in \{\text{RZF}, \text{TZF}, \text{RTZF}\}$, $c_0 = a_2\frac{\lambda_{h_2}}{\lambda_{BP}}$, $c_1 = a_1\frac{\lambda_{h_2}}{\lambda_{BP}}$, $c_2 = 2\lambda_{PR}\frac{\rho_U}{\rho_{\text{th}}}$ with $\rho_U = \frac{P_U}{\sigma_n^2}$, and $c_3 = \frac{\rho_{\text{th}}}{2}\frac{\lambda_{f_2}}{\lambda_{RP}}$. Moreover, N_i and M_i for different beamforming schemes are summarized in Table 7.2.

Table 7.2 Parameters of the asymptotic outage probability expressions of different ZF-based schemes

Scheme	N_i	M_i	$\Psi_i(N_R)$	$\Phi_i(N_R)$	Diversity order
TZF	$N_T - 1$	N_R	$N_R(N_R+1)$	$\frac{N_R(N_R+1)^2(N_R+2)}{2}$	$\min(N_R, N_T - 1)$
RZF	N_T	$N_R - 1$	$N_R(N_R-1)$	$\frac{(N_R-1)N_R^2(N_R+1)}{2}$	$\min(N_R - 1, N_T)$
RTZF	$N_T - 1$	$N_R - 1$	$N_R(N_R-1)$	$\frac{(N_R-1)N_R^2(N_R+1)}{2}$	$\min(N_R-1, N_T-1)$

We now look into the high SNR region, i.e., $\rho_{th} \to \infty$, to provide some insights on the diversity order of the ZF-based schemes. The asymptotic outage probability expression for the TZF, RZF, and RTZF schemes can be derived as [43]

$$\tilde{P}^i_{out,2} \approx \left(\frac{2\lambda_{RP}\theta_2}{\lambda_{f_2}}\right)^{N_i}\left(\frac{1}{\rho_{th}}\right)^{N_i} + e^{\frac{1}{\rho_U\lambda_{PR}}}\,\Gamma(N_R+1)\left(\frac{2\rho_U\lambda_{PR}\theta_2}{c_0-\theta_2 c_1}\right)^{M_i}\left(\frac{1}{\rho_{th}}\right)^{M_i}$$

$$\times\left(1-\Psi_i(N_R)\left(\frac{2\rho_U\lambda_{PR}\theta_2}{c_0-\theta_2 c_1}\right)\left(\frac{1}{\rho_{th}}\right)+\Phi_i(N_R)\left(\frac{2\rho_U\lambda_{PR}\theta_2}{c_0-\theta_2 c_1}\right)^2\left(\frac{1}{\rho_{th}}\right)^2\right),$$

$$(7.56)$$

where $i \in \{RZF, TZF, RTZF\}$, $\Gamma(a)$ is the Gamma function [48, Eq. (8.310.1)], and $\Psi_i(N_R)$, and $\Phi_i(N_R)$ for different scheme are given in Table 7.2. We report the diversity order of the ZF-based schemes in Table 7.2.

Figure 7.12 illustrates the outage performance of the far with ZF-based beamforming schemes and for different antenna setups. It can be observed that the TZF, RZF, and RTZF schemes achieve the intended diversity orders, reported in Table 7.2. Moreover, it is noted that with different antenna setups, diversity order of RTZF and TZF/RZF is identical, however, in some cases due to the higher array gain, the TZF/RZF scheme provides a superior performance in the low SNR region.

7.6 Future Research Directions

As of now, research community and industry face several important issues for the design of full-duplex NOMA systems. These issues can be grouped into key areas as:

- **Interference management in full-duplex NOMA systems:** In general, due to the full-duplex operation, several interference links are created. For example, in a multi-cellular setting, uplink-downlink operation creates SI at the BS, inter-user interference at the downlink user as well as inter-cell interference. Such increased levels of interference can lower the ability of a NOMA user to decode message using SIC. Hence, in order to understand the NOMA performance degradation, interference due to full-duplex operation should be carefully studied. Moreover,

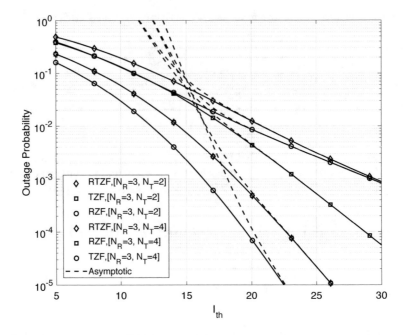

Fig. 7.12 Outage probability of CU_2 versus ρ_{th} ($\mathscr{R}_1 = \mathscr{R}_2 = 0.5\,\text{bps/Hz}$)

in order to establish a specific level of QoS, effective interference mitigation techniques should be adopted network wise.

- **Multiple-antenna processing and user pairing:** Spatial SI cancellation is an attractive solution to control the input-output coupling at a full-duplex transceiver. However, such schemes cannot be applied directly in NOMA systems. This is because, MIMO processing would also affect the effective link conditions at the NOMA users. Hence, new beamforming techniques, antenna selection, and user pairing schemes should be developed such that a balance between SI suppression and link conditions could be achieved. Moreover, massive MIMO technology is attractive to implement full-duplex NOMA systems as it can suppress the effect of SI using simple linear processing. To this end, how issues of massive MIMO such as pilot contamination, channel aging, imperfect RF chain calibration affect full-duplex NOMA performance forms a body of interesting future research work.

- **Massive access with machine-type communications:** NOMA stands out as an attractive solution to implement massive access in machine-type devices. However, it is well known that such devices are subject to energy constraints. Therefore, how available limited energy can be efficiently managed for operations such as SIC and SI cancellation seems a challenging problem to solve. Moreover, in order to implement massive access, full-duplex medium access control protocols that allow monitoring of the channel and backoff when a collision is detected in combination with NOMA and power control should be developed.

- **Wireless security for full-duplex NOMA systems:** NOMA systems are vulnerable to various security threats, yet at the same time physical layer security techniques can be implemented with NOMA. Use of NOMA ensures mixing of signals, hence common messages along with the ability of full-duplex to transmit while receiving can be exploited create jamming signals that can overwhelm an eavesdropper's receiver. As such, it is important to research on schemes that show promise to augment the security of full-duplex NOMA systems.

There has been an upsurge of interest recently to use machine learning and artificial intelligence techniques in wireless systems. Use of such algorithms also holds promise in full-duplex NOMA systems, for example, to learn interference characteristics, user mobility, and time varying channel conditions so that efficient power control techniques, user pairing methods, massive access protocols, and secure transmission schemes of dynamic nature can be developed.

7.7 Conclusions

In this chapter, applications of NOMA in full-duplex wireless communication systems were investigated. We first considered downlink NOMA transmission helped by a full-duplex multi-antenna relay. We derived optimum and suboptimal beamforming schemes at the relay and provided the outage probability of the RNRF and NNNF user selection. Then, we studied the antenna selection problem for a full-duplex multi-antenna equipped cooperative NOMA system. Two specific antenna selection schemes to improve the near/far user end-to-end SINR values were presented. The performance of both antenna selection schemes was derived in terms of the ergodic sum rate and outage probability. We next explored a beamformer design and power allocation problem for a full-duplex relay-assisted NOMA cognitive radio system. In particular, our optimization goal was to maximize the NOMA near user rate under a NOMA far user rate threshold.

In all systems studied, use of full-duplex communications can deliver improved performance in comparison to half-duplex communications. These gains depend on the number of antennas of the system as well as on channel parameters. In order to translate the performance benefits highlighted in the chapter for future practical implementation, careful selection of their values is therefore recommended.

Acknowledgements The work of Z. Ding was supported by the UK EPSRC under grant number EP/L025272/1, NSFC under grant number 61728101 and H2020-MSCA-RISE-2015 under grant number 690750.

References

1. A. Sabharwal, P. Schniter, D. Guo, D. W. Bliss, S. Rangarajan, and R. Wichman, "In-band full-duplex wireless: Challenges and opportunities," *IEEE J. Sel. Areas Commun.*, vol. 32, no. 9, pp. 1637–1652, Sep. 2014.
2. Z. Zhang, X. Chai, K. Long, A. V. Vasilakos, and L. Hanzo, "Full duplex techniques for 5G networks: Self-interference cancellation, protocol design, and relay selection," *IEEE Commun. Mag.*, vol. 53, pp. 128–137, May 2015.
3. Y. Saito, Y. Kishiyama, A. Benjebbour, T. Nakamura, A. Li, and K. Higuchi, "Non-orthogonal multiple access (NOMA) for cellular future radio access," in *Proc. IEEE Veh. Technol. Conf. (VTC'13)*, Dresden, Germany, June 2013, pp. 1–5.
4. L. Dai, B. Wang, Z. Ding, Z. Wang, S. Chen, and L. Hanzo, "A survey of non-orthogonal multiple access for 5G," *IEEE Commun. Surveys Tuts.*, vol. 20, pp. 2294–2323, thirdquarter 2018.
5. M. Mohammadi, X. Shi, B. K. Chalise, Z. Ding, H. A. Suraweera, C. Zhong, and J. S. Thompson, "Full-duplex non-orthogonal multiple access for next generation wireless systems," *IEEE Commun. Mag.*, vol. 57, no. 5, pp. 110–116, May 2019.
6. M. Duarte, "Full-duplex wireless: Design, implementation and characterization," Ph.D. dissertation, Dept. Elect. and Computer Eng., Rice University, Houston, TX, 2012.
7. S. S. Hong, J. Brand, J. I. Choi, M. Jain, J. Mehlman, S. Katti, and P. Levis, "Applications of self-interference cancellation in 5G and beyond," *IEEE Commun. Mag.*, vol. 52, no. 2, pp. 114–121, Feb. 2014.
8. E. Everett, A. Sahai, and A. Sabharwal, "Passive self-interference suppression for full-duplex infrastructure nodes," vol. 13, no. 2, pp. 680–694, Feb. 2014.
9. C. Psomas, M. Mohammadi, I. Krikidis, and H. A. Suraweera, "Impact of directionality on interference mitigation in full-duplex cellular networks," *IEEE Trans. Wireless Commun.*, vol. 16, no. 1, pp. 487–502, Jan 2017.
10. T. Riihonen, S. Werner, and R. Wichman, "Mitigation of loopback self-interference in full-duplex MIMO relays," *IEEE Trans. Signal Process.*, vol. 59, no. 12, pp. 5983–5993, Dec. 2011.
11. L. Laughlin, M. A. Beach, K. A. Morris, and J. L. Haine, "Optimum single antenna full duplex using hybrid junctions," *IEEE J. Sel. Areas Commun.*, vol. 32, pp. 1653–1661, Sep. 2014.
12. H. A. Suraweera, I. Krikidis, G. Zheng, C. Yuen, and P. J. Smith, "Low-complexity end-to-end performance optimization in MIMO full-duplex relay systems," *IEEE Trans. Wireless Commun.*, vol. 13, no. 2, pp. 913–927, Jan. 2014.
13. H. Q. Ngo, H. A. Suraweera, M. Matthaiou, and E. G. Larsson, "Multipair full-duplex relaying with massive arrays and linear processing," *IEEE J. Sel. Areas Commun.*, vol. 32, pp. 1721–1737, Sep. 2014.
14. Y. Cai, Z. Qin, F. Cui, G. Y. Li, and J. A. McCann, "Modulation and multiple access for 5G networks," *IEEE Commun. Surveys Tuts.*, vol. 20, pp. 629–646, Firstquarter 2018.
15. Z. Zhang, H. Sun, and R. Q. Hu, "Downlink and uplink non-orthogonal multiple access in a dense wireless network," *IEEE J. Sel. Areas Commun.*, vol. 35, pp. 2771–2784, Dec. 2017.
16. Z. Ding, M. Peng, and H. V. Poor, "Cooperative non-orthogonal multiple access in 5G systems," *IEEE Commun. Lett.*, vol. 19, pp. 1462–1465, Aug. 2015.
17. L. Lv, J. Chen, Q. Ni, Z. Ding, and H. Jiang, "Cognitive non-orthogonal multiple access with cooperative relaying: A new wireless frontier for 5G spectrum sharing," *IEEE Communications Magazine*, vol. 56, pp. 188–195, Apr. 2018.
18. Z. Yang, W. Xu, H. Xu, J. Shi, and M. Chen, "Energy efficient non-orthogonal multiple access for machine-to-machine communications," *IEEE Commun. Lett.*, vol. 21, pp. 817–820, Apr. 2017.
19. F. Zhou, Y. Wu, R. Q. Hu, Y. Wang, and K. K. Wong, "Energy-efficient NOMA enabled heterogeneous cloud radio access networks," *IEEE Netw.*, vol. 32, pp. 152–160, Mar. 2018.

20. Z. Ding, P. Fan, and H. V. Poor, "On the coexistence between full-duplex and NOMA," *IEEE Wireless Commun. Lett.*, vol. 7, pp. 692–695, Oct. 2018.
21. C. Zhong and Z. Zhang, "Non-orthogonal multiple access with cooperative full-duplex relaying," *IEEE Commun. Lett.*, vol. 20, pp. 2478–2481, Dec. 2016.
22. Z. Mobini, M. Mohammadi, B. K. Chalise, H. A. Suraweera, and Z. Ding, "Beamforming design and performance analysis of full-duplex cooperative NOMA systems," *IEEE Trans. Wireless Commun.*, vol. 18, no. 6, pp. 3295–3311, June 2019.
23. M. Mohammadi, Z. Mobini, H. A. Suraweera, and Z. Ding, "Antenna selection in full-duplex cooperative NOMA systems," in *Proc. IEEE Intl. Conf. Commun. (ICC'18)*, Kansas City, Missouri, USA, May 2018, pp. 1–6.
24. X. Yue, Y. Liu, S. Kang, A. Nallanathan, and Z. Ding, "Exploiting full/half-duplex user relaying in NOMA systems," *IEEE Trans. Commun.*, vol. 66, pp. 560–575, Feb. 2017.
25. B. Zheng, M. Wen, C. Wang, X. Wang, F. Chen, J. Tang, and F. Ji, "Secure NOMA based two-way relay networks using artificial noise and full duplex," *IEEE J. Sel. Areas Commun.*, vol. 36, pp. 1426–1440, Jul. 2018.
26. M. F. Kader, S. Y. Shin, and V. C. M. Leung, "Full-duplex non-orthogonal multiple access in cooperative relay sharing for 5G systems," *IEEE Trans. Veh. Technol.*, vol. 67, pp. 5831–5840, July 2018.
27. Q. Y. Liau, C. Y. Leow, and Z. Ding, "Amplify-and-forward virtual full-duplex relaying-based cooperative NOMA," *IEEE Wireless Commun. Lett.*, vol. 7, no. 3, pp. 464–467, June 2018.
28. Q. Y. Liau and C. Y. Leow, "Cooperative NOMA system with virtual full duplex user relaying," *IEEE Access*, vol. 7, pp. 2502–2511, 2019.
29. Z. Zhang, Z. Ma, M. Xiao, Z. Ding, and P. Fan, "Full-duplex device-to-device aided cooperative non-orthogonal multiple access," *IEEE Trans. Veh. Technol.*, vol. 66, pp. 4467–4471, May 2017.
30. L. Zhang, J. Liu, M. Xiao, G. Wu, Y. C. Liang, and S. Li, "Performance analysis and optimization in downlink NOMA systems with cooperative full-duplex relaying," *IEEE J. Sel. Areas Commun.*, vol. 35, pp. 2398–2412, Oct. 2017.
31. T. N. Do, D. B. da Costa, T. Q. Duong, and B. An, "Improving the performance of cell-edge users in NOMA systems using cooperative relaying," *IEEE Trans. Commun.*, vol. 66, pp. 1883–1901, May 2018.
32. Y. Sun, D. W. K. Ng, Z. Ding, and R. Schober, "Optimal joint power and subcarrier allocation for full-duplex multicarrier non-orthogonal multiple access systems," *IEEE Trans. Commun.*, vol. 65, pp. 1077–1091, Mar. 2017.
33. M. S. Elbamby, M. Bennis, W. Saad, M. Debbah, and M. Latva-aho, "Resource optimization and power allocation in in-band full duplex-enabled non-orthogonal multiple access networks," vol. 35, no. 12, pp. 2860–2873, Dec. 2017.
34. L. Lei, E. Lagunas, S. Chatzinotas, and B. Ottersten, "NOMA aided interference management for full-duplex self-backhauling HetNets," *IEEE Communications Letters*, vol. 22, pp. 1696–1699, Aug. 2018.
35. Y. Sun, D. W. K. Ng, J. Zhu, and R. Schober, "Robust and secure resource allocation for full-duplex MISO multicarrier NOMA systems," *IEEE Trans. Commun.*, vol. 66, pp. 4119–4137, Sep. 2018.
36. R. Tang, H. Qu, J. Zhao, J. Cheng, and Z. Cao, "Distributed resource allocation for IBFD-enabled NOMA systems," *IEEE Commun. Lett.*, vol. 22, pp. 2318–2321, Nov. 2018.
37. X. Zhang and F. Wang, "Resource allocation for wireless power transmission over full-duplex OFDMA/NOMA mobile wireless networks," *IEEE J. Sel. Areas Commun.*, vol. 37, pp. 327–344, Feb. 2019.
38. Y. Alsaba, C. Y. Leow, and S. K. A. Rahim, "Full-duplex cooperative non-orthogonal multiple access with beamforming and energy harvesting," *IEEE Access*, vol. 6, pp. 19 726–19 738, 2018.
39. Z. Wang, X. Yue, and Z. Peng, "Full-duplex user relaying for NOMA system with self-energy recycling," *IEEE Access*, vol. 6, pp. 67 057–67 069, 2018.

40. P. Deng, B. Wang, W. Wu, and T. Guo, "Transmitter design in MISO-NOMA system with wireless-power supply," *IEEE Commun. Lett.*, vol. 22, pp. 844–847, Apr. 2018.
41. C. Guo, L. Zhao, C. Feng, Z. Ding, and H. Chen, "Energy harvesting enabled NOMA systems with full-duplex relaying," *IEEE Trans. Veh. Technol.*, pp. 1–1, 2019.
42. Y. Yuan, Y. Xu, Z. Yang, P. Xu, and Z. Ding, "Energy efficiency optimization in full-duplex user-aided cooperative SWIPT NOMA systems," *IEEE Trans. Commun.*, pp. 1–1, 2019.
43. M. Mohammadi, B. K. Chalise, A. Hakimi, Z. Mobini, H. A. Suraweera, and Z. Ding, "Beamforming design and power allocation for full-duplex non-orthogonal multiple access cognitive relaying," *IEEE Trans. Commun.*, vol. 66, pp. 5952–5965, Dec. 2018.
44. L. Lei, E. Lagunas, S. Maleki, Q. He, S. Chatzinotas, and B. Ottersten, "Energy optimization for full-duplex self-backhauled HetNet with non-orthogonal multiple access," in *Proc. 18th Intl. Workshop on Signal Process. Advances in Wireless Commun. (SPAWC'17)*, Sapporo, Japan, July 2017, pp. 1–5.
45. Y. Liu, Z. Ding, M. Elkashlan, and H. V. Poor, "Cooperative non-orthogonal multiple access with simultaneous wireless information and power transfer," *IEEE J. Sel. Areas Commun.*, vol. 34, pp. 938–953, Apr. 2016.
46. Z. Ding, H. Dai, and H. V. Poor, "Relay selection for cooperative NOMA," *IEEE Wireless Commun. Lett.*, vol. 5, pp. 416–419, Aug. 2016.
47. M. Mohammadi, B. K. Chalise, H. A. Suraweera, C. Zhong, G. Zheng, and I. Krikidis, "Throughput analysis and optimization of wireless-powered multiple antenna full-duplex relay systems," *IEEE Trans. Commun.*, vol. 64, no. 4, pp. 1769–1785, Apr. 2016.
48. I. S. Gradshteyn and I. M. Ryzhik, *Table of Integrals, Series and Products*, 7th ed. San Diego, CA: Academic Press, 2007.
49. Y. Yu, H. Chen, Y. Li, Z. Ding, and B. Vucetic, "Antenna selection for MIMO-NOMA networks," in *Proc. IEEE ICC 2017*, Paris, France, May 2017, pp. 1–6.
50. N. T. Do, D. B. da Costa, T. Q. Duong, and B. An, "Transmit antenna selection schemes for MISO-NOMA cooperative downlink transmissions with hybrid SWIPT protocol," in *Proc. IEEE ICC 2017*, Paris, France, May 2017, pp. 1–6.
51. Y. Zhang, J. Ge, and E. Serpedin, "Performance analysis of nonorthogonal multiple access for downlink networks with antenna selection over Nakagami-*m* fading channels," *IEEE Trans. Veh. Technol.*, vol. 66, pp. 10 590–10 594, Nov. 2017.
52. Y. Deng, K. J. Kim, T. Q. Duong, M. Elkashlan, G. K. Karagiannidis, and A. Nallanathan, "Full-duplex spectrum sharing in cooperative single carrier systems," *IEEE Trans. Cognit. Commun. Netw.*, vol. 2, pp. 68–82, Mar. 2016.
53. Z. Ding, P. Fan, and H. V. Poor, "Impact of user pairing on 5G nonorthogonal multiple-access downlink transmissions," *IEEE Trans. Veh. Technol.*, vol. 65, pp. 6010–6023, Aug. 2016.
54. H. Kim, S. Lim, H. Wang, and D. Hong, "Optimal power allocation and outage analysis for cognitive full-duplex relay systems," *IEEE Trans. Wireless Commun.*, vol. 11, pp. 3754–3765, Oct. 2012.

Chapter 8
Full Duplex and Wireless-Powered Communications

Onel L. Alcaraz López and Hirley Alves

Abstract Wireless energy transfer (WET) is an attractive energy-efficient technology that has been identified as a potential enabler for the Internet of Things (IoT) era. Recently, there is an increasing interest in combining full duplex (FD) and WET to achieve greater system performance, and in this chapter we overview the main characteristics of the wireless-powered communication networks (WPCNs), the approaches for modeling the energy-harvesting (EH) conversion process, and the main FD architectures, namely (1) FD bidirectional communications, (2) FD relay communications, and (3) FD hybrid access point (AP), along with their particularities. We also discuss two example setups related with architectures (1) and (3), while demonstrating the suitable regions for FD operation in each case. For the former we demonstrate the gains in energy efficiency (EE) when the base station (BS) operates with FD for self-energy (SEg) recycling, while for the latter we show that the linear EH model is optimistic and that the FD design is not only simpler than half duplex's (HD) but it also can offer significant performance gains in terms of system reliability when the successive interference cancellation (SIC) hardware performs not too bad. Finally, some possible research directions for the design and deployment of FD wireless-powered systems are identified.

8.1 Introduction

With the advent of the Internet of Things (IoT) era, where communication systems should support a huge number of connected devices, there is an increasing interest in energy-efficient technologies. This is because (1) IoT devices are mostly low power and (2) powering and uninterrupted operation of such potential massive number of IoT nodes is a major challenge. Energy-harvesting (EH) techniques have recently drawn significant attention as a potential solution, and a variety of energy

O. L. A. López (✉) · H. Alves
University of Oulu, Centre for Wireless Communications, Oulu, Finland
e-mail: onel.alcarazlopez@oulu.fi; hirley.alves@oulu.fi

© Springer Nature Singapore Pte Ltd. 2020   219
H. Alves et al. (eds.), *Full-Duplex Communications for Future Wireless Networks*,
https://doi.org/10.1007/978-981-15-2969-6_8

sources such as heat, light, and wind have been considered for EH in wireless networks [1]. These natural energy sources are usually location, weather, or climate dependent and may not always be available in enclosed/indoor environments or suitable for mobile devices. In that sense, wireless energy transfer (WET) [2], as a particular EH technique that allows the devices to harvest energy from radio-frequency (RF) signals, is very attractive because of the coverage advantages of RF signals, especially for IoT scenarios where replacing or recharging batteries requires high cost and/or can be inconvenient or hazardous (e.g., in toxic environments), or highly undesirable (e.g., for sensors embedded in building structures or inside the human body) [3].[1]

Recently, there is an increasing interest in combining full duplex (FD) and WET to achieve greater system performance [4–31]. The reasoning behind is based on the fact that FD radios should not operate with very high transmit power due to the negative effect of the self-interference (SI), while path loss significantly limits the EH performance. Additionally, the idea of self-energy (SEg) recycling from the SI [14, 25, 26] has been shown to be beneficial by providing a secondary energy source, and consequently SI from the data transmission is no longer a nuisance to be strongly suppressed by an expensive hardware. Also, multiple-antenna setups are appropriate for both FD and WET operation since they help the receivers to accumulate more energy, while at the same time spatial SI cancellation techniques can be deployed [11, 16]. Therefore, incorporating FD operation in many WET systems will bring benefits in terms of increased spectral and energy efficiency (EE); of course, as long as the systems are properly designed.

8.1.1 Wireless-Powered Networks: An Overview

The idea of WET was first conceived and experimented by Nicola Tesla in 1899. However, the area did not pick up until the 1960s when microwave technologies rapidly advanced [2]. Nowadays, systems incorporating WET are becoming feasible due to the further advances in technology, the miniaturization of form factors, and availability of hardware with extremely low-power requirements, e.g., low-power sensors nodes. Actually, commercialization already began and one can find, for instance, PowerCast [32] and Cota system [33] products in the market.

Three main WET approaches have been identified in the literature:

1. Wireless EH, which refers to harvesting energy from the ambient RF signals, e.g., TV broadcasting, Wi-Fi, and GSM signals, and imposes some challenges due to the variable nature of the ambient RF signals, the path loss and shadowing vulnerability, and the choice of the rectifier since usually a range of frequencies must be scanned in order to harvest sufficient amounts of energy;

---

[1]Note that we are referring to far-field RF energy transmission, which is different from (close contact) inductive RF energy transmission or nonradiative RF energy transmission.

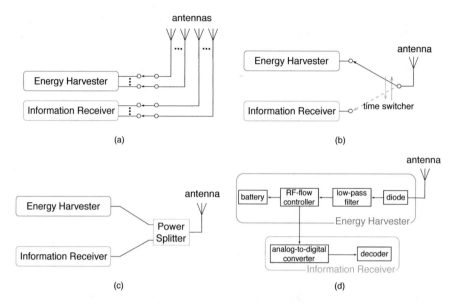

Fig. 8.1 Receiver architecture designs for SWIPT. (**a**) Separated receiver architecture; (**b**) time-switching architecture; (**c**) power splitting architecture; (**d**) integrated receiver architecture

2. Dedicated WET, which refers to harvesting energy using dedicated external sources such as a power beacon (PB). The gains depend on the placement of PBs, number of users to be served, the ability of the antenna array to focus the radiated power in the desired direction(s), and others; and

3. Simultaneous wireless information and power transfer (SWIPT), where both, information and RF energy, are conveyed from the source to the destination [34] using the same waveform, thus, saving spectrum.

With respect to SWIPT, four receiver architecture designs have been proposed and analyzed last years [35] (see Fig. 8.1):

- Separated receiver or antenna switching (AS) architecture is shown in Fig. 8.1a. The antenna array is divided into two sets with each connected to the EH circuitry or the information receiver. Consequently, this architecture allows to perform EH and information decoding independently and concurrently, and can be used to optimize the performance of separated receiver architecture as in [3].

- Co-located receiver architecture permits the EH and information receiver to share the same antenna(s). This architecture can be categorized into two models, e.g., time switching (TS) and power-splitting (PS) architectures, which are shown in Fig. 8.1b and c, respectively. When operating with TS, the network node switches and uses either the information receiver or the RF EH for the received RF signals at a time, while when operating with PS, the received RF signals are split into two streams for the information receiver and RF energy harvester with different power levels.

- Integrated receiver architecture [36] is shown in Fig. 8.1d. The implementation of RF-to-baseband conversion for information decoding is integrated with the energy harvester via the rectifier. The RF flow controller can also adopt a switcher or power splitter, like in the co-located receiver architecture, but the difference is that the switcher and power splitter are adopted in the integrated receiver architecture.

In general, the incident RF power at the EH receiver can be written as

$$P_{\mathrm{RF}} = \left| \sum_{i=1}^{T} P_i [\mathbf{w}_i]_{1 \times N_i} [\mathbf{H}_i]_{N_i \times M} \mathbf{1}_{M \times 1} \right|^2, \tag{8.1}$$

or

$$P_{\mathrm{RF}} = \sum_{i=1}^{T} \left| P_i [\mathbf{w}_i]_{1 \times N_i} [\mathbf{H}_i]_{N_i \times M} \mathbf{1}_{M \times 1} \right|^2, \tag{8.2}$$

where T is the number of energy transmitters, P_i, N_i, and $\mathbf{w}_i$ are the overall transmit power, the number of transmit antennas, and the normalized complex precoding vector, e.g., $||\mathbf{w}_i||_2^2 = 1$, at the i-th energy transmitter, respectively, M is the number of EH antennas at the receiver, while $\mathbf{H}_i$ is the complex channel matrix between the i-th energy transmitter and the EH receiver. Notice that (8.1) holds when the energy transmitters are fully synchronized and transmitting the same energy signals, while (8.2) holds when signals are fully independent. In both cases it is assumed a common EH circuitry for all receive antennas.

In case that the PS architecture is used, only a portion $\rho \in [0, 1]$ of this power goes to the RF-to-DC converter, while the remaining portion, $1 - \rho$, goes to the information decoding circuitry. Similarly, if TS scheme is utilized, the energy is harvested during a portion $\tau \in [0, 1]$ of the time, and the remaining portion, $1 - \tau$, is used for information decoding. Therefore, at the input of the RF-to-DC converter the RF power is $P_{\mathrm{RF}}^* = \rho P_{\mathrm{RF}}$ under PS, and $P_{\mathrm{RF}}^* = P_{\mathrm{RF}}$ during a portion τ of the time and $P_{\mathrm{RF}}^* = 0$ the remaining portion under TS. Now, the harvested energy,[2] P_{DC}, can be written as a function of P_{RF}^*, as

$$P_{\mathrm{DC}} = f\left(P_{\mathrm{RF}}^*\right), \tag{8.3}$$

where $f : \mathcal{R} \to \mathcal{R}$ is a non-decreasing function of its argument. One important factor that limits the performance of a WET receiver is the RF-to-DC energy conversion efficiency. In fact, in most of the works f is assumed to be linear for

---

[2]We use the terms energy and power indistinctly, which can be interpreted as if harvesting time is normalized.

Table 8.1 Nonlinear EH models

$f(x)$	Reference	$f(x)$	Reference
$ce^{-ab}\left(e^{ab} - e^{-a(x-b)}\right)\left(1 + e^{-a(x-b)}\right)^{-1}$	[38]	$\frac{ax^3 + bx^2 + cx}{p_3 x^3 + p_2 x^2 + p_1 x + p_0}$	[37]
$\frac{ax+b}{x+c} - \frac{b}{c}$	[16]	$ax^2 + bx + c$	[39]

analytical tractability, which implies that

$$f(x) = \eta x, \tag{8.4}$$

where η is a RF-to-DC energy conversion efficiency constant, thus, independent of the input power. However, many measurement data have revealed that the conversion efficiency actually depends on the input power [37] and consequently, the relationship between the input power and the output power is nonlinear. Table 8.1 summarizes the main nonlinear EH models in the literature where $a, b, c, p_3, p_2, p_1, p_0$ are constants determined by standard curve-fitting and are related to the detailed circuit specifications such as the resistance, capacitance, and diode turn-on voltage. The accuracy of the models depends on the fitting regions and on the specific characteristics of the EH circuits.

Finally, by just considering three factors such as (1) the sensitivity, which is the minimum RF input power required for energy harvesting, (2) the saturation level, which is the RF input power for which the diode starts working in the breakdown region and from that point onwards the output DC power keeps practically constant, and (3) a constant energy efficiency η between the sensitivity and saturation points; one can come up with a simple but accurate piece-wise model as illustrated in [40].

8.1.2 State of the Art on FD Wireless-Powered Networks

In the following we overview the recent results in FD wireless-powered communications separately according to three main topologies: FD bidirectional, FD relay, and FD hybrid access point (AP) communications, while other works are also discussed. Usually, the authors consider the harvest-use (HU) protocol, where the harvested energy cannot be stored and immediately must be consumed in order to maintain operability. A summary of the discussed results is shown in Table 8.2.

8.1.2.1 FD Bidirectional Communications

This kind of topology involves two-way information flow and one or two-way energy flow between the devices [4–7]. Specifically, the authors in [4] consider the PS architecture at EH receiver (one-way energy flow) and minimize the weighted sum transmit power by jointly designing the transmit beamforming vector of the

Table 8.2 Summary of the results on FD wireless-powered communications

Literature	Topology	WET approach and receiver architecture	Tuning parameter for optimization	Optimization goal
[4]	Bidirectional	SWIPT-PS	Transmit beamforming vector, ρ and the transmit power	Minimizing the weighted sum transmit power
[5]	Bidirectional	SWIPT-PS	Transmit power and ρ	Maximizing the sum-rate
[6]	Bidirectional	SWIPT-TS	Transmit beamforming vector, τ	Maximizing the UL throughput
[7]	Bidirectional	SWIPT-TS	τ, beamforming and transmit power	Maximizing the EE
[8]	Relay (AF/DF)	SWIPT-TS	τ	Maximizing the throughput
[9]	Relay (AF)	SWIPT-TS	τ	Minimizing outage probability
[10]	Relay (DF)	SWIPT-PS	ρ	Maximizing the end-to-end signal-to-interference-plus-noise ratio (SINR)
[11]	Relay (DF)	SWIPT-TS	Transmit and receive beamforming	Maximizing the end-to-end SINR
[12]	Relay (AF)	SWIPT-TS ($\tau = 1/2$)	Power allocation and transmit beamforming	Maximizing the throughput
[13]	Relay (DF)	SWIPT-TS	Receive and transmit beamformers, and τ	Maximizing the instantaneous throughput
[14]	Relay (AF)	SWIPT-TS ($\tau = 1/2$)	Beamforming	Maximizing signal-to-noise ratio (SNR)
[15]	Relay (AF/DF)	SWIPT-TS	Transmit beamforming (AF,DF) and τ (DF)	Maximizing the physical-layer security
[16]	Relay (AF)	SWIPT-TS/PS	Relay node	Minimizing outage probability
[17]	Hybrid AP	Dedicated WET	Antenna role	Ergodic capacity

(continued)

Table 8.2 (continued)

Literature	Topology	WET approach and receiver architecture	Tuning parameter for optimization	Optimization goal
[18]	Hybrid AP	Dedicated WET	DL and UL transmit powers and time slots duration	Maximizing the long-term weighted throughput
[19]	Hybrid AP	Dedicated WET	Time slots duration	Maximizing the throughput, and minimizing the total time
[20]	Hybrid AP	Dedicated WET	Transmit powers	Minimizing the aggregate power and maximizing the throughput
[21]	Hybrid AP	Dedicated WET	Subcarrier scheduling and transmit powers	Maximizing the sum-rate
[22]	Hybrid AP	Dedicated WET	Time slots duration and transmit power	Maximizing the throughput
[23]	Hybrid AP	Dedicated WET	Time slots duration	Maximizing the sum-throughput
[24]	Hybrid AP	Dedicated WET	DL beamforming and UL transmit power	Minimizing DL/UL transmit power and maximizing the total harvested energy
[25]	Hybrid AP	Dedicated WET	Duration of each phase, DL beamformers, and the UL transmit power	Maximizing the sum-rate, and EE
[26]	Hybrid FD/HD 4-node setup	SWIPT-PS	–	–
[27]	Relay (AF) with EH destination	SWIPT-PS	Beamforming	Minimizing MSE
[28]	Co-channel energy and information transfer	Dedicated WET	Energy and information beamformings	Maximizing the achievable data rate
[29]	FD friendly jammer	Dedicated WET	–	–

energy transmitter, the receive PS ratio, ρ, and the transmit power value of the EH device. The main limitation is that perfect successive interference cancellation (SIC) is assumed, which is not a practical assumption. Meanwhile, a point-to-point system where each two-antenna node houses identical transmitter–receiver pair is investigated in [5]. Each receiver intends to simultaneously transmit & decode information and harvests energy from the received signal. Therein, the authors propose transmit power and received PS ratio optimization algorithms that maximize the sum-rate subject to the transmit power and EH constraints. According to simulation results the residual SI may inhibit the system performance when it is not properly handled. On the other hand, some techniques for optimizing transmit beamforming in a FD multiple-input multiple-output (MIMO) setup, where the TS architecture is implemented at the EH receiver, are proposed in [6]. The problem is addressed when either complete instantaneous channel state information (CSI) or only its second-order statistics are available at the transmitter, while the authors demonstrate the advantages of the proposed methods over the sub-optimum and HD ones.

Finally, the potential of harvesting energy from the SI of a FD base station (BS) is further investigated in [7]. The BS is equipped with a SIC switch, which is turned off for a fraction of the transmission period for harvesting the energy from the SI that arises due to the downlink (DL) transmission. For the remaining transmission period, the switch is on such that the uplink (UL) transmission takes place simultaneously with the DL transmission. The authors explore the optimal time-splitting factor, τ, that maximizes the EE of the system along with the optimal beamforming and power allocation design for the DL and UL, respectively.

8.1.2.2 FD Relay Communications

A low-power FD relay node that assists the communications between the source and destination is powered by a WET process. The three-node FD relay wireless-powered communication network has been investigated in [8–15].

Both amplify and forward (AF) and decode and forward (DF) relaying protocols are studied in [8] using the TS architecture. An analytical characterization of the achievable throughput of three different communication modes, instantaneous transmission, delay-constrained transmission, and delay-tolerant transmission, is provided, while the optimal τ is investigated. It is shown that the FD relaying could substantially boost the system throughput compared to the conventional half duplex (HD) relaying architecture for the three transmission modes. An analysis of the outage probability in a TS AF FD setup is presented in [9] under imperfect channel state information (CSI). Numerical results provide some insights into the effect of various system parameters, such as τ, η, the noise power, and the channel estimation error on the performance of this network. Different from previous works, [10] explores both, the HU and harvest-use-store models, where the latter is implemented by switching between two batteries for charging and discharging with the aid of a PS architecture. A greedy switching policy is implemented with energy accumulation across transmission blocks in the harvest-use-store model. Also, the optimal ρ is presented and the corresponding outage probability is derived by modeling the relay's energy levels as a Markov chain with a two-stage state transition.

The above works consider single-antenna source and destination nodes, while the relay is equipped with two antennas, one receive antenna for EH and information reception and one transmit antenna, which enables the FD operation. The case of multiple-input single-output (MISO) or multiple-input multiple-output (MIMO) relaying is considered in [11–15]. Three different interference mitigation schemes are studied in [11], namely (1) optimal, (2) zero-forcing (ZF), and (3) maximum ratio combining (MRC)/maximum ratio transmission (MRT), for a FD DF MIMO relaying system, and the authors attain outage probability expressions to investigate delay-constrained transmission throughput. In [12], the source is equipped with multiple antennas, while the FD MISO relay node operates with the AF protocol, and the authors find the optimal power allocation and beamforming design. The instantaneous throughput is maximized in [13], while results reveal how the relay beamforming, using MRC, ZF, and MRT, increases both the EH and SI suppression capabilities at the MIMO FD DF relay. The possibilities and limitations of SEg recycling for RF powered MIMO relay channels is studied in [14]. Therein, a geometric geodesic approach rendered a simple beamformer optimization for both DL only and joint up-down link protocols, providing efficient strategies for power allocation between data transmission and WET. The authors show that the hardware design for a high recycling ratio is the prerequisite for the proposed system to operate with EE. Different from previous literatures, in [15] an eavesdropper is threatening the information confidentiality in the last hop and the goal is to maximize the physical-layer security under harvested energy constraints. The authors show the trade-off between AF and DF protocols according to the occurrence probability of non-zero secrecy rate.

On a different note, [16] investigates a FD wireless-powered two-way communication networks, where two hybrid APs and multiple AF relays operate all in FD mode. Both, TS and PS, architectures are considered, the new time division duplexing static power splitting (TDD SPS) and the full duplex static power splitting (FDSPS) schemes as well as a simple relay selection strategy are proposed to improve the system performance.

8.1.2.3 FD Hybrid AP

Herein, data/energy from/to users in the UL/DL channel are transmitted and received simultaneously. The single user case is investigated in [17], where the serving user is also FD. The AP and user are both equipped with two antennas, one for WET from the AP to user and the other for UL WIT from the user to AP, where WET and WIT are performed simultaneously through the same frequency band. The role of each antenna (i.e., transmission or reception) is not predefined and the authors propose an antenna pair selection scheme to improve the system performance. The two-HD user scenario is studied in [18] where the focus is on the long-term weighted throughput maximization problem. To that end, the analysis takes into account CSI variations over future slots and the evolution of the batteries when deciding the optimal resource allocation.

Different from above, the works in [19–25] analyze scenarios with an arbitrary number of serving users. In [19], the authors investigate the sum-throughput maximization problem and the total time minimization problem under perfect SIC for a setup where each user can continuously harvest wireless power from the AP until it transmits. Two distributed power control schemes for controlling the UL transmit power by the FD user equipments (UEs) and the DL energy-harvesting signal power by the FD AP are proposed in [20]. The objectives are minimizing the aggregate power subject to the quality of service requirement constraint and maximizing the aggregate throughput. Meanwhile, maximizing the sum-rate is the goal of the work in [21] where a joint subcarrier scheduling and power allocation algorithms for OFDM systems are investigated to that end. In [22], the authors design a protocol to support simultaneous WET in the DL and WIT in the UL by jointly optimizing the time allocations for WET and different users for UL WIT, and the transmit power allocations over time at the AP, such that the users' weighted sum-rate is maximized. With a similar setup, but now considering FD operation also at the UEs and with perfect SIC, the authors in [23] derive the optimal UL time allocation to users to maximize the network sum-throughput.

A new kind of systems comprising a FD AP, multiple single-antenna HD users and multiple energy harvesters equipped with multiple antennas, is considered in [24]. Therein, a multi-objective optimization taking into account heterogeneous quality of service requirements for UL and DL communication and WET is proposed, and the authors study the trade-off between UL and DL transmit power minimization and total harvested energy maximization. Finally, a system considering a FD multiple-antenna AP, multiple single-antenna DL users, and single-antenna UL users, where the latter need to harvest energy for transmitting information to the AP, is investigated in [25]. The communication is divided into two phases such that in the first one, the AP uses all available antennas for conveying information to DL users and wireless energy to UL users via information and energy beamforming, respectively, while in the second phase, UL users send their independent information to the AP using their harvested energy while the AP transmits the information to the DL users. The aim of the work is to maximize the sum rate and EE under UL user's achievable information throughput constraints by jointly optimizing beamforming and time allocation. In all the above works, the authors utilize the HD setup as benchmark while showing the benefits and limitations of the investigated FD operation.

8.1.2.4 Others

Herein we briefly discuss some works that do not fit, or at least not well, into the previous topologies. In [26], an energy-recycling single-antenna FD radio is designed by including a power divider and an energy harvester between the circulator and the receiver chain. This brings advantages in terms of both spectral efficiency and energy consumption since it allows performing both, an arbitrary attenuation of the incoming signal and the recycling of a non-negligible portion of the energy leaked through the nonideal circulator. The authors analyze the performance gains of this architecture in a four-node scenario in which two nodes operate in FD and two nodes in HD, while they also provide valuable numerical results obtained under practical parameter assumptions. In [27], the authors consider a FD MIMO AF relay network but different from the works in Sect. 8.1.2.2 herein the destination node is the energy-limited node and performs SWIPT with the PS architecture. A joint optimization problem of source and relay beamformers under their transmit powers' constraints and the user's EH constraint is formulated and solved such that the mean-square-error (MSE) in the detection is minimized. In [28], the authors refer to a setup where the EH node harvests energy from a PB and at the same time and frequency it transmits its information to the destination, thus, a FD operation different than in all the previous works. The achievable data rate is maximized by jointly optimizing the energy beamforming at the PB and the information beamforming at the EH node subject to their individual transmit power constraints. Finally, a cooperative jamming protocol, termed as accumulate-and-jam (AnJ) is proposed in [29] to improve physical-layer security. This is done by deploying a FD EH friendly jammer to secure the direct communication between source and destination in the presence of a passive eavesdropper and under imperfect CSI.

In the next two sections we present and analyze two example setups related with the FD bidirectional (Sect. 8.2) and FD hybrid AP (Sect. 8.3) architectures discussed previously. The suitable regions for FD operation are investigated in each case.

8.2 SEg Recycling for EE

Herein we consider a FD small cell BS (SBS) equipped with one transmit (T_x) and one receive (R_x) antenna, which serves one DL and one UL single-antenna UE, denoted as D and U, respectively. The SBS is powered by a regular grid source that can supply at most a power P_{max}, and is also equipped with an RF EH device and a rechargeable battery for energy storage. The setup is represented in Fig. 8.2 where UD, TD, TR, UR denote the links $U \to D$, $T_x \to D$, $T_x \to R_x$, and $U \to R_x$, respectively. Notice that the system model corresponds to a FD bidirectional topology as described in Sect. 8.1.2.1. We assume quasi-static Rayleigh fading such that $h_{UD}, h_{UR}, h_{TD} \sim Exp(1)$ are the normalized power channel coefficients of the far-field links, which remain unchanged for a time block of duration T. Meanwhile, $g = h_{TR}$ is treated as a constant since this channel exhibits the near-field property different from the far-field fading [14, 41].

The transmission time is divided into phases of duration τT and $(1 - \tau)T$, and without loss of generality we set $T = 1$. In the first phase, the SBS transmits the information-bearing to D, while U is silent and the SIC switch in this phase is turned

Fig. 8.2 System model. FD
SBS serving one DL and UL
devices

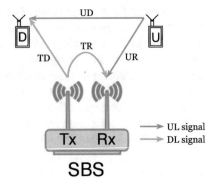

off for SEg harvesting.[3] Then, the received signal at D is given by

$$y_{d,1} = \sqrt{h_{TD}\varphi_{TD}(p_F + p_{EH})}s_d + w_d, \tag{8.5}$$

where s_d, $\mathbb{E}[|s_d|^2] = 1$, is the data signal from SBS and p_F and p_{EH} are the power
that the SBS draws from the power grid and the power coming from the EH process.
Thus, $p_F + p_{EH}$ constitutes the transmit power of the SBS. Additionally, φ_j denotes
the path loss in the link $j \in \{UD, UR, TD\}$, and w_i is the complex additive white
Gaussian noise (AWGN) at node $i \in \{D, U, SBS\}$ and we assume them with the
same variance σ^2. Since the EH process is also carried out in this phase when the
SBS harvests the energy from the SI, the signal at R_x in the SBS is given by

$$y_{sbs,1} = \sqrt{g(p_F + p_{EH})}s_{sbs} + w_{sbs}, \tag{8.6}$$

where $\mathbb{E}[|s_{sbs}|^2] = 1$. Consequently, using the linear EH model in (8.4) the SEg
harvested at the SBS in the first phase is given by

$$E = \eta\tau\mathbb{E}[|y_{sbs,1}|^2] = \eta\tau g(p_F + p_{EH}), \tag{8.7}$$

where the noise term is ignored in the last step as its contribution is negligible. Since
the energy E in (8.7) is going to be used during the entire block (of unit duration)
for transmission by assisting the power-grid consumption, we have that $E = p_{EH}$
and there is an energy loop described by

$$p_{EH} = \eta\tau g(p_F + p_{EH}) = \frac{\eta\tau g p_F}{1 - \eta\tau g}, \tag{8.8}$$

where the last step comes from isolating p_{EH} in the first line. Notice that since
$\eta, \tau, g < 1$ we have that $1 - \eta\tau g > 0$.

---

[3]U keeps silent in this phase since channel UR is not sufficiently strong for WET, thus, the SBS
only relies on the SEg for EH purpose.

In the second phase, the SBS turns the SIC switch on, which causes an attenuation $\zeta \ll 1$ of the SI signal, thus, the signal received at R_x in the SBS is given as

$$y_{\text{sbs},2} = \sqrt{h_{\text{UR}}\varphi_{\text{UR}}p_{\text{U}}}s_u + \sqrt{\zeta g(p_{\text{F}} + p_{\text{EH}})}s_{\text{sbs}} + w_{\text{sbs}}, \tag{8.9}$$

with $\mathbb{E}[|s_u|^2] = 1$, and the instantaneous SINR is

$$\gamma_{\text{sbs}} = \frac{h_{\text{UR}}\varphi_{\text{UR}}p_{\text{U}}}{\zeta g(p_{\text{F}} + p_{\text{EH}}) + \sigma^2} = \frac{h_{\text{UR}}\varphi_{\text{UR}}p_{\text{U}}}{\frac{\zeta g p_{\text{F}}}{1 - \eta\tau g} + \sigma^2}, \tag{8.10}$$

where the last step comes from using (8.8). Meanwhile, the signal received at D in this phase is given by

$$y_{\text{d},2} = \sqrt{h_{\text{TD}}\varphi_{\text{TD}}(p_{\text{F}} + p_{\text{EH}})}s_d + \sqrt{h_{\text{UD}}\varphi_{\text{UD}}p_{\text{U}}}s_u + w_d. \tag{8.11}$$

Notice that different from (8.5), in (8.11) it is reflected the impact of the interference coming from U. The impact of this interference is for $1 - \tau$ portion of the time, thus, we can write the instantaneous SINR at D for the whole block duration as

$$\begin{aligned}
\gamma_{\text{d}} &= \frac{h_{\text{TD}}\varphi_{\text{TD}}(p_{\text{F}} + p_{\text{EH}})}{(1 - \tau)h_{\text{UD}}\varphi_{\text{UD}}p_{\text{U}} + \sigma^2} \\
&= \frac{h_{\text{TD}}\varphi_{\text{TD}}p_{\text{F}}}{(1 - \eta\tau g)\big((1 - \tau)h_{\text{UD}}\varphi_{\text{UD}}p_{\text{U}} + \sigma^2\big)},
\end{aligned} \tag{8.12}$$

where the last step comes from using (8.8).

8.2.1 Problem Formulation

We assume delay-limited transmission for both DL and UL links such that each destination node has to decode the received signal block by block, and the transmit rates (r_d and r_{sbs}, respectively) are assumed fixed for many consecutive blocks and established according to transmitter and/or receiver requirements. Therefore, the outage event in the link TD, UR is defined as $\mathbb{O}_d \overset{\triangle}{=} \{\log_2(1 + \gamma_d) < r_d\}$ and $\mathbb{O}_{\text{sbs}} \overset{\triangle}{=} \{(1 - \tau)\log_2(1 + \gamma_{\text{sbs}}) < r_{\text{sbs}}\}$, respectively. Meanwhile, the power consumed from the grid at the SBS and from the battery or grid source at U and D is given by

$$P_{\text{con}} = \frac{p_{\text{F}} + p_{\text{U}}(1 - \tau)}{\epsilon} + P_{\text{cir},1} + P_{\text{cir},2}, \tag{8.13}$$

where ϵ is the amplifier efficiency at the SBS and U; and $P_{\text{cir},1}$, $P_{\text{cir},2}$ is the circuit consumption at the SBS, and the total circuit consumption at D and U, respectively.

In compliance with 5G networks, we maximize the system's EE by jointly designing p_F and τ. Notice that τ^* (optimum τ) might be equal to zero, and in such case SEg is not beneficial for the system performance. We measure the EE as the ratio between the throughput and the aggregated energies drawn from any source but from the SEg recycling process. Therefore,

$$\text{EE} = \frac{r_d(1 - \mathbb{P}[\mathbb{O}_d]) + r_{\text{sbs}}(1 - \mathbb{P}[\mathbb{O}_{\text{sbs}}])}{P_{\text{con}}}, \tag{8.14}$$

and the optimization problem is stated as follows:

$$\textbf{P1}: \quad \underset{p_F, \tau}{\arg \max} \quad \text{EE} \tag{8.15a}$$

$$\text{s.t.} \quad 0 \le \tau \le 1, \tag{8.15b}$$

$$\frac{p_F}{\epsilon} + P_{\text{cir},1} \le P_{\text{max}}. \tag{8.15c}$$

In the next subsection we characterize the outage performance in TD, UR links as it is required for evaluating (8.14).

8.2.2 Outage Analysis

We proceed as follows:

$$\mathbb{P}[\mathbb{O}_d] = \mathbb{P}[\log_2(1 + \gamma_d) < r_d] \overset{(a)}{=} \mathbb{P}[\gamma_d < 2^{r_d} - 1]$$

$$\overset{(b)}{=} \mathbb{P}\left[\frac{h_{\text{TD}}\varphi_{\text{TD}}p_F}{(1 - \eta\tau g)((1 - \tau)h_{\text{UD}}\varphi_{\text{UD}}p_U + \sigma^2)} < 2^{r_d} - 1 \right]$$

$$\overset{(c)}{=} \mathbb{P}\left[h_{\text{TD}} < \frac{(2^{r_d} - 1)(1 - \eta\tau g)((1 - \tau)h_{\text{UD}}\varphi_{\text{UD}}p_U + \sigma^2)}{\varphi_{\text{TD}}p_F} \right]$$

$$\overset{(d)}{=} 1 - \mathbb{E}\left[e^{-(b_1 + b_2 h_{\text{UD}})}\right] \overset{(e)}{=} 1 - \frac{e^{-b_1}}{1 + b_2}, \tag{8.16}$$

where (a) comes from isolating γ_d, (b) comes from using (8.12), (c) is attained by isolating h_{TD}, while (d) follows by using its cumulative distribution function (CDF) along with

$$b_1 = (2^{r_d} - 1)(1 - \eta\tau g)\frac{\sigma^2}{\varphi_{\text{TD}}p_F}, \tag{8.17}$$

$$b_2 = (2^{r_d} - 1)(1 - \eta\tau g)(1 - \tau)\frac{\varphi_{\text{UD}}p_U}{\varphi_{\text{TD}}p_F}, \tag{8.18}$$

and finally, (e) is the result of computing the expectation with respect to the exponential random variable (RV) h_{UD}.

In the case of $\mathbb{P}[\mathbb{O}_{\mathrm{sbs}}]$ we proceed as follows:

$$\mathbb{P}[\mathbb{O}_{\mathrm{sbs}}] = \mathbb{P}[(1-\tau)\log_2(1+\gamma_{\mathrm{sbs}}) < r_{\mathrm{sbs}}] \stackrel{(a)}{=} \mathbb{P}\left[\gamma_{\mathrm{sbs}} < 2^{\frac{r_{\mathrm{sbs}}}{1-\tau}} - 1\right]$$

$$\stackrel{(b)}{=} \mathbb{P}\left[\frac{h_{\mathrm{UR}}\varphi_{\mathrm{UR}}p_{\mathrm{U}}}{\frac{\zeta g p_{\mathrm{F}}}{1-\eta\tau g} + \sigma^2} < 2^{\frac{r_{\mathrm{sbs}}}{1-\tau}} - 1\right] \stackrel{(c)}{=} \mathbb{P}\left[h_{\mathrm{UR}} < \frac{\left(2^{\frac{r_{\mathrm{sbs}}}{1-\tau}} - 1\right)\left(\frac{\zeta g p_{\mathrm{F}}}{1-\eta\tau g} + \sigma^2\right)}{\varphi_{\mathrm{UR}}p_{\mathrm{U}}}\right]$$

$$\stackrel{(d)}{=} 1 - e^{-\frac{\left(2^{\frac{r_{\mathrm{sbs}}}{1-\tau}} - 1\right)\left(\frac{\zeta g p_{\mathrm{F}}}{1-\eta\tau g} + \sigma^2\right)}{\varphi_{\mathrm{UR}}p_{\mathrm{U}}}}, \tag{8.19}$$

where (a) comes from isolating γ_{sbs}, (b) comes from using (8.10) along with (8.8), (c) is attained by isolating h_{UR}, while (d) follows by using its CDF.

8.2.3 Numerical Solution

Based on the results in the previous subsection, **P1** in (8.15) can be stated as

$$\textbf{P2}: \quad \underset{p_{\mathrm{F}},\tau}{\arg\max} \quad \frac{r_d\frac{e^{-b_1}}{1+b_2} + r_{\mathrm{sbs}}e^{-\frac{\left(2^{\frac{r_{\mathrm{sbs}}}{1-\tau}} - 1\right)\left(\frac{\zeta g p_{\mathrm{F}}}{1-\eta\tau g} + \sigma^2\right)}{\varphi_{\mathrm{UR}}p_{\mathrm{U}}}}}{\frac{p_{\mathrm{F}}+p_{\mathrm{U}}(1-\tau)}{\epsilon} + P_{\mathrm{cir},1} + P_{\mathrm{cir},2}} \tag{8.20a}$$

$$\text{s.t.} \quad 0 \leq \tau \leq 1, \tag{8.20b}$$

$$p_{\mathrm{F}} \leq \epsilon(P_{\max} - P_{\mathrm{cir},1}). \tag{8.20c}$$

Notice that **P2** would be cumbersome to solve in closed-form because of the tangled dependence of the objective function on p_{F} and τ. Since convexity analysis seems also a cumbersome task (in fact, numerical results show that the objective function is not concave), herein we resort to numerical methods to solve **P2** while drawing some insights from the numerical results. We assume $\varphi_{\mathrm{TD}} = \varphi_{\mathrm{UR}} = \varphi_{\mathrm{UD}} = \varphi$ and unless stated otherwise the parameter values are those presented in Table 8.3.

Figure 8.3 shows a contour plot of EE vs τ, p_{F}. As clearly shown in the figure, the objective function in **P2** is not concave because of the presence of two local peaks: one maximum and one minimum, and any algorithm for solving the optimization problem has to deal with that issue. Notice that $p_{\mathrm{F}}^* = 25$ dBm and most interesting $\tau^* \approx 0.42$ which is much larger than 0; thus, SEg harvesting results are advantageous for improving the EE of the system. The success probabilities at that operation point are 0.63 and 0.99 for TD and UR links, respectively.

In Fig. 8.4, we show the optimization results as functions of r_{sbs} for different transmit power levels at U and transmit rates in the DL. As shown in Fig. 8.4a,

Table 8.3 System
parameters for setup in
Fig. 8.2

Parameter	Value	
p_U	25 dBm	
P_{max}	40 dBm	
η	0.5	[4, 7, 16]
ε	0.69	[7]
r_d, r_{sbs}	1, 5 bps/Hz	
φ	10^{-5}	
σ^2	-70 dBm	[4]
g, ζ	$-15, -110$ dB	[7, 12]
$P_{cir,1}, P_{cir,2}$	30, 40 dBm	[7]

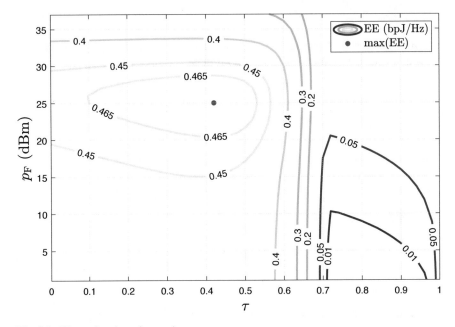

Fig. 8.3 EE as a function of p_F and τ

only for relatively large data rates in the UL, the SEg harvesting is not beneficial
since $\tau \to 0$. A smaller p_U and/or DL transmit rate makes SEg less beneficial. In
the first case because the success probability in the UL becomes more dependent
on the transmission time since the U's transmit power decreases; while in the
second case, when the DL transmit rate is small, less power resources at the SBS
are necessary, which is also shown in Fig. 8.4b, thus, the need of SEg decreases.
Interesting, Fig. 8.4b shows that as p_U increases, p_F^* increases as well in almost
all the region. The region for which this holds is particularly dependent on to the
specific values of r_{sbs} and r_d, and notice that for $r_{sbs} \geq 7$ bps/Hz and $r_d = 1$
bps/Hz, such relation no longer holds. The abrupt changes in the red lines in that
region reflect the interplay of both, UL and DL throughputs, where the r_{sbs} values at

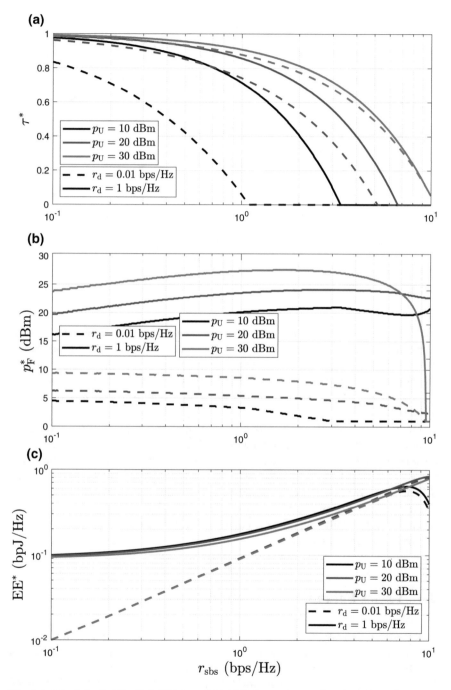

Fig. 8.4 (a) τ^* (top), (b) p_F^* (middle), and (c) EE* (bottom), as a function of r_{sbs} for $p_U \in \{10, 20, 30\}$ dBm and $r_d \in \{0.01, 1\}$ bps/Hz

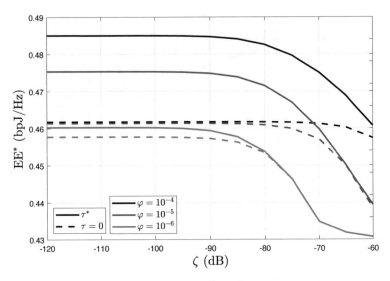

Fig. 8.5 EE* as a function of ζ for $\varphi \in \{10^{-4}, 10^{-5}, 10^{-6}\}$ with/without SEg recycling

the changing points determine when UL or DL throughputs become more relevant between each other. The EE values attained with the optimum allocation are shown in Fig. 8.4c and notice that the system is more energy efficient when p_U is around 20 dBm. Also, from $r_{sbs} = 7.5$ bps/Hz onwards the EE is no longer increasing on r_{sbs} for $p_U = 10$ dBm because the UL success probability affects heavily the UL throughput.

Finally, in Fig. 8.5 we show the optimum EE as a function of the SIC performance parameter for the case with and without SEg harvesting and for different path-loss attenuations. As expected, as the path loss becomes more critical, the EE decreases, but also the SEg harvesting becomes less beneficial. Additionally, the better the SIC performance (smaller ζ) is, more gain from SEg harvesting can be obtained. This is because, SIC performance affects directly the UL success probabilities, and when SIC is relatively high, the impact of the UL transmission time becomes more important and τ^* decreases.

8.3   FD for Sporadic IoT Transmissions

In this subsection we consider a FD SBS serving k single-antennas IoT devices, referred as UEs. The SBS is assumed to be equipped with two antennas. One is for the DL WET, while the other one is used to receive the UL information transmission from the users (WIT). Thus, the normalized channel power gains for the DL channel, h_i, and UL channel, g_i, are different, and the topology matches the FD Hybrid AP model described in Sect. 8.1.2.3. The system setup is illustrated in Fig. 8.6, and we

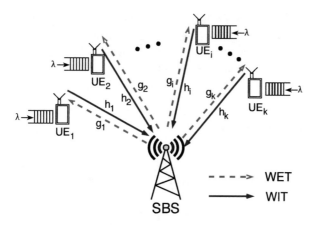

Fig. 8.6 System model. FD SBS serving k UEs with Poisson traffic

FD Operation	HD Operation
1. The entire time block is divided in δ slots. Thus, each time slot has duration of $1/\delta$.	1. The entire time block is divided in two parts. The first part, of duration $\tau < 1$, is used only for WET from the SBS to the UEs, while the second part, of duration $1-\tau$, is divided in δ time slots. Thus, each time slot has duration of $(1-\tau)/\delta$.
2. UEs are continuously harvesting energy until a message arrives to its output buffer.	
3. After a data message arrives, the UE keeps harvesting energy until the current time slot ends, and then it transmits the message in the following time slot using all the harvested energy.	2. When a message arrives to the UE's output buffer, it selects randomly one transmit time slot from those that are still to come and transmits over it.

Fig. 8.7 Operation scheme

assume quasi-static fading, e.g., such that channel remains unchanged during the duration T of a block, and varies independently from one block to another one. We assume that all the links are subject to the same path-loss attenuation φ, and without loss of generality we set $T = 1$.

We assume sporadic and short transmissions, which are typical characteristics of future IoT networks [42]. Hence, we model the arrivals of data messages of fixed length d (in bits/Hz) to the first-in first-out (FIFO) queue buffer before transmission, as a Poisson RV with rate $\lambda \ll 1$ (in mssg/block—messages per block). Thus, the inter-arrivals time, β (in blocks), is exponentially distributed with mean $1/\lambda$.

8.3.1 Slotted Operation

The operation procedure is illustrated in Fig. 8.7 for the FD scheme and the HD operation counterpart which is used as benchmark. Notice that, for the HD scheme there is only one antenna available at the SBS. Additionally, since $\lambda \ll 1$, the chances of two (or more) consecutive messages arriving to be transmitted in the same time slot is small, and because it is a rare event and for simplicity, we assume that the second message cannot be delivered.

8.3.2 FD Performance

By denoting p_T as the SBS transmit power, the energy available for harvesting at the i-th UE in the j-th time slot, is approximately given by

$$E_i^{(j)} = \frac{p_T \varphi g_i^{(j)}}{\delta}, \tag{8.21}$$

where the contribution of the noise energy and UL signals from other UEs in the same slot was ignored. Notice that the transmit power of the low-power UEs that are concurrently transmitting in the j-th slot is much smaller than the powering signal coming from the SBS, therefore its contribution would be negligible unless they are very close from the harvesting device. Herein, we assume its energy contribution is negligible. On the other hand, because of the quasi-static property we have that $g_i^{(n\delta+1)} = g_i^{(n\delta+2)} = \cdots = g_i^{((n+1)\delta)}$, where $i = 1, \ldots, k$ and $n = 0, 1, \ldots$, which also holds for the UL channels, h_i.

We consider the two EH models discussed at the end of Sect. 8.1.1. Therefore, after the WET phase, which lasts for $\lceil \delta \beta_i \rceil$ consecutive time slots for UE_i, the transmit power is a RV that can be expressed as

$$p_i = \sum_{j=1}^{\lceil \delta \beta_i \rceil} f(\delta E_i^{(j)}) = \sum_{j=1}^{\lceil \delta \beta_i \rceil} f(p_T \varphi g_i^{(j)}), \tag{8.22}$$

which for the case of the linear model in (8.4) reduces to $p_i = \eta p_T \varphi \sum_{j=1}^{\lceil \delta \beta_i \rceil} g_i^{(j)}$.

At the SBS side, the received signal in a given time slot j is given by

$$y_j = \sum_{i=1}^{k} \sqrt{h_i^{(j)} \varphi p_i} \chi_i^{(j)} s_i + \sqrt{\psi p_T} s_{ue} + \omega_{sbs}, \tag{8.23}$$

where s_i, $\mathbb{E}[|s_i|^2] = 1$, is the data signal from UE_i and $\chi_i^{(j)}$ denotes the activity of that UE in the time slot j, this is, $\chi_i^{(j)} = 1$ or $\chi_i^{(j)} = 0$ if UE_i is active or inactive, respectively. More formally,

$$\chi_i^{(j)} = \mathbb{I}(\mathrm{mod}(\lceil \delta \beta_i \rceil, \delta) + 1 = j), \tag{8.24}$$

where mod$(\cdot, \cdot)$ is the modulo operation and $\mathbb{I}(\cdot)$ is the indicator function, which is equal to the unity if the argument is true; otherwise, its output is zero. Finally, ψ accounts for the combined effect of the SI channel attenuation and the SIC performance, which similarly to the scenario in the previous subsection is assumed as a constant; and ω_{sbs} is the complex AWGN at the SBS with variance σ^2. In case of several UEs transmitting in the same time slot, successive interference cancellation (also referred as SIC) is carried out when decoding. Therefore, herein we are interested in SINR after SIC in order to evaluate the system performance. For ease of exposition let us sort the UEs in descending order according to the value of $h_i^{(j)} \varphi p_i \chi_i^{(j)}$, thus, UE$_i$ is the i-th UE in the list, and assuming it is active and the first $l = 1, \ldots, i-1$ UEs were correctly decoded, its SINR is given by

$$\text{SINR}_i^{(j)} = \frac{h_i^{(j)} \varphi p_i}{\varphi \sum_{l=i+1}^{k} h_l^{(j)} p_l \chi_l^{(j)} + \psi p_T + \sigma^2}. \tag{8.25}$$

Notice that $\sum_{l=i+1}^{k} h_l^{(j)} p_l \chi_l^{(j)} = 0$ when $i = k$.

There is success in decoding the information from the i-th UE in the j-th time slot when $\text{SINR}_i^{(j)} > 2^{d\delta} - 1$ and the data from the l-th UEs ($l = 1, \ldots, i-1$) were also decoded correctly. Thus, the overall system reliability is defined as

$$P_{\text{succ}}^{\text{fd}} = \mathbb{E}\left[\frac{\sum_{i=1}^{k} \mathbb{I}(\text{SINR}_1^{(j)} > 2^{d\delta} - 1) \times \cdots \times \mathbb{I}(\text{SINR}_i^{(j)} > 2^{d\delta} - 1)}{\sum_{i=1}^{k} \chi_i^{(j)}} \right], \tag{8.26}$$

where the expectation is taken with respect to the fading realizations and user activity, which depends on the traffic characteristics. Notice that outage events for the concurrent transmitting UEs are correlated, thus, numerator in (8.26) cannot be further efficiently simplified.

8.3.3   HD Performance

In the HD scenario, the UEs are harvesting energy during a portion τ of the block time. Therefore, the energy harvested in the j-th block by UE$_i$ is given by

$$E_i^{(j)} = \tau p_T \varphi g_i^{(j)}, \tag{8.27}$$

whereas in the previous case, the noise energy is ignored. Notice that for FD j identifies a time slot, while here it identifies a whole block. Now, after the WET procedure, which lasts for $\lceil \beta_i \rceil$ consecutive blocks but without including the last

$(1 - \tau)$ portion of each block, the transmit power of UE$_i$ in its transmit slot is

$$p_i = \frac{\tau\delta}{1-\tau} \sum_{j=1}^{\lceil \beta_i \rceil} f\left(\frac{E_i^{(j)}}{\tau}\right), \tag{8.28}$$

which for the case of the linear model in (8.4) reduces to $p_i = \frac{\tau\delta\eta p_T\varphi}{1-\tau} \sum_{j=1}^{\lceil \beta_i \rceil} g_i^{(j)}$.

At the SBS side, the received signal in a given time slot j is given by

$$y_j = \sum_{i=1}^{k} \sqrt{h_i^{(j)} \varphi p_i} \chi_i^{(j)} s_i + \omega_{\text{sbs}}, \tag{8.29}$$

and again $\chi_i^{(j)}$ denotes the activity of UE$_i$ in the time slot j, which according to the procedure described in Fig. 8.7 for HD is now given by

$$\chi_i^{(j)} = \begin{cases} \mathbb{I}(v = j), & \text{if } \beta_i - \lfloor \beta_i \rfloor \leq \tau \text{ or } \beta_i - \lfloor \beta_i \rfloor > \tau + \frac{(\delta-1)(1-\tau)}{\delta} \\ \mathbb{I}(\upsilon = j), & \text{otherwise} \end{cases}, \tag{8.30}$$

where v, υ are discrete RVs with PMFs given by $\mathbb{P}(v = j) = 1/\delta$ for $j = 1, \ldots, \delta$ and $\mathbb{P}(\upsilon = j) = 1/(\delta - \rho)$ for $j = \rho + 1, \rho + 2, \ldots, \delta$, respectively. In the latter expression ρ is the smallest non-negative integer that satisfies $\beta_i - \lfloor \beta_i \rfloor - \tau < \rho(1 - \tau)/\delta$ for $\rho < \delta$, thus, $\rho = \lceil \delta(\beta_i - \lfloor \beta_i \rfloor - \tau)/(1 - \tau) \rceil$. In a simplified form we can state $\mathbb{P}(\upsilon = j) = 1/\left(\delta - \text{mod}\left(\lceil \delta(\beta_i - \lfloor \beta_i \rfloor - \tau)/(1 - \tau) \rceil, \delta\right)\right)$.

Once again, as in the FD case, in case of several UEs transmitting in the same time slot, SIC is carried out when decoding. By sorting the UEs in descending order according to the value of $h_i^{(j)} \varphi p_i \chi_i^{(j)}$, and assuming UE$_i$ is active and the first $l = 1, \ldots, i - 1$ UEs were correctly decoded, its SINR is given by

$$\text{SINR}_i^{(j)} = \frac{h_i^{(j)} \varphi p_i}{\varphi \sum_{l=i+1}^{k} h_l^{(j)} p_l \chi_l^{(j)} + \sigma^2}. \tag{8.31}$$

and again $\sum_{l=i+1}^{k} h_l^{(j)} p_l \chi_l^{(j)} = 0$ when $i = k$.

There is success in decoding the information from the i-th UE in the j-th time slot when $\text{SINR}_i^{(j)} > 2^{\frac{d\delta}{1-\tau}} - 1$ and the data from the l-th UEs ($l = 1, \ldots, i - 1$) were also decoded correctly. Thus, the overall system reliability is defined as

$$P_{\text{succ}}^{\text{hd}} = \mathbb{E}\left[\frac{\sum_{i=1}^{k} \mathbb{I}\left(\text{SINR}_1^{(j)} > 2^{\frac{d\delta}{1-\tau}} - 1\right) \times \cdots \times \mathbb{I}\left(\text{SINR}_i^{(j)} > 2^{\frac{d\delta}{1-\tau}} - 1\right)}{\sum_{i=1}^{k} \chi_i^{(j)}} \right], \tag{8.32}$$

where the expectation is taken with respect to the fading realizations and user activity, which depends on the traffic characteristics. Notice that outage events for the concurrent transmitting UEs are correlated, thus, numerator in (8.26) cannot be further efficiently simplified.

8.3.4 Performance Analysis

In this subsection we use Monte Carlo simulations to analyze the system performance under different settings. We obtain the results for the ideal linear EH model presented in (8.4) and a suitable nonlinear EH model.

To that end we assume that the UEs are equipped with the EH hardware proposed in [43]. The input–output power relation for the measurement data offered in [43] is shown in Fig. 8.8, and we compare the different EH models discussed in Sect. 8.1.1 for that specific hardware. Notice that the linear model is suitable for the range of small input power but it fails mimicking the saturation region. As shown in the figure, the nonlinear model in [38] performs the best, and the saturation is ensured for values greater than 1000 μW since it is based on a logistic function, different from the other nonlinear models. Therefore, we adopt the nonlinear model in [38] as well as the linear alternative with $\eta = 0.07$. Finally, channels are assumed i.i.d with Nakagami-m fading, such that $h_i, g_i \sim \Gamma(m, 1/m)$, and unless stated otherwise the parameter values are those presented in Table 8.4.

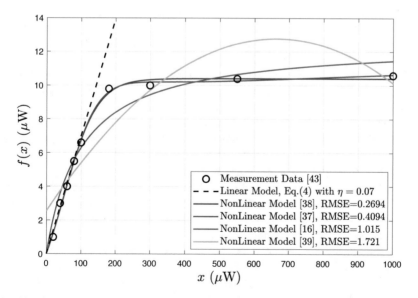

Fig. 8.8 Comparison of measurement [43] and the existing linear and nonlinear EH models

Table 8.4 System
parameters for setup in
Fig. 8.6

Parameter	Value	
k	20	
m	3	
δ	1	
d	0.5 bps/Hz	
λ	0.1 mssg/block	
σ^2	-70 dBm	[4]
φ	10^{-5}	
p_T	45 dBm	
ψ	-110 dB	[7]

As shown in (8.26) and (8.32), a small δ is convenient in the sense that it increases the chances of successful individual transmissions but at the same time is prejudicial by increasing the chances of concurrent transmissions. Therefore, selecting an appropriate δ is crucial for optimizing the system reliability performance. Figure 8.9 shows how the selection of δ impacts on the system reliability for different number of UEs, $k \in \{10, 100\}$, and using both, linear and nonlinear, EH models. First to notice is that there is a significant gap in the performance when using the linear vs the nonlinear EH model, thus, reinforcing the idea that the linear EH model, although suitable for analytical tractability most of the time, is idealistic and should be used with caution. For the chosen system parameter values, the FD scheme overcomes its HD counterparts using $\tau \in \{1/7, 1/4, 1/2\}$, and in all the cases as k increases, more slots are required for an efficient communication. The results shown in this figure raise an important disadvantage of the HD scheme, and it is that besides optimizing δ, the optimization over τ is also required, therefore, the HD design is even more complicated and its performance may not even reach the one achieves by the FD scheme.

Figure 8.10 shows the performance of FD and HD operating with the optimum τ as a function of the SIC parameter for $\delta \in \{1, 4\}$ and $d \in \{0.5, 1\}$ bps/Hz. Since the HD configuration does not depend on this parameter, its performance appears as a straight line. Notice that the gap in performance of HD and FD setups can be substantial when SI is efficiently eliminated, while FD is no longer suitable when SIC performance is poor, e.g., $\psi \geq -106$ dB for $\delta = 4$ and $d = 1$ bps/Hz. Even more interesting are the results shown in Fig. 8.11 where the performance appears as a function of the message arrival rate for different Line-of-Sight conditions, $m \in \{1, 4, 10\}$. As λ increases, the FD scheme becomes more suitable than HD with optimum τ as shown in Fig. 8.11a. For very small λ, HD becomes more suitable for the chosen parameter values, but notice that (1) there is still the problem of finding the optimum τ and (2) when operating with relatively large m, e.g., $m \geq 4$, the gain from using HD with optimum τ is no longer significant. Figure 8.11b shows that as λ increases, the optimum time required for WET in the HD setup decreases, and more time for information transmission is required in order to resolve the concurrent transmissions. Notice that the number of Monte Carlo samples used

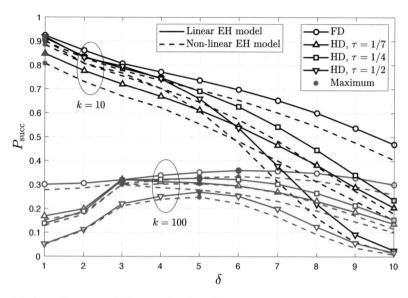

Fig. 8.9 Overall system reliability as a function of the number of transmit slots for $k \in \{10, 100\}$ and using both, linear and nonlinear, EH models. For the HD scheme we evaluated setups with $\tau = \{1/7, 1/4, 1/2\}$

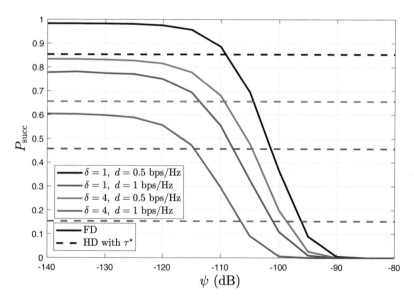

Fig. 8.10 Overall system reliability as a function of SIC parameter for $\delta \in \{1, 4\}$, $d \in \{0.5, 1\}$ bps/Hz and using nonlinear EH model. The HD performance is evaluated with the optimum τ

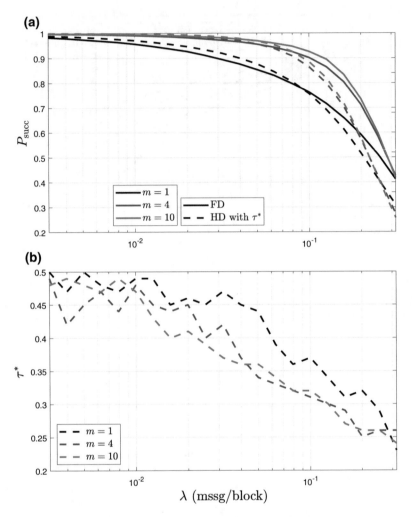

Fig. 8.11 Performance as a function of the message arrival rate for $m \in \{1, 4, 10\}$ and using nonlinear EH model. (**a**) Overall system reliability for FD and HD with the optimum τ (top) and (**b**) Optimum τ for HD scheme (bottom)

was insufficient for reaching smooth curves, but increasing such number crashes with the software and hardware limitations. Even so, the tendency can be easily appreciated, and the irregularities do not impact on the P_{succ} performance as its curves are already smooth as shown in Fig. 8.11a.

8.4   Conclusions and Outlook

In this chapter, we presented an overview of the main characteristics of the wireless-powered communication networks and the main approaches for modeling the EH conversion process. We have discussed the main FD architectures investigated in the literature, namely (1) FD bidirectional communications, (2) FD relay communications, and (3) FD hybrid AP, along with their particularities. Additionally, we have provided and analyzed two example setups related with architectures (1) and (3), while demonstrating the suitable regions for FD operation in each case. We first considered a small cell BS serving one DL and one UL users, and we demonstrated the gains in EE when the BS operates with FD for SEg recycling, especially for relatively small data rates and large transmit power in the UL and/or large transmit rate in the DL. In the second example we have considered a BS serving multiple IoT devises with a Poisson sporadic traffic pattern. For this case, we have shown that the linear EH model is optimistic and a more realistic model should be used for getting practical results. Additionally, we have compared the FD and HD schemes, and shown that the FD design is not only simpler but it also can offer significant performance gains in terms of system reliability when the SIC hardware performs not too bad.

Many possible research directions for the design and deployment of FD wireless-powered systems can be still identified. For instance: (1) Efficient MIMO implementations. MIMO techniques can be used for both, spatial-domain SIC and to harvest more energy. However, optimal solutions in FD wireless-powered systems are complex and require significant power consumption for computation purpose, even more if nonlinear EH models are used; hence, it is necessary looking at low-complexity MIMO schemes suitable for efficient FD MIMO designs. Additionally, exploiting the benefits of both, massive MIMO and FD operation in the context of wireless-powered systems, seems as an interesting direction to pursue. (2) Practical and energy-efficient FD transceiver design. Research on SI mitigation techniques with low-power consumption is essential since circuitry required in the cancellation step could drain harvested energy significantly. The nonlinear behavior of the EH circuit components needs to be taken into account for an efficient design. (3) Medium access control (MAC) layer performance. Research work on MAC layer issues is on infancy since majority of research has focused on physical-layer aspects, thus, energy-efficient FD MAC algorithms are still required. (4) Interference exploitation. In wireless-powered networks, interference may not be a completely detrimental factor, since it can be also exploited as an extra source of energy. Investigating how much interference can be exploited to improve the performance of FD wireless-powered systems is of paramount importance. Therefore, transmission schemes, scheduling, and interference cancellation algorithms that optimize the system performance taking into account these trade-offs are required.

Acknowledgements This research has been financially supported by Academy of Finland, 6Genesis Flagship (Grant no318927), and EE-IoT (no319008).

References

1. S. Priya and D. J. Inman, *Energy harvesting technologies*. Springer, 2009, vol. 21.
2. N. Shinohara, *Wireless power transfer via radiowaves*. John Wiley & Sons, 2014.
3. R. Zhang and C. K. Ho, "MIMO broadcasting for simultaneous wireless information and power transfer," *IEEE Transactions on Wireless Communications*, vol. 12, no. 5, pp. 1989–2001, May 2013.
4. Z. Hu, C. Yuan, F. Zhu, and F. Gao, "Weighted sum transmit power minimization for full-duplex system with SWIPT and self-energy recycling," *IEEE Access*, vol. 4, pp. 4874–4881, 2016.
5. A. A. Okandeji, M. R. A. Khandaker, and K. Wong, "Wireless information and power transfer in full-duplex communication systems," in *IEEE International Conference on Communications (ICC)*, May 2016, pp. 1–6.
6. B. K. Chalise, H. A. Suraweera, G. Zheng, and G. K. Karagiannidis, "Beamforming optimization for full-duplex wireless-powered MIMO systems," *IEEE Transactions on Communications*, vol. 65, no. 9, pp. 3750–3764, Sep. 2017.
7. A. Yadav, O. A. Dobre, and H. V. Poor, "Is self-interference in full-duplex communications a foe or a friend?" *IEEE Signal Processing Letters*, vol. 25, no. 7, pp. 951–955, July 2018.
8. C. Zhong, H. A. Suraweera, G. Zheng, I. Krikidis, and Z. Zhang, "Wireless information and power transfer with full duplex relaying," *IEEE Transactions on Communications*, vol. 62, no. 10, pp. 3447–3461, Oct 2014.
9. T. N. Nguyen, D.-T. Do, P. T. Tran, and M. Vozňák, "Time switching for wireless communications with full-duplex relaying in imperfect CSI condition," 2016.
10. H. Liu, K. J. Kim, K. S. Kwak, and H. V. Poor, "Power splitting-based SWIPT with decode-and-forward full-duplex relaying," *IEEE Transactions on Wireless Communications*, vol. 15, no. 11, pp. 7561–7577, Nov 2016.
11. M. Mohammadi, H. A. Suraweera, G. Zheng, C. Zhong, and I. Krikidis, "Full-duplex MIMO relaying powered by wireless energy transfer," in *IEEE International Workshop on Signal Processing Advances in Wireless Communications (SPAWC)*, June 2015, pp. 296–300.
12. Y. Zeng and R. Zhang, "Full-duplex wireless-powered relay with self-energy recycling." *IEEE Wireless Commun. Letters*, vol. 4, no. 2, pp. 201–204, 2015.
13. M. Mohammadi, B. K. Chalise, H. A. Suraweera, C. Zhong, G. Zheng, and I. Krikidis, "Throughput analysis and optimization of wireless-powered multiple antenna full-duplex relay systems," *IEEE Transactions on Communications*, vol. 64, no. 4, pp. 1769–1785, April 2016.
14. D. Hwang, K. C. Hwang, D. I. Kim, and T. Lee, "Self-energy recycling for RF powered multi-antenna relay channels," *IEEE Transactions on Wireless Communications*, vol. 16, no. 2, pp. 812–824, Feb 2017.
15. H. Kim, J. Kang, S. Jeong, K. E. Lee, and J. Kang, "Secure beamforming and self-energy recycling with full-duplex wireless-powered relay," in *IEEE Annual Consumer Communications Networking Conference (CCNC)*, Jan 2016, pp. 662–667.
16. G. Chen, P. Xiao, J. R. Kelly, B. Li, and R. Tafazolli, "Full-duplex wireless-powered relay in two way cooperative networks," *IEEE Access*, vol. 5, pp. 1548–1558, 2017.
17. M. Gao, H. H. Chen, Y. Li, M. Shirvanimoghaddam, and J. Shi, "Full-duplex wireless-powered communication with antenna pair selection," in *IEEE Wireless Communications and Networking Conference (WCNC)*, March 2015, pp. 693–698.
18. M. A. Abd-Elmagid, A. Biason, T. ElBatt, K. G. Seddik, and M. Zorzi, "On optimal policies in full-duplex wireless powered communication networks," in *2016 14th International Symposium on Modeling and Optimization in Mobile, Ad Hoc, and Wireless Networks (WiOpt)*, May 2016, pp. 1–7.
19. X. Kang, C. K. Ho, and S. Sun, "Full-duplex wireless-powered communication network with energy causality," *IEEE Transactions on Wireless Communications*, vol. 14, no. 10, pp. 5539–5551, Oct 2015.

20. R. Aslani and M. Rasti, "Distributed power control schemes for in-band full-duplex energy harvesting wireless networks," *IEEE Transactions on Wireless Communications*, vol. 16, no. 8, pp. 5233–5243, Aug 2017.
21. H. Kim, H. Lee, M. Ahn, H. Kong, and I. Lee, "Joint subcarrier and power allocation methods in full duplex wireless powered communication networks for OFDM systems," *IEEE Transactions on Wireless Communications*, vol. 15, no. 7, pp. 4745–4753, July 2016.
22. H. Ju and R. Zhang, "Optimal resource allocation in full-duplex wireless-powered communication network," *IEEE Transactions on Communications*, vol. 62, no. 10, pp. 3528–3540, Oct 2014.
23. H. Ju, K. Chang, and M. Lee, "In-band full-duplex wireless powered communication networks," in *International Conference on Advanced Communication Technology (ICACT)*, July 2015, pp. 23–27.
24. S. Leng, D. W. K. Ng, N. Zlatanov, and R. Schober, "Multi-objective resource allocation in full-duplex SWIPT systems," in *IEEE International Conference on Communications (ICC)*, May 2016, pp. 1–7.
25. V. Nguyen, T. Q. Duong, H. D. Tuan, O. Shin, and H. V. Poor, "Spectral and energy efficiencies in full-duplex wireless information and power transfer," *IEEE Transactions on Communications*, vol. 65, no. 5, pp. 2220–2233, May 2017.
26. M. Maso, C. Liu, C. Lee, T. Q. S. Quek, and L. S. Cardoso, "Energy-recycling full-duplex radios for next-generation networks," *IEEE Journal on Selected Areas in Communications*, vol. 33, no. 12, pp. 2948–2962, Dec 2015.
27. Z. Wen, X. Liu, N. C. Beaulieu, R. Wang, and S. Wang, "Joint source and relay beamforming design for full-duplex MIMO AF relay SWIPT systems," *IEEE Communications Letters*, vol. 20, no. 2, pp. 320–323, Feb 2016.
28. Y. L. Che, J. Xu, L. Duan, and R. Zhang, "Multiantenna wireless powered communication with cochannel energy and information transfer." *IEEE Communications letters*, vol. 19, no. 12, pp. 2266–2269, 2015.
29. Y. Bi and H. Chen, "Accumulate and jam: Towards secure communication via a wireless-powered full-duplex jammer," *IEEE Journal of Selected Topics in Signal Processing*, vol. 10, no. 8, pp. 1538–1550, Dec 2016.
30. A. A. Okandeji, M. R. A. Khandaker, K. Wong, and Z. Zheng, "Joint transmit power and relay two-way beamforming optimization for energy-harvesting full-duplex communications," in *IEEE GLOBECOM Workshops (GC Wkshps)*, Dec 2016, pp. 1–6.
31. H. Kim, J. Kang, S. Jeong, K. E. Lee, and J. Kang, "Secure beamforming and self-energy recycling with full-duplex wireless-powered relay," in *IEEE Annual Consumer Communications Networking Conference (CCNC)*, Jan 2016, pp. 662–667.
32. PowerCast. [Online]. Available: www.powercastco.com
33. C. System. [Online]. Available: http://www.ossia.com/cota/
34. L. R. Varshney, "Transporting information and energy simultaneously," in *2008 IEEE International Symposium on Information Theory*, July 2008, pp. 1612–1616.
35. X. Lu, P. Wang, D. Niyato, D. I. Kim, and Z. Han, "Wireless networks with RF energy harvesting: A contemporary survey," *IEEE Communications Surveys Tutorials*, vol. 17, no. 2, pp. 757–789, Secondquarter 2015.
36. X. Zhou, R. Zhang, and C. K. Ho, "Wireless information and power transfer: Architecture design and rate-energy tradeoff," *IEEE Transactions on Communications*, vol. 61, no. 11, pp. 4754–4767, November 2013.
37. Y. Chen, K. T. Sabnis, and R. A. Abd-Alhameed, "New formula for conversion efficiency of RF EH and its wireless applications," *IEEE Transactions on Vehicular Technology*, vol. 65, no. 11, pp. 9410–9414, Nov 2016.
38. E. Boshkovska, D. W. K. Ng, N. Zlatanov, and R. Schober, "Practical non-linear energy harvesting model and resource allocation for SWIPT systems," *IEEE Communications Letters*, vol. 19, no. 12, pp. 2082–2085, Dec 2015.
39. X. Xu, A. Özçelikkale, T. McKelvey, and M. Viberg, "Simultaneous information and power transfer under a non-linear RF energy harvesting model," in *IEEE International Conference on Communications (ICC)*, May 2017, pp. 179–184.

40. O. L. A. López, H. Alves, R. D. Souza, and S. Montejo-Śńchez, "Statistical analysis of multiple antenna strategies for wireless energy transfer," *arXiv preprint arXiv:1811.10308*, 2018, https://ieeexplore.ieee.org/abstract/document/8760520.
41. H. G. Schantz, "Near field propagation law & A novel fundamental limit to antenna gain versus size," in *IEEE Antennas and Propagation Society International Symposium*, vol. 3A, July 2005, pp. 237–240 vol. 3A.
42. F. Boccardi, R. W. Heath, A. Lozano, T. L. Marzetta, and P. Popovski, "Five disruptive technology directions for 5G," *IEEE Communications Magazine*, vol. 52, no. 2, pp. 74–80, February 2014.
43. T. Le, K. Mayaram, and T. Fiez, "Efficient far-field radio frequency energy harvesting for passively powered sensor networks," *IEEE Journal of Solid-State Circuits*, vol. 43, no. 5, pp. 1287–1302, May 2008.

Chapter 9
Full-Duplex Transceivers for Defense and Security Applications

Karel Pärlin and Taneli Riihonen

Abstract The full-duplex (FD) radio technology that promises to improve the spectral efficiency of wireless communications was, however, initially used in continuous-wave (CW) radars by means of same-frequency simultaneous transmission and reception (SF-STAR). In this chapter, we explore how the recent advances in the FD technology, which have been mainly motivated by higher throughput in commercial networks, could in turn be used in defense and security applications, including CW radars and also electronic warfare (EW) systems. We suggest that, by integrating tactical communications with EW operations such as signals intelligence and jamming, multifunction military full-duplex radios (MFDRs) could provide a significant technical advantage to armed forces over an adversary that does not possess comparable technology. Similarly in the civilian domain, we examine the prospective benefits of SF-STAR concepts in security critical applications in the form of a radio shield.

9.1 Introduction

In contrast to classical half-duplex (HD) wireless communication models that divide transmission and reception in either time or frequency domain, full-duplex (FD), or otherwise referred to as same-frequency simultaneous transmit and receive (SF-STAR), has the potential to double the spectral efficiency of wireless communications by not requiring such division. In addition to the significant benefits that SF-STAR is capable of delivering in terms of increased throughput in commercial wireless networks, it also has potential uses in defense and security applications [1, 2]. Indeed, the first use of SF-STAR actually emerged from the defense domain in

K. Pärlin (✉)
Rantelon, Tallinn, Estonia
e-mail: karel.parlin@rantelon.ee

T. Riihonen
Unit of Electrical Engineering, Tampere University, Tampere, Finland
e-mail: taneli.riihonen@tuni.fi

© Springer Nature Singapore Pte Ltd. 2020
H. Alves et al. (eds.), *Full-Duplex Communications for Future Wireless Networks*,
https://doi.org/10.1007/978-981-15-2969-6_9

the form of continuous-wave (CW) radars, which have been studied since at least the 1940s [3].

In order to receive echoes from targets simultaneously to transmitting, CW radars require the near-end local leakage, i.e., self-interference (SI), to be reduced similarly to FD wireless communication systems. This had initially been achieved by using separate antennas or circulators in single-antenna systems [3]. Such passive methods, however, provide only moderate isolation which consequently restricts the usable transmission power. In order to increase the radar's working range by amplifying the output power while also limiting the SI, active SI cancellation methods using analog circuitry were developed based on feed-through nulling which attenuated the SI by as much as 60 dB [4]. To potentially double the spectral efficiency in wireless networks, FD radio technology has from thereon evolved to yield wideband SI suppression of up to 100 dB through combination of passive and active methods.

These advances have been recognized by NATO's Science and Technology Organization as its exploratory team has recently completed its report that focuses on how the FD technology can alleviate spectral congestion issues in tactical communications [5, 6]. The report also identifies possible applications in electronic warfare (EW). Most notably, SF-STAR could deliver a paradigm shift in military communications by merging tactical communications with simultaneous electronic attack and defense capabilities, therefore enabling the spectrum resources to be used based on operational circumstances rather than technological limitations. However, a different set of requirements, such as operating frequencies and transmission powers, needs to be considered when designing military radios as opposed to commercial applications, for which the FD radio prototypes have been mostly developed.

Similarly to the potential paradigm shift in military communications, the FD technology can also become central to the security of civilian wireless communications. For example, in the form of a radio shield, simultaneous wireless reception and jamming could be used to prevent eavesdropping on wireless corporate or body area networks. Moreover, the radio shield could be used to prevent unauthorized usage of the radio spectrum to, e.g., restrict remotely controlled unmanned aerial vehicles (UAVs) from entering the airspace covered by the shield. In the security domain, an outstanding challenge is to introduce new capabilities while not requiring any changes to the legacy communication standards. Transferring the FD radio technology from its current state to the military and security domains therefore requires careful planning on how to benefit from SF-STAR operation but also on what are the technical prerequisites for applying FD technology in these domains.

The remainder of this chapter is organized as follows. In Sect. 9.2, we discuss the challenges in transferring the FD technology from its current civilian/commercial state to the military domain and the prospective applications of multifunction military full-duplex radios (MFDRs) in both communication and non-communication systems. In Sect. 9.3, we identify possible security applications of the FD radio technology in commercial systems in the form of a radio shield and briefly reflect on the relation to the information-theoretic physical layer security aspects. Finally, Sect. 9.4 concludes the chapter.

9.2 Applications for Full-Duplex Radios in Military Communications

Most of the ongoing FD research focuses on improving SI cancellation methods, studying the physical layer security aspects from an information-theoretic viewpoint, or developing scheduling and routing algorithms that can leverage the SI cancellation for commercial applications by improving spectral efficiency. Unlike commercial systems, however, their military counterparts are required to perform in adverse propagation environments and hostile conditions. Such circumstances place generally more rigorous requirements on the radios but also present new applications for the FD technology in the form of MFDRs, including those illustrated and categorized in Figs. 9.1 and 9.2, respectively.

In the following, we first discuss the requirements for military radios in general and also from the viewpoint of the FD radio technology in particular. We then consider the advantages of MFDR radios over conventional HD military radios in combinations of tactical communications with EW and also in tactical communica-

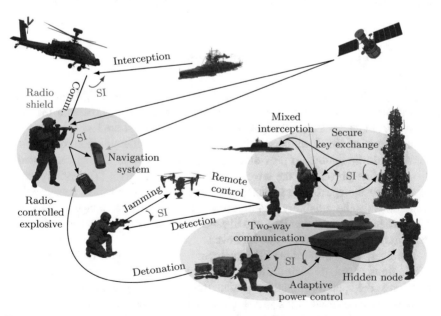

Fig. 9.1 Conceptual use of military full-duplex radios in the battlefield for communications and electronic battle. Self-interference is abbreviated as SI. © [2018] IEEE. Reprinted, with permission, from [2]

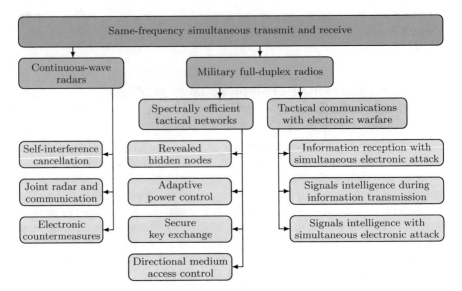

Fig. 9.2 Same-frequency simultaneous transmit and receive applications in the military domain

tion networks. We also present an overview of CW radars and multifunction radios together with potential uses of the FD radio technology in those applications.

9.2.1 Requirements for Military Radios

Typically military radios share the physical and electromagnetic (EM) environment with radars, EW applications, and navigation systems. Not to mention the interference from adversarial radio systems that further congest the EM spectrum. The environment, in which military radios are required to operate, therefore imposes considerable limitations to providing host forces the use of EM spectrum and at the same time preventing the adversary from doing likewise [7]. Military radios need to use the spectrum efficiently to fulfill the communication needs without compromising the reliability requirements [8]. By taking advantage of the recent advances in FD radio technology and SI cancellation in particular, spectral efficiency in military radios can possibly be improved.

However, so far most of the FD prototypes have been designed with commercial applications in mind. Main differences between military and civilian radios, in addition to the operating conditions, arise from the used frequency bands. Typically military radios operate in the very high frequency (VHF) or high frequency (HF) bands whereas nearly all academic FD prototypes demonstrate SI cancellation in the upper ultra high frequency (UHF) bands only. Additional studies are needed to confirm the feasibility of FD radios at military frequencies, but also at higher transmission powers. Military radios can require much higher output powers than what has been proven usable in laboratory environments so far. Moreover, the inherent mobility of tactical units requires the radio's size, power consumption, and weight to be kept at minimum while other requirements include the need for higher bandwidth, lower latency, and security [8].

The security considerations in military radios are of paramount importance not only to their operation, but also to the integrity and survival of the physical systems that they support [9]. Hence it is desirable for military radios to have low probability of detection (LPD), low probability of interception (LPI), good jamming resistance, and means to obfuscate the communicated information from potential eavesdroppers. Classically, LPD, LPI, and jamming resistance have been achieved by the use of spread spectrum techniques and adaptive power control [10] while intelligence is typically obfuscated through the use of encryption, which relies on secure key exchange protocols and the adversary's limited computing power [11]. In addition to protecting one's own communications, hindering the enemy's radio correspondence is an important aspect to consider in the electronic battlefield.

Hostile operating conditions not only affect the point-to-point links between military radios but also impose stringent requirements on the networks in which those radios operate. Tactical networks have highly time-variant topologies and are expected to work in a self-forming, self-healing, infrastructure-less manner without sacrificing data rate, latency, nor node mobility. Such requirements have motivated the design of decentralized routing and scheduling protocols, which can in turn be enhanced by the FD radio technology. Still, developing cognitive algorithms that comprise of power control, spectrum management, electronic combat tasks, and network topology adjustment for tactical networks is one the most challenging aspects of designing radios for future military communications [12].

9.2.2 Tactical Communications with Electronic Warfare

In the military domain, EW provides means to oppose and resist hostile actions that involve the EM spectrum in all battle stages. It is an important avenue in advancing desired military objectives or, on the contrary, hindering undesired ones and improving the survivability of the host force [7]. Effective use of EW countermeasures relies on signals intelligence and reconnaissance while EW as a whole consists of the following interrelated operational functions:

- electronic attack (EA), which involves the offensive use of EM energy to reduce the enemy's battle capabilities;
- electronic protection (EP), which protects the host forces from the opponent's EAs through EM countermeasures;
- electronic support (ES), which combines surveillance and reconnaissance of the EM environment in order to provide information for EA and EP.

Classically, EW functions have been separated from tactical communications in time or frequency domain, so that the host forces' use of the EM spectrum for tactical communication is not obstructed. However, use of the SF-STAR capability in military radios would not only enable spectrally efficient two-way information exchange but also allow armed forces to merge tactical communications with EW and so introduce novel combat tactics. Through such combinations, the radios could either receive or transmit communication signals while at the same time conducting EW tasks in the opposite direction. As the pioneering works dedicated to exploring the potential benefits of MFDRs, [1, 2] provide insight into such combinations and how they could present armed forces with a significant technical advantage over an adversary that does not possess comparable technology. In the following, we consider those combinations in detail.

9.2.2.1 Simultaneous Communication and Jamming

Deploying EW systems, such as jammers, against radio-controlled (RC) improvised explosive devices (IEDs) or UAVs, can significantly help in protecting the personnel and platforms from those threats. However, jammers can inadvertently interfere with the host's communication systems that operate in the close vicinity [13]. Suppressing the EM interference in the communication systems caused by jamming is therefore a crucial challenge with high technical complexity and operational significance. Ordinarily, frequency-based separation with fixed filters or time division is used to alleviate the EM interference. Such methods, however, limit the spectral efficiency and, in case of frequency-division duplexing, require duplex filters to be changed in accordance to the environment and threats. Whenever jamming is carried out alternately in time with tactical communications, it presents the opponent with similar possibilities to use the EM spectrum. This results in inefficient use of the EM spectrum and can severely limit the efficiency of the EA.

It is therefore desirable to enable simultaneous same-frequency communication and jamming [13], which is exactly what recent advances in SI cancellation facilitate. The cancellation techniques allow a FD transceiver to simultaneously transmit a jamming signal and receive tactical communication signals on the exact same frequency, therefore preventing opponents in the FD transceiver's proximity from using the frequency band. Numerical results in [1] illustrate the gain margins which tactical forces could therefore achieve. Mitigation of in-band interference from co-located jammers through SI cancellation techniques at the communication system's receiver has been demonstrated to enhance the reception of signals of interest [14]. Such jamming could be used to block the enemy from detonating radio-controlled IEDs or operating UAVs while the host could still receive communications from allied forces. Another conceivable use case in the battlefield would be to jam or spoof the adversaries' reception of navigation satellite system signals while itself retaining the ability to receive such signals and consequently the positioning capabilities.

9.2.2.2 Simultaneous Interception and Communication

Similar to the above case, FD radio technology makes it also possible to combine signals intelligence with tactical communications. Compared to the combination of communication with jamming, this is a somewhat different task because communication systems' transmitters usually do not use as high output power as jammers. Therefore, the integration of current SI cancellation techniques to MFDRs could already suffice to achieve simultaneous interception and communication, given that those techniques can be transferred from the UHF to HF and VHF bands. Such combination would facilitate devices which perform spectrum monitoring and signal surveillance to, e.g., transmit the gathered intelligence to other tactical units without compromising the surveillance capabilities during transmission. Otherwise, when considering conventional HD radios that carry out surveillance and communications at the same frequency in an alternating pattern in time, the opponent would have a chance of hiding its communications by transmitting at the same time as the signal intelligence unit. It has been highlighted in [1] that performing simultaneous interception with information transmission does not degrade the host's communication link and therefore the interception comes almost at no cost if the transceiver has effective SF-STAR capability.

9.2.2.3 Simultaneous Interception and Jamming

Although not strictly a combination of tactical communications and EW, simultaneous interception and jamming can, e.g., be used to degrade the quality of a communication link between adversaries which is at the same time being intercepted. Reduction in communication link quality can lead the opponents to inadvertently increasing their transmission power in order to sustain the communication link. By carefully choosing the jamming power, it is therefore probable that the interception quality becomes better with simultaneous jamming despite the residual SI as a result of the opponent's countermove [1]. The feasibility of such strategy has already been demonstrated in a laboratory environment by successfully degrading the opponent's reception quality while retaining the ability to intercept it [15].

On the other hand, being able to receive and analyze the targeted communication link under jamming allows one to adapt the jamming waveform to the targeted signals. For example, *a priori* knowledge about UAV remote control systems has been shown to aid in designing effective jamming signals against those systems [16]. Instead of requiring the jammer to have the knowledge beforehand, similar effect could be achieved by gathering such knowledge while jamming through the use of SF-STAR. This would be especially beneficial against systems for which the reaction to jamming cannot be anticipated or known in advance. Thus, replacing conventional jammers which either transmit a wideband jamming signal or alternate between monitoring and jamming stages. Furthermore, by using such target aware jamming, it can become much more difficult for the opponent to detect that it is being jammed [17].

9.2.3   Tactical Communication Networks

Tactical communications in the battlefield result in highly time-varying topologies and typically ad hoc networks, such as the packet radio network (PRN) and mobile ad hoc network (MANET), are considered suitable for connecting tactical units. Ad hoc networking aims to provide a flexible method for establishing communications in scenarios that require rapid deployment of survivable and efficient dynamic networking [18]. Furthermore, ad hoc networks are attractive because they do not require infrastructure and tactical operations often take place in locations where infrastructure is lacking [19], or rendered inaccessible. Tactical MANETs are expected to provide completely self-forming, self-healing, and decentralized platforms for tactical units to join and leave swiftly.

Aside from the dynamic topologies, tactical networks typically also require LPD and LPI. To achieve that, impulse PRNs have been considered because impulse radios' ultra-wideband spectrum usage offers potentially covert operation [20]. Even before the recent advances in SI cancellation techniques, the idea of FD impulse PRNs was studied to combine the covertness of impulse radios with the increased network throughput of FD radios [21]. In order to allow bidirectional information transfer, the FD impulse PRN technology proposes to blank the receiving front-end during transmissions at the expense of some degradation in the received signals. However, due to the nature of impulse radios, as long as the transmitted and received pulses do not completely overlap, information can be exchanged.

Although the concept of FD impulse PRNs does not rely on the true FD radio technology as considered herein, the idea already emphasized the benefit that the true FD radio technology can bring in tactical networks in terms of improved throughput [22]. However, due to the typically asymmetrical data flow, imperfect SI cancellation, and increased inter-node interference, the improvement in throughput may not always be remarkable. Nevertheless, as discussed next, the FD radio technology also has the potential to improve several other aspects of tactical networks which in turn can enhance situational awareness and network security.

9.2.3.1   Hidden Node

One of the most prominent challenges in tactical and also commercial ad hoc networks is the hidden node issue since it is a major source of collisions. The hidden node, or sometimes referred to as the hidden terminal, issue arises when a node is not aware that the recipient, to whom it is about to start transmitting, is already receiving signals because those signals are not reaching the node which intends to transmit. In this case, the two nodes that have information to transmit to a common node are hidden from each other. If the second node were to also start transmitting then the recipient would receive mixed signals and not be able to make sense of either of those transmissions, which in turn would result in decreased network throughput and increased latency as information has to be retransmitted.

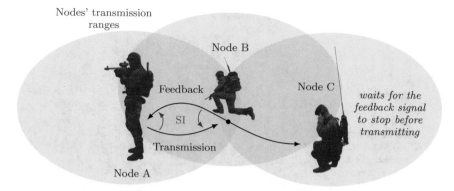

Fig. 9.3 Application of military full-duplex radios to prevent the hidden node problem from occurring in tactical ad hoc networks. Self-interference is abbreviated as SI

To solve this problem, a busy-tone scheme which uses a separate wireless channel to acknowledge the ongoing transmission was initially proposed [23, 24]. This scheme is able to eliminate collisions, but the requirement of allocating a separate wireless channel for collision avoidance only makes it impractical in real ad hoc networks. A pragmatic and widely accepted solution is the use of request to send/clear to send (RTS/CTS) mechanism before data transmission [25]. This way, both parties, the transmitter and the receiver, acknowledge to all nodes in their transmission range that they are about to start communicating. This results in performance increase by reducing the number of collisions and required retransmissions, while on the other hand, this method also introduces considerable overhead in the form of the RTS/CTS exchange. In case the network does not have any hidden nodes, such prior exchange is redundant and prevents the network from achieving the otherwise highest possible throughput.

Similarly to the busy-tone scheme which uses a second frequency to acknowledge the reception with a feedback signal, the FD radio technology enables the recipient to acknowledge the reception with simultaneous transmission but on the exact same frequency. As illustrated in Fig. 9.3, the recipient can consequently inform any nodes in its range about ongoing communications and therefore prevent the hidden node issue from occurring [26, 27]. Furthermore, since simultaneous listening and sensing is being performed on a frequency band while the signals are being transmitted, each node can decide whether or not the other nodes have simultaneously started transmitting and thus prevent multiple access collisions [28].

9.2.3.2 Adaptive Power Control

By facilitating simultaneous two-way information exchange, the FD radio technology significantly reduces latency and end-to-end delays in wireless networks [27]. Lower latency enables tactical networks to employ faster adaptive power control

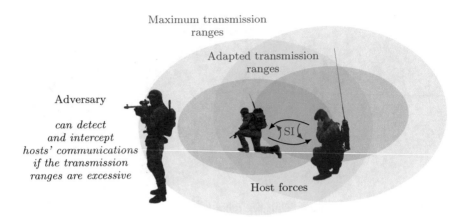

Fig. 9.4 Application of low latency military full-duplex radios with adaptive power control towards low probability of detection and interception. Self-interference is abbreviated as SI

so that the radio links do not use excessive output powers for extended periods of time. This could possibly improve battery life and reduce inter-node interference in multi-hop networks [29, 30]. More importantly in the context of military wireless communications, however, fast adaptive power control can help keep the transmission range as small as possible and therefore lower the probabilities of detection and interception as illustrated in Fig. 9.4. Adapting the transmit power can also reduce the SI in FD radios and therefore improve the reception quality in some cases [31].

9.2.3.3   Secure Key Exchange

As was stressed when discussing requirements for military radios, a prerequisite for securely encrypted communications in wireless networks is secure key exchange. However, due to the broadcast nature of the wireless medium it is not trivial to achieve wirelessly. If an adversary intercepts a wireless key exchange, then it can decrypt the following communications encrypted with that key. Works on secret key extraction from radio channel measurements have demonstrated that two devices can generate shared keys based on the channel variations between the devices [32, 33]. The key generation rate with such methods, however, depends on the rate of channel variations and can be low in static environments. Furthermore, methods which rely on channel variations are susceptible to disagreements about the generated keys between the two devices. An alternative method relies on sending the key twice, each time jamming different parts of the key by the receiver and assuming that the eavesdropper cannot discern which parts have been jammed during either transmission [34].

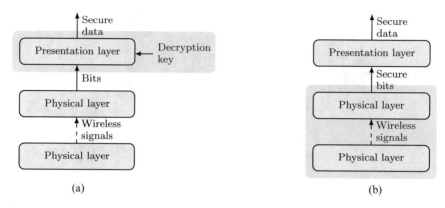

Fig. 9.5 Cryptographic and physical layer security approaches to securing wireless communications. (**a**) Encryption based security. (**b**) Physical layer security

Simultaneous reception and jamming that is facilitated by the FD technology simplifies such key exchange methods to require the key to be transmitted only once [35]. Adversary then receives superposed signals that are difficult to separate and consequently is prevented from intercepting the key. Incorporation of such key exchange schemes in military networks could enable secure wireless key exchange with reduced risk of enemy's signals intelligence decrypting the host's communications should they successfully intercept any. Figure 9.5 illustrates how the FD radio technology enables exchanging secure messages by shifting the security focus from the upper communication layers to the physical layer. A significant benefit of physical layer security compared to cryptographic methods is that physical layer security does not rely on the opponent's limited computational capabilities and therefore the applications for such methods go beyond secure key exchange [36].

9.2.3.4 Directional Medium Access Control

To increase jamming resistance and lower the detection and interception probability in tactical networks, directional medium access control (MAC) protocols have been proposed [37]. This approach aims to concentrate the transmission power towards the intended recipient through beamforming [38]. A significant challenge in applying directional protocols in dynamic topologies is to keep a good estimate of the direction of the intended receiver. To that end, several solutions have been proposed, mostly using a variation of the RTS/CTS exchange to let both the source and destination nodes determine each other's directions [39–41]. However, the performance of such schemes can be expected to degrade as the node mobility increases [41].

It is reasonable to envision that SF-STAR is used so that the transmitter processes the feedback signal from the intended receiver to update the estimated direction

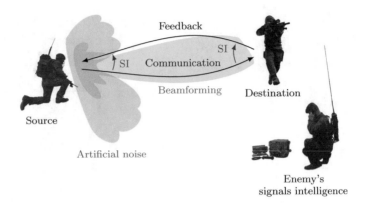

Fig. 9.6 Applying the military full-duplex radio technology to improve direction estimation in directional medium access control protocols for improved throughput and security. Self-interference is abbreviated as SI

of the recipient simultaneously to transmitting as illustrated in Fig. 9.6. Similar concept has been evaluated based on a retrodirective array system that enables FD communication and high-speed beam tracking [42]. Therefore, the node mobility issue that has been of a concern in directional MAC protocols so far can possibly be solved by enabling SF-STAR operation. Additionally, artificial noise can be transmitted in the surrounding directions to further ensure LPI [38, 43]. In static environments and network topologies, the combination of directional antennas with the FD technology has been analytically shown to increase network throughput [44], while beamforming improves the secrecy rate of FD point-to-point links [45].

9.2.4 Continuous-Wave Radars

Radars use high-power radio frequency (RF) transmissions ranging from HF to millimeter-waves (mmWaves) in order to illuminate targets by collecting the reflected echoes in either pulsed or CW modes. The received echoes are used to determine each target's location and velocity, which can be used in both offensive and defensive weapon systems to control and direct the weapon at the target [7]. In pulsed radars, the RF front-end is switched from transmission to reception mode to transmit and then receive the pulse without interfering with itself. In CW radars, echoes from the targets are received simultaneously to transmitting, which causes direct leakage from the radar's transmitter to its receiver that needs to be suppressed by some form of SI cancellation [46]. In that sense, CW radars are quite similar to FD radios.

9.2.4.1   Self-Interference Cancellation

Even though military radars typically operate with much higher frequencies [7] than the currently reported academic FD radio prototypes, many of the SI cancellation solutions could be potentially applied also in low-power CW military radars [1]. More so because typically radar systems require less isolation than FD data transfer applications. However, efficient SI cancellation is not the only challenge in military radars. Radar and data communications are often opposing one another and compete for the same spectral resources, which can result in degradation of sensitivity in the radar or communication systems.

Recent results suggest that by co-designing the radar and communication systems from the ground up, the scarce RF resources could be shared by those seemingly conflicting applications [47]. Based on the advances in SI cancellation, a method for cancelling the radar-induced interference to enable spectrum sensing has been presented in [48]. Classically such coexistent systems could only operate in a time-multiplexed manner, preventing either system from continuously carrying out its task. However, by using the cancellation methods, the known radar signal can be sufficiently suppressed in adjacent receivers. It is reasonable to envision that not only spectrum sensing can be achieved simultaneously to the radar operation but also receiving wireless communications as illustrated in Fig. 9.7.

Besides suppressing the interference caused by radars in co-located receivers, there is significant interest in using the radar waveforms for both object detection and information transmission [49–51]. Such joint radar and communication systems typically study the use of waveforms, such as direct-sequence spread spectrum (DSSS) and orthogonal frequency-division multiplexing (OFDM), which are similar to those used in experiments with research prototype FD transceivers. Such joint radar and communication platforms could therefore take advantage of the SI cancellation techniques to improve near-end local leakage suppression in the radar to improve the radar performance but also to suppress the reflected radar signals

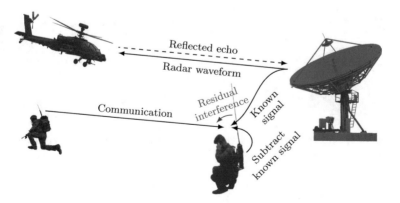

Fig. 9.7 Concurrent radar and communication operation by using the same-frequency simultaneous transmit and receive methods to suppress known radar signals

in order to receive communication signals in the same frequency band. The latter combination is essentially the same as the FD technology used in wireless communications to improve the spectral efficiency.

9.2.4.2 Electronic Countermeasures

In order to evade an opponent's radars, electronic countermeasures (ECMs) such as suppression jamming and deception jamming are often used. Suppression jamming is exercised to impair the opponent's ability to detect objects in the operational environment [52], while deception jamming, which is arguably more difficult to perform, is used to mislead the enemy about the operational environment [53]. For example, through false target generation or delayed radar signal replaying, by use of the digital radio frequency memory (DRFM), the target could be shown to be at a different distance altogether [54]. Through velocity or angle deception, the target could be shown to be moving with a different speed than it actually is or prevent the correct angle from being detected.

To circumvent and detect ECMs in radars, electronic counter-countermeasures (ECCMs) such as frequency agility, frequency diversity, and jamming cancellation through various signal processing techniques are employed [53, 55]. These methods rely to some extent on the jammer's incapability to quickly respond to changes in the radar signal. By integrating the SF-STAR capabilities into radar ECM systems, those systems could simultaneously receive the radar signal and transmit a spoofed echo back. Given adequate signal processing abilities, SF-STAR therefore enables the ECM systems to adapt to the radar signal in real time and possibly evade the aforementioned ECCMs as illustrated in Fig. 9.8.

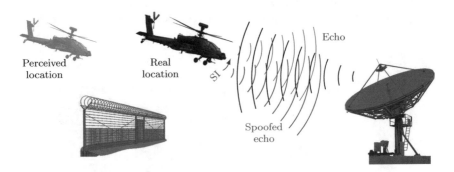

Fig. 9.8 Use of same-frequency simultaneous transmit and receive in electronic countermeasures against radars Self-interference is abbreviated as SI

9.2.5 Multifunction Radios

The military domain is characterized by long-term acquisitions, while missions and technical requirements change at quicker rates, therefore the ability to upgrade and reconfigure radio systems through software rather than hardware is highly sought after [56]. Concepts like the joint tactical radio system (JTRS) have focused on replacing aging legacy radios with a single, versatile system based on software defined radio (SDR) [57, 58]. Thus, enabling the radio to be upgraded or modified to operate with other communications systems by the addition or reconfiguration of software as opposed to redesigning or changing hardware. Depending on the mission requirements, each JTRS is envisioned to be capable of executing different waveforms or communication standards, therefore enabling collaboration between otherwise incompatible systems [59].

Furthermore, integrating multiple communication and non-communication tasks simultaneously in the form of advanced multifuncion radio frequency concept (AMRFC) and subsequently integrated topside (INTOP) have been proposed [60–62]. Those concepts encompass the integration of RF functions, such as radar, EW operations, and communications, into a single system utilizing a common set of hardware (as illustrated in Fig. 9.9) for which the functionality is programmed as necessary. The potential benefits of such multifunction systems include reduced number of antennas, increased potential for future growth without adding new aperture therefore resulting in significantly lower upgrade costs, and better control over EM interference through agile and intelligent frequency management. However, the ultimate power of multifunction military radios lies in the ability to adapt the functionality together with key parameters of the equipment to the current tactical operations [61].

Conventional HD single-function systems are able to operate at peak performance by applying various isolation techniques that are tailored to each individual system. However, most of those techniques cannot be directly applied when a single aperture performs multiple functions [63]. So far, multifunction military RF

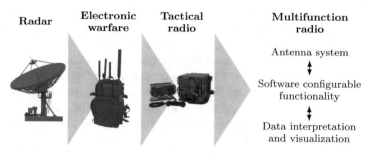

Fig. 9.9 Integration of multiple functions, including radar, electronic warfare, and communications, into a shared set of antennas and signal processing hardware to provide radio functionality depending on the operational needs

systems have mostly relied on separation of transmit and receive antennas to provide moderate isolation between those paths and consequently a key topic for further refinement of multifunction RF systems is to employ improved transmit-to-receive isolation techniques [61]. Therefore FD radio technology can become an elemental part of the multifunction radio vision because it potentially allows transmit and receive functions, whatever they are, to operate simultaneously.

9.3 Applications for Full-Duplex Radios in Civilian Security

When considering the civilian security domain instead of the electronic battlefield, defensive applications rather than offensive ones are paramount. Another significant difference is the fact that many military communication systems operate in the HF and VHF bands while their commercial counterparts work in the UHF band. In that sense, the existing FD prototypes can be more readily applied in the civilian security domain rather than in the military. The malicious wireless communications to be considered in the civilian security domain are, e.g., unauthorized use of remotely controlled UAVs near restricted areas and eavesdropping on or tampering with private wireless communications.

9.3.1 Radio Shield

In order to counter the aforementioned threats, the FD radio technology can be exploited through jamming to propagate a protective electromagnetic field, i.e., a "radio shield," around the transceiver. The jamming prevents any third party within the shield from successfully receiving wireless transmissions while the transceiver's own reception of any other transmissions is unaffected. Moreover, if using a known pseudo-random jamming signal, any other authorized device can also cancel the jamming signal and thereby be capable of transmission and reception inside the radio shield. A conventional HD jammer on the other hand cannot receive at the same frequencies while transmitting and this leads to potentially dangerous situations, e.g., when the malicious wireless communications use the same frequencies as the law enforcement. Using conventional jammers, law enforcement then has to decide whether to block or allow all communications, including their own.

The radio shield could be useful for any common wireless device, including mobile terminals and network infrastructure. For example, the radio shield could be useful in a corporate environment to prevent unintentional information leakage, decreasing the risk of improper or lacking use of encryption. Such wireless physical layer firewalls have been previously proposed on the basis of reactive jammers, which rely on first analyzing the wireless communications and begin to jam when the communication is deemed obtrusive [64]. In case of FD jamming transceivers, it is also possible to carry out simultaneous spectrum surveillance. The transceiver would therefore be able to detect and identify malicious users who attempt to communicate within the radio shield despite being prohibited from doing so.

However, the predominant challenge in implementing a radio shield is maintaining backwards compatibility with the existing communication systems. This means that the radio shield blocks unwarranted communications while at the same time allowing authorized users to continue using legacy communication standards as if there was no SI. Furthermore, colluding eavesdroppers present a security risk [65] as the radio shield requires the number of antennas transmitting artificial noise or jamming signal to exceed the number of eavesdropper antennas [43].

Several works have already been published regarding the radio shield and they mainly divide into two separate categories: the information-theoretical works, where the secrecy rate under jamming is formally investigated and signal processing works which provide results with high practical value. The latter is mainly focused on the following topics that exemplify the potential value of a FD radio shield in civilian wireless security.

9.3.1.1 Drones

Due to the increased availability of consumer-grade UAVs, it has become necessary to restrict their unauthorized use in areas where they might cause accidents or be used for malicious purposes. Disabling UAV remote control links by wideband jamming while simultaneously retaining the ability to receive communications [66] or detect such links [67, 68] has been shown feasible with the FD radio technology. Consequently, the FD transceiver is also able to detect and identify malicious users who attempt to remotely control UAVs within the radio shield despite being prohibited from doing so. Ideally such restrictions should not prevent authorized UAVs from operating in the same space and if the radio shield used pseudo-random jamming signals, then authorized UAVs could cancel its effect using co-located interference cancellation methods [69] as envisioned in Fig. 9.10. From a non-security perspective, the FD radio technology enables UAVs to form efficient ad hoc networks [70].

9.3.1.2 Wireless Energy Transfer

The fundamental challenge in enabling the ever-growing number of wireless devices part of the Internet of Things (IoT) to communicate is in developing protocols that enable energy-efficient communications between devices without interfering with one another. Acquiring energy from RF signals has opened the way for unified wireless power transmission and communication since those signals carry energy and information simultaneously. Combining such energy harvesting with the FD radio technology potentially enables nodes to power simultaneous reception and transmission from the received signal [71], while at the same time reducing multiple access collisions and improving transmission throughput [72, 73]. The nodes could

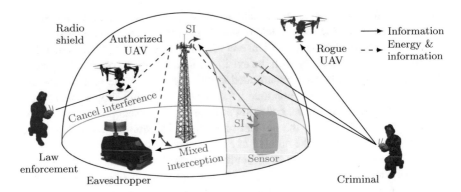

Fig. 9.10 Conceptual use of full-duplex radio shield for wireless power transfer and restricting unauthorized use of the radio frequency spectrum. Self-interference is abbreviated as SI. © [2018] IEEE. Reprinted, with permission, from [2]

be powered from base stations or even from UAVs that could act as FD relays [74]. Therefore, the radio shield can conceivably prevent unauthorized spectrum usage or eavesdropping inside the protective dome, while authorized devices can harvest energy and communicate as illustrated in Fig. 9.10. By adopting beamforming instead of omnidirectional methods, both energy harvesting and SI cancellation capabilities can be increased at FD transceivers [75].

9.3.1.3 Medical Devices

Wearable medical sensors and implanted medical devices (IMDs) are also going through rapid development as they promise to revolutionize healthcare in the form of wireless body area networks (WBANs). However, among other challenges, such as power consumption and aesthetic issues, WBANs face the need to secure the wireless communications from eavesdropping and tampering. Typically, encryption is being considered as a solution [76], yet, concerned by the lack of encryption in existing devices, methods based on FD and reactive jamming have been presented [77, 78]. In such methods, the IMD user wears an additional device—the radio shield generator, which acts as a secure gateway for external devices that want to communicate with the IMD. The radio shield, as an external device, can establish secure connection to a legitimate reader more conveniently than the IMD. The shield jams unauthorized transmission to the IMD or transmissions from the IMD, preventing perpetrators from gaining access to the IMD as illustrated in Fig. 9.11. However, such radio shield does lend itself to attacks from adversaries with multiple reception antennas, as the single-antenna radio shield cannot provide strong confidentiality guarantees in all settings where the attacker can be freely positioned [79].

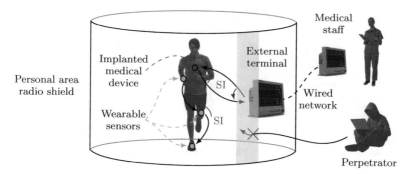

Fig. 9.11 Conceptual use of full-duplex radio shield to provide physical layer security in wireless body area networks and for medical implanted devices. Self-interference is abbreviated as SI. © [2018] IEEE. Reprinted, with permission, from [2]

9.3.1.4   Automotive Radars and Vehicle-to-Vehicle Communications

The automotive industry is also seeking to take advantage of the RF spectrum as the industry is edging towards self-driving cars. To that end, two technologies in particular are essential: automotive radars and vehicle-to-vehicle communications. Radars have been already deployed on consumer vehicles to avoid collisions and provide some self-driving features [47], while vehicular ad hoc network (VANET) protocols are being developed by the automotive industry to provide vehicle operators a better overview of the environment [80, 81]. For example, such communication methods could be used to warn the driver of an accident ahead.

Compared to the previous topics, confidentiality of wireless communications in VANETs is not as important as the authenticity of the information and therefore the physical layer security is typically not considered [82]. However, spectrum congestion and multiple access collisions are as significant issues as they are elsewhere. The proposed VANETs are based on the exchange of periodic cooperative awareness messages (CAMs) and transmitting such messages in highly dynamic network topologies can result in collisions which in turn makes the data transmission unreliable. The use of the FD radio technology can considerably improve the reliability of CAM delivery [83, 84] as simulated results indicate improvements compared to HD broadcasting techniques and cancellation of SI has been successfully demonstrated in a realistic multipath environment on a moving vehicle [85].

Spectrum congestion has motivated studies on the coexistence of automotive radar and communication technologies since they are so closely related. Consequently, radar waveforms can be coded with information without negative influences on the radar performance [50, 86]. Since the feasibility of such waveforms in FD radios has already been demonstrated with numerous prototypes, the combined radio in vehicles could be transmitting the CAMs and use the echoes for object detection or suppress the echoes and receive messages from other vehicles. In

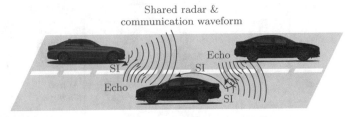

Fig. 9.12 Application of the full-duplex radio technology to enable simultaneous radar and communication capabilities for enhanced spectrum reuse and public safety in the automotive domain. Self-interference is abbreviated as SI

automotive applications, the radio shield could therefore consist of shared radar and communication waveforms that the vehicle uses to detect and track objects inside the shield, while at the same time communicating with other vehicles in the close vicinity as illustrated in Fig. 9.12.

9.3.2 Physical Layer Security

Practicality of the FD radio shield concept has already been demonstrated through experimental results as covered in the previous section. However, these studies have been complemented to a great extent by the physical layer security research incorporating the FD radio technology from an information-theoretic viewpoint. Information-theoretic studies on physical layer security in general have existed long before the emergence of FD radio technology. Most notably the introduction of the wiretap channel and subsequently the Gaussian wiretap channel sparked interest in this field [87, 88]. The fundamental principle behind physical layer security that resulted from these works is that the secrecy capacity of a wireless communications system is inherent in the difference between the channel capacities of the intended and wiretap channels. Non-zero secrecy capacity can only be achieved if the wiretap channel is of lower quality than the channel between the transmitter and the intended receiver. Furthermore, the emergence of multiple-input multiple-output (MIMO) systems led to the realization that the secrecy capabilities of wireless systems could be enhanced by taking advantage of the available spatial dimensions [89].

Assuming that the receiver operates in the HD mode, solutions against eavesdropping have been proposed, e.g., through the use of cooperating jammer nodes that confuse the eavesdropper [90]. Although cooperation has been shown to significantly improve the system security as compared to transmission without cooperation, then in order to effectively use cooperative jammers, challenges such as external node mobility, synchronization, and trustworthiness need to be addressed. By making use of the FD mode at the receiver, i.e., the possibility to transmit jamming noise simultaneously to receiving data as in case of the radio shield, the need for external cooperating nodes together with the respective challenges is eliminated while still degrading the eavesdropper channel. Even more, simultaneous data reception and jamming possibly allows to hide the existence of the communication and thus provide physical layer privacy, something that is not typically considered in the information-theoretic physical layer security works but is emphasized in the signal processing-specific research.

Applications for which physical layer security through SF-STAR operation has been considered include increasing the security against eavesdroppers between point-to-point links [91], in relay networks [92], and in cellular base stations [93]. That being said, the use of the FD technology with regard to physical layer security has also been explored for offensive scenarios in the form of active eavesdroppers [94, 95]. The idea being that an active eavesdropper with FD capabilities can degrade the channel between the transmitter and receiver, therefore also reducing the secrecy rate of the system. Thus, active eavesdropping imposes a more significant challenge as compared to conventional passive eavesdropping from the wireless communications security perspective.

Herein we have given only a brief introduction to the physical layer security research problem and to how the information-theoretic research involving the FD technology in that sense relates to the signal processing research efforts. The information-theoretic research with regard to FD technology is considered in more detail in Chap. 10, which specifically focuses on resource allocation within multiuser FD communication systems in order to secure simultaneous downlink and uplink transmission.

9.4 Conclusion

The importance of electronic warfare (EW) is on the rise and it further establishes the electromagnetic (EM) spectrum as an operational environment, in which tasks must be coordinated and collaborated to enhance the capability to advance tactical and strategical aims. As the sophistication of EW increases so does the importance of the underlying technologies. Consequently, the radio frequency (RF) technology community is challenged with the task of delivering the technological base for EW systems to form a solid framework for conducting operations. Encouraged by the recent advances of the full-duplex (FD) radio technology in the wireless networking domain, we anticipate this technology not only to award spectrally efficient wireless communications but also to pave the way for combinations of non-communication and communication tasks in the military domain. Thus, this chapter surveyed the perspectives of military full-duplex radios (MFDRs) in electronic battlefields, combining tactical communications with EW operations. We have also reviewed possible related defensive applications in the civilian security field. Arguably the FD radio technology can provide a key technical advantage in either domain over an opponent or a perpetrator that is limited to employ the conventional half-duplex (HD) radio technology.

References

1. T. Riihonen, D. Korpi, O. Rantula, H. Rantanen, T. Saarelainen, and M. Valkama, "Inband full-duplex radio transceivers: A paradigm shift in tactical communications and electronic warfare?" *IEEE Communications Magazine*, vol. 55, no. 10, pp. 30–36, Oct. 2017.
2. K. Pärlin, T. Riihonen, R. Wichman, and D. Korpi, "Transferring the full-duplex radio technology from wireless networking to defense and security," in *Proc. 52nd Asilomar Conference on Signals, Systems and Computers*, Oct. 2018.
3. A. Sabharwal, P. Schniter, D. Guo, D. W. Bliss, S. Rangarajan, and R. Wichman, "Inband full-duplex wireless: Challenges and opportunities," *IEEE Journal on Selected Areas in Communications*, vol. 32, no. 9, pp. 1637–1652, Sep. 2014.
4. F. O'Hara and G. Moore, "A high performance CW receiver using feedthrough nulling," *Microwave Journal*, vol. 6, pp. 63–71, Sep. 1963.
5. M. Adrat, R. Keller, S. Wilden, V. L. Nir, T. Riihonen, M. Bowyer, and K. Pärlin, "Full duplex radio: Increasing the spectral efficiency for military applications," NATO, Tech. Rep., Jan. 2020.
6. M. Adrat, R. Keller, M. Tschauner, S. Wilden, V. L. Nir, T. Riihonen, M. Bowyer, and K. Pärlin, "Full-duplex radio technology – Increasing the spectral efficiency for military applications," in *Proc. International Conference on Military Communications and Information Systems*, May 2019.
7. A. E. Spezio, "Electronic warfare systems," *IEEE Transactions on Microwave Theory and Techniques*, vol. 50, no. 3, pp. 633–644, Mar. 2002.
8. N. Suri, G. Benincasa, M. Tortonesi, C. Stefanelli, J. Kovach, R. Winkler, U. S. R. Kohler, J. Hanna, L. Pochet, and S. Watson, "Peer-to-peer communications for tactical environments: Observations, requirements, and experiences," *IEEE Communications Magazine*, vol. 48, no. 10, pp. 60–69, Oct. 2010.
9. S. M. Al-Shehri, P. Loskot, T. Numanoğlu, and M. Mert, "Comparing tactical and commercial MANETs design strategies and performance evaluations," in *Proc. IEEE Military Communications Conference*, Oct. 2017, pp. 599–604.
10. R. Poisel, *Modern Communications Jamming: Principles and Techniques*. Artech House, 2011, pp. 5–10.
11. A. Mukherjee, S. A. A. Fakoorian, J. Huang, and A. L. Swindlehurst, "Principles of physical layer security in multiuser wireless networks: A survey," *IEEE Communications Surveys & Tutorials*, vol. 16, no. 3, pp. 1550–1573, Aug. 2014.
12. H. Saarnisaari and T. Bräysy, "Future military mobile radio communication systems from electronic warfare perspective," in *Proc. International Conference on Military Communications and Information Systems*, May 2017.
13. G. Karawas, K. Goverdhanam, and J. Koh, "Wideband active interference cancellation techniques for military applications," in *Proc. 5th European Conference on Antennas and Propagation*, Apr. 2011, pp. 390–392.
14. S. Enserink, M. P. Fitz, K. Goverdhanam, C. Gu, T. R. Halford, I. Hossain, G. Karawasy, and O. Y. Takeshita, "Joint analog and digital interference cancellation," in *Proc. IEEE MTT-S International Microwave Symposium*, Jun. 2014.
15. T. Riihonen, D. Korpi, M. Turunen, T. Peltola, J. Saikanmäki, M. Valkama, and R. Wichman, "Tactical communication link under joint jamming and interception by same-frequency simultaneous transmit and receive radio," in *Proc. IEEE Military Communications Conference*, Oct. 2018.
16. K. Pärlin, M. M. Alam, and Y. Le Moullec, "Jamming of UAV remote control systems using software defined radio," in *Proc. International Conference on Military Communications and Information Systems*, May 2018.
17. M. Lichtman, J. D. Poston, S. Amuru, C. Shahriar, T. C. Clancy, R. M. Buehrer, and J. H. Reed, "A communications jamming taxonomy," *IEEE Security & Privacy*, vol. 14, no. 1, pp. 47–54, Jan. 2016.

18. S. Corson and J. Macker, "Mobile ad hoc networking (MANET): Routing protocol performance issues and evaluation considerations," Jan. 1999.
19. J. L. Burbank, P. F. Chimento, B. K. Haberman, and W. T. Kasch, "Key challenges of military tactical networking and the elusive promise of MANET technology," *IEEE Communications Magazine*, vol. 44, no. 11, Nov. 2006.
20. R. J. Fontana, "Recent system applications of short-pulse ultra-wideband (UWB) technology," *IEEE Transactions on Microwave Theory and Techniques*, vol. 52, no. 9, pp. 2087–2104, Sep. 2004.
21. S. S. Kolenchery, J. K. Townsend, and J. A. Freebersyser, "A novel impulse radio network for tactical military wireless communications," in *Proc. IEEE Military Communications Conference*, vol. 1, Oct. 1998, pp. 59–65.
22. M. A. Alim, M. Kobayashi, S. Saruwatari, and T. Watanabe, "In-band full-duplex medium access control design for heterogeneous wireless LAN," *EURASIP Journal on Wireless Communications and Networking*, vol. 2017, no. 1, p. 83, May 2017.
23. F. Tobagi and L. Kleinrock, "Packet switching in radio channels: part II–the hidden terminal problem in carrier sense multiple-access and the busy-tone solution," *IEEE Transactions on Communications*, vol. 23, no. 12, pp. 1417–1433, Dec. 1975.
24. C.-s. Wu and V. Li, "Receiver-initiated busy-tone multiple access in packet radio networks," in *ACM SIGCOMM Computer Communication Review*, vol. 17, no. 5, Aug. 1987, pp. 336–342.
25. K. Xu, M. Gerla, and S. Bae, "How effective is the IEEE 802.11 RTS/CTS handshake in ad hoc networks," in *Proc. IEEE Global Telecommunications Conference*, vol. 1, Nov. 2002, pp. 72–76.
26. N. Singh, D. Gunawardena, A. Proutiere, B. Radunovi, H. V. Balan, and P. Key, "Efficient and fair MAC for wireless networks with self-interference cancellation," in *International Symposium on Modeling and Optimization in Mobile, Ad Hoc and Wireless Networks*, May 2011, pp. 94–101.
27. K. M. Thilina, H. Tabassum, E. Hossain, and D. I. Kim, "Medium access control design for full duplex wireless systems: challenges and approaches," *IEEE Communications Magazine*, vol. 53, no. 5, pp. 112–120, May 2015.
28. D. Kim, H. Lee, and D. Hong, "A survey of in-band full-duplex transmission: From the perspective of PHY and MAC layers," *IEEE Communication Surveys and Tutorials*, vol. 17, no. 4, pp. 2017–2046, Q4 2015.
29. J. P. Monks, J.-P. Ebert, A. Wolisz, and W.-M. W. Hwu, "A study of the energy saving and capacity improvement potential of power control in multi-hop wireless networks," in *Proc. 26th Annual IEEE Conference on Local Computer Networks*, Nov. 2001, pp. 550–559.
30. W. Choi, H. Lim, and A. Sabharwal, "Power-controlled medium access control protocol for full-duplex WiFi networks," *IEEE Transactions on Wireless Communications*, vol. 14, no. 7, pp. 3601–3613, Jul. 2015.
31. T. Riihonen, S. Werner, and R. Wichman, "Hybrid full-duplex/half-duplex relaying with transmit power adaptation," *IEEE Transactions on Wireless Communications*, vol. 10, no. 9, pp. 3074–3085, Sep. 2011.
32. N. Patwari, J. Croft, S. Jana, and S. K. Kasera, "High-rate uncorrelated bit extraction for shared secret key generation from channel measurements," *IEEE Transactions on Mobile Computing*, vol. 9, no. 1, p. 17, Jan. 2010.
33. S. N. Premnath, S. Jana, J. Croft, P. L. Gowda, M. Clark, S. K. Kasera, N. Patwari, and S. V. Krishnamurthy, "Secret key extraction from wireless signal strength in real environments," *IEEE Transactions on Mobile Computing*, vol. 12, no. 5, pp. 917–930, May 2013.
34. S. Gollakota and D. Katabi, "Physical layer wireless security made fast and channel independent," in *Proc. IEEE INFOCOM*, Apr. 2011.
35. R. Jin, X. Du, Z. Deng, K. Zeng, and J. Xu, "Practical secret key agreement for full-duplex near field communications," *IEEE Transactions on Mobile Computing*, vol. 15, no. 4, pp. 938–951, Apr. 2016.
36. R. Bassily, E. Ekrem, X. He, E. Tekin, J. Xie, M. R. Bloch, S. Ulukus, and A. Yener, "Cooperative security at the physical layer: A summary of recent advances," *IEEE Signal Processing Magazine*, vol. 30, no. 5, pp. 16–28, Sep. 2013.

37. S. L. Cotton, W. G. Scanlon, and B. K. Madahar, "Millimeter-wave soldier-to-soldier communications for covert battlefield operations," *IEEE Communications Magazine*, vol. 47, no. 10, pp. 72–81, Oct. 2009.

38. Y.-W. P. Hong, P.-C. Lan, and C.-C. J. Kuo, "Enhancing physical-layer secrecy in multiantenna wireless systems: An overview of signal processing approaches," *IEEE Signal Processing Magazine*, vol. 30, no. 5, pp. 29–40, Sep. 2013.

39. A. Nasipuri, S. Ye, J. You, and R. E. Hiromoto, "A MAC protocol for mobile ad hoc networks using directional antennas," in *IEEE Wireless Communications and Networking Conference*, vol. 3, Sep. 2000, pp. 1214–1219.

40. R. R. Choudhury, X. Yang, R. Ramanathan, and N. H. Vaidya, "On designing MAC protocols for wireless networks using directional antennas," *IEEE Transactions on Mobile Computing*, vol. 5, no. 5, pp. 477–491, May 2006.

41. T. Korakis, G. Jakllari, and L. Tassiulas, "CDR-MAC: A protocol for full exploitation of directional antennas in ad hoc wireless networks," *IEEE Transactions on Mobile Computing*, vol. 7, no. 2, pp. 145–155, Feb. 2008.

42. K. M. Leong, Y. Wang, and T. Itoh, "A full duplex capable retrodirective array system for high-speed beam tracking and pointing applications," *IEEE Transactions on Microwave Theory and Techniques*, vol. 52, no. 5, pp. 1479–1489, May 2004.

43. S. Goel and R. Negi, "Guaranteeing secrecy using artificial noise," *IEEE Transactions on Wireless Communications*, vol. 7, no. 6, pp. 2180–2189, Jun. 2008.

44. K. Miura and M. Bandai, "Node architecture and MAC protocol for full duplex wireless and directional antennas," in *Proc. 23rd International Symposium on Personal, Indoor and Mobile Radio Communications*, Sep. 2012, pp. 369–374.

45. F. Zhu, F. Gao, M. Yao, and H. Zou, "Joint information- and jamming-beamforming for physical layer security with full duplex base station," *IEEE Transactions on Signal Processing*, vol. 62, no. 24, pp. 6391–6401, Dec. 2014.

46. M. A. Richards, J. Scheer, W. A. Holm, and W. L. Melvin, *Principles of modern radar*. Institution of Engineering and Technology, 2010.

47. B. Paul, A. Chiriyath, and D. Bliss, "Survey of RF communications and sensing convergence research," *IEEE Access*, vol. 5, no. 99, pp. 252–270, Dec. 2016.

48. M. P. Fitz, T. R. Halford, I. Hossain, and S. W. Enserink, "Towards simultaneous radar and spectral sensing," in *Proc. IEEE International Symposium on Dynamic Spectrum Access Networks*, Apr. 2014, pp. 15–19.

49. M. Jamil, H.-J. Zepernick, and M. I. Pettersson, "On integrated radar and communication systems using Oppermann sequences," in *Proc. IEEE Military Communications Conference*, Oct. 2008.

50. C. Sturm and W. Wiesbeck, "Waveform design and signal processing aspects for fusion of wireless communications and radar sensing," *Proceedings of the IEEE*, vol. 99, no. 7, pp. 1236–1259, Jul. 2011.

51. L. Han and K. Wu, "Joint wireless communication and radar sensing systems–state of the art and future prospects," *IET Microwaves, Antennas & Propagation*, vol. 7, no. 11, pp. 876–885, Aug. 2013.

52. R. N. Lothes, M. B. Szymanski, and R. G. Wiley, *Radar vulnerability to jamming*. Artech House, 1990.

53. L. Neng-Jing and Z. Yi-Ting, "A survey of radar ECM and ECCM," *IEEE Transactions on Aerospace and Electronic Systems*, vol. 31, no. 3, pp. 1110–1120, Jul. 1995.

54. S. Roome, "Digital radio frequency memory," *Electronics & Communication Engineering Journal*, vol. 2, no. 4, pp. 147–153, Aug. 1990.

55. M. Greco, F. Gini, and A. Farina, "Radar detection and classification of jamming signals belonging to a cone class," *IEEE Transactions on Signal Processing*, vol. 56, no. 5, pp. 1984–1993, May 2008.

56. T. Ulversoy, "Software defined radio: Challenges and opportunities," *IEEE Communications Surveys & Tutorials*, vol. 12, no. 4, pp. 531–550, Oct. 2010.

57. A. Feickert, "The joint tactical radio system (JTRS) and the army's future combat system (FCS): Issues for congress," Congressional Research Service, Tech. Rep. RL33161, 2005.
58. R. North, N. Browne, and L. Schiavone, "Joint tactical radio system – connecting the GIG to the tactical edge," in *Proc. IEEE Military Communications Conference*, Oct. 2006.
59. B. Perlman, J. Laskar, and K. Lim, "Fine-tuning commercial and military radio design," *IEEE Microwave Magazine*, vol. 9, no. 4, Aug. 2008.
60. P. K. Hughes and J. Y. Choe, "Overview of advanced multifunction RF system (AMRFS)," in *Proc. IEEE International Conference on Phased Array Systems and Technology*, May 2000, pp. 21–24.
61. G. C. Tavik, C. L. Hilterbrick, J. B. Evins, J. J. Alter, J. G. Crnkovich, J. W. de Graaf, W. Habicht, G. P. Hrin, S. A. Lessin, D. C. Wu *et al.*, "The advanced multifunction RF concept," *IEEE Transactions on Microwave Theory and Techniques*, vol. 53, no. 3, pp. 1009–1020, Mar. 2005.
62. J. A. Molnar, I. Corretjer, and G. Tavik, "Integrated topside-integration of narrowband and wideband array antennas for shipboard communications," in *Proc. IEEE Military Communications Conference*, Oct. 2011, pp. 1802–1807.
63. M. Parent, D. Taylor, G. Tavik, M. Kluskens, and J. Valenzi, "RF isolation of separate transmit and receive phased array antennas in a multifunction environment," in *Proc. Antenna Application Symposium*, vol. 2, Sep. 2001, pp. 413–442.
64. M. Wilhelm, I. Martinovic, J. B. Schmitt, and V. Lenders, "WiFire: a firewall for wireless networks," in *ACM SIGCOMM Computer Communication Review*, vol. 41, no. 4, Aug. 2011, pp. 456–457.
65. P. C. Pinto, J. Barros, and M. Z. Win, "Secure communication in stochastic wireless networks– part II: Maximum rate and collusion," *IEEE Transactions on Information Forensics and Security*, vol. 7, no. 1, pp. 139–147, Feb. 2012.
66. T. Riihonen, D. Korpi, M. Turunen, T. Peltola, J. Saikanmäki, M. Valkama, and R. Wichman, "Military full-duplex radio shield for protection against adversary receivers," in *Proc. International Conference on Military Communications and Information Systems*, May 2019.
67. T. Riihonen, D. Korpi, M. Turunen, and M. Valkama, "Full-duplex radio technology for simultaneously detecting and preventing improvised explosive device activation," in *Proc. International Conference on Military Communications and Information Systems*, May 2018.
68. J. Saikanmäki, M. Turunen, M. Mäenpää, A.-P. Saarinen, and T. Riihonen, "Simultaneous jamming and RC system detection by using full-duplex radio technology," in *Proc. International Conference on Military Communications and Information Systems*, May 2019.
69. K. Pärlin, T. Riihonen, and M. Turunen, "Sweep jamming mitigation using adaptive filtering for detecting frequency agile systems," in *Proc. International Conference on Military Communications and Information Systems*, May 2019.
70. Y. Cai, F. R. Yu, J. Li, Y. Zhou, and L. Lamont, "Medium access control for unmanned aerial vehicle (UAV) ad-hoc networks with full-duplex radios and multipacket reception capability," *IEEE Transactions on Vehicular Technology*, vol. 62, no. 1, pp. 390–394, Jan. 2013.
71. I. Krikidis, S. Timotheou, S. Nikolaou, G. Zheng, D. W. K. Ng, and R. Schober, "Simultaneous wireless information and power transfer in modern communication systems," *IEEE Communications Magazine*, vol. 52, no. 11, pp. 104–110, Nov. 2014.
72. C. Zhong, H. A. Suraweera, G. Zheng, I. Krikidis, and Z. Zhang, "Wireless information and power transfer with full duplex relaying," *IEEE Transactions on Communications*, vol. 62, no. 10, pp. 3447–3461, Oct. 2014.
73. L. Zhao, X. Wang, and T. Riihonen, "Transmission rate optimization of full-duplex relay systems powered by wireless energy transfer," *IEEE Transactions on Wireless Communications*, vol. 16, no. 10, pp. 6438–6450, Oct. 2017.
74. Y. Ma, N. Selby, and F. Adib, "Drone relays for battery-free networks," in *Proc. Conference of the ACM Special Interest Group on Data Communication*, Aug. 2017, pp. 335–347.
75. M. Mohammadi, B. K. Chalise, H. A. Suraweera, C. Zhong, G. Zheng, and I. Krikidis, "Throughput analysis and optimization of wireless-powered multiple antenna full-duplex relay systems," *IEEE Transactions on Communications*, vol. 64, no. 4, pp. 1769–1785, Apr. 2016.

76. M. Li, W. Lou, and K. Ren, "Data security and privacy in wireless body area networks," *IEEE Wireless Communications*, vol. 17, no. 1, Feb. 2010.
77. S. Gollakota, H. Hassanieh, B. Ransford, D. Katabi, and K. Fu, "They can hear your heartbeats: Non-invasive security for implantable medical devices," in *ACM SIGCOMM Computer Communication Review*, vol. 41, no. 4, Aug. 2011, pp. 2–13.
78. F. Xu, Z. Qin, C. C. Tan, B. Wang, and Q. Li, "IMDGuard: Securing implantable medical devices with the external wearable guardian," in *Proc. IEEE INFOCOM*, Apr. 2011, pp. 1862–1870.
79. N. O. Tippenhauer, L. Malisa, A. Ranganathan, and S. Capkun, "On limitations of friendly jamming for confidentiality," in *IEEE Symposium on Security and Privacy*, May 2013, pp. 160–173.
80. J. B. Kenney, "Dedicated short-range communications (DSRC) standards in the United States," *Proceedings of the IEEE*, vol. 99, no. 7, pp. 1162–1182, Jul. 2011.
81. K. Sjöberg, P. Andres, T. Buburuzan, and A. Brakemeier, "Cooperative intelligent transport systems in Europe: current deployment status and outlook," *IEEE Vehicular Technology Magazine*, vol. 12, no. 2, pp. 89–97, Jun. 2017.
82. M. Raya and J.-P. Hubaux, "Securing vehicular ad hoc networks," *Journal of computer security*, vol. 15, no. 1, pp. 39–68, Jan. 2007.
83. A. Bazzi, B. M. Masini, and A. Zanella, "Performance analysis of V2V beaconing using LTE in direct mode with full duplex radios," *IEEE Wireless Communications Letters*, vol. 4, no. 6, pp. 685–688, Dec. 2015.
84. C. Campolo, A. Molinaro, and A. O. Berthet, "Improving CAMs broadcasting in VANETs through full-duplex radios," in *IEEE 27th Annual International Symposium on Personal, Indoor, and Mobile Radio Communications*, Sep. 2016, pp. 1–6.
85. K. E. Kolodziej, B. T. Perry, and J. S. Herd, "Simultaneous transmit and receive (STAR) system architecture using multiple analog cancellation layers," in *IEEE MTT-S International Microwave Symposium*, May 2015, pp. 1–4.
86. P. Kumari, J. Choi, N. González-Prelcic, and R. W. Heath, "IEEE 802.11 ad-based radar: An approach to joint vehicular communication-radar system," *IEEE Transactions on Vehicular Technology*, vol. 67, no. 4, pp. 3012–3027, Apr. 2018.
87. A. D. Wyner, "The wire-tap channel," *Bell System Technical Journal*, vol. 54, no. 8, pp. 1355–1387, 1975.
88. S. Leung-Yan-Cheong and M. Hellman, "The Gaussian wire-tap channel," *IEEE Transactions on Information Theory*, vol. 24, no. 4, pp. 451–456, Jul. 1978.
89. A. O. Hero, "Secure space-time communication," *IEEE Transactions on Information Theory*, vol. 49, no. 12, pp. 3235–3249, Dec. 2003.
90. L. Dong, Z. Han, A. P. Petropulu, and H. V. Poor, "Improving wireless physical layer security via cooperating relays," *IEEE Transactions on Signal Processing*, vol. 58, no. 3, pp. 1875–1888, Mar. 2010.
91. G. Zheng, I. Krikidis, J. Li, A. P. Petropulu, and B. Ottersten, "Improving physical layer secrecy using full-duplex jamming receivers," *IEEE Transactions on Signal Processing*, vol. 61, no. 20, pp. 4962–4974, Oct. 2013.
92. G. Chen, Y. Gong, P. Xiao, and J. A. Chambers, "Physical layer network security in the full-duplex relay system," *IEEE Transactions on Information Forensics and Security*, vol. 10, no. 3, pp. 574–583, 2015.
93. F. Zhu, F. Gao, T. Zhang, K. Sun, and M. Yao, "Physical-layer security for full duplex communications with self-interference mitigation," *IEEE Transactions on Wireless Communications*, vol. 15, no. 1, pp. 329–340, Jan. 2016.
94. A. Mukherjee and A. L. Swindlehurst, "A full-duplex active eavesdropper in MIMO wiretap channels: Construction and countermeasures," in *Proc. 45th Asilomar Conference on Signals, Systems and Computers*, 2011, pp. 265–269.
95. X. Tang, P. Ren, Y. Wang, and Z. Han, "Combating full-duplex active eavesdropper: A hierarchical game perspective," *IEEE Transactions on Communications*, vol. 65, no. 3, pp. 1379–1395, Mar. 2016.

Chapter 10
Multi-Objective Optimization for Secure Full-Duplex Wireless Communication Systems

Yan Sun, Derrick Wing Kwan Ng, and Robert Schober

Abstract In traditional half-duplex (HD) communication systems, the HD base station (BS) can transmit artificial noise (AN) to jam the eavesdroppers for securing the downlink (DL) communication. However, guaranteeing uplink (UL) transmission is not possible with an HD BS because HD BSs cannot jam the eavesdroppers during UL transmission. In this chapter, we investigate the resource allocation algorithm design for secure multiuser systems employing a full-duplex (FD) BS for serving multiple HD DL and UL users simultaneously. In particular, the FD BS transmits AN to guarantee the concurrent DL and UL communication security. We propose a multi-objective optimization framework to study two conflicting yet desirable design objectives, namely total DL transmit power minimization and total UL transmit power minimization. To this end, the weighted Tchebycheff method is adopted to formulate the resource allocation algorithm design as a multi-objective optimization problem (MOOP). The considered MOOP takes into account the quality-of-service (QoS) requirements of all legitimate users for guaranteeing secure DL and UL transmission in the presence of potential eavesdroppers. Thereby, secure UL transmission is enabled by the FD BS, which would not be possible with an HD BS. Although the considered MOOP is non-convex, we solve it optimally by semidefinite programming (SDP) relaxation. Simulation results not only unveil the trade-off between the total DL transmit power and the total UL transmit power, but also confirm that the proposed secure FD system can guarantee concurrent secure DL and UL transmission and provide substantial power savings over a baseline system.

©Portions of this chapter are reprinted from [19], with permission from IEEE.

Y. Sun (✉) · R. Schober
Friedrich-Alexander-University Erlangen-Nürnberg, Erlangen, Germany
e-mail: yan.sun@fau.de; robert.schober@fau.de

D. W. K. Ng
The University of New South Wales, Sydney, NSW, Australia
e-mail: w.k.ng@unsw.edu.au

© Springer Nature Singapore Pte Ltd. 2020
H. Alves et al. (eds.), *Full-Duplex Communications for Future Wireless Networks*,
https://doi.org/10.1007/978-981-15-2969-6_10

10.1 Introduction

In the past two decades, the telecommunication industry has developed rapidly thanks to the advance of signal processing. The success of the third-generation (3G) and the fourth-generation (4G) wireless communication systems has promoted numerous innovative mobile applications and has led to an explosive and continuing growth in mobile data traffic [23]. According to the Cisco visual network index report [10], the global mobile data traffic will increase sevenfold between 2016 and 2021. As a result, it is foreseen that the demand for data traffic will exceed the capacity of the existing wireless communication systems in the near future. Besides, the exponential growth in high-data rate communications has triggered a tremendous demand for radio resources such as bandwidth and energy [22–24]. An important technique for reducing the energy and bandwidth consumption of wireless systems while satisfying quality-of-service (QoS) requirements is multiple-input multiple-output (MIMO), as it offers extra spatial degrees of freedom (DoF) facilitating the design of efficient resource allocation. However, the MIMO gain may be difficult to achieve in practice due to the high computational complexity of MIMO receivers. As an alternative, multiuser MIMO (MU-MIMO) has been proposed as an effective technique for realizing MIMO performance gains. In particular, in MU-MIMO systems, a transmitter equipped with multiple antennas (e.g., a base station (BS)) serves multiple single-antenna users, which shifts the computational complexity from the receivers to the transmitter [8]. Yet, the spectral resource is still underutilized even if MU-MIMO is employed as long as the BS adopts the traditional half-duplex (HD) protocol, where uplink (UL) and downlink (DL) communication are separated orthogonally in either time or frequency which leads to a significant waste of in-system resources.

On the other hand, security is a crucial issue for wireless communication due to the broadcast nature of the wireless medium. Traditionally, secure communication is achieved by cryptographic encryption performed at the application layer and is based on the assumption of limited computational capabilities of the eavesdroppers [4]. However, new computing technologies (e.g., quantum computers) may make this assumption invalid which results in a potential vulnerability of traditional approaches to secure communication. The pioneering work in [21] proposed an alternative approach for providing perfectly secure communication by utilizing the nature of the channel in the physical layer. Specifically, Wyner [21] revealed that secure communication can be achieved whenever the information receiver enjoys better channel conditions than the eavesdropper. Motivated by this finding, physical layer security has received significant attention for preventing eavesdropping in wireless communication systems [4, 14, 17, 26]. An important technique to ensure communication security via physical layer security is multiple-antenna transmission which utilizes the spatial DoF for degrading the quality of the eavesdroppers' channels. In particular, artificial noise (AN) transmission is an effective approach to deliberately impair the information reception at the eavesdroppers. For instance, in [17], a power allocation algorithm was designed for maximizing the secrecy outage capacity via AN generation. In [26], the authors investigated the secrecy performance of DL massive MIMO systems and derived a lower bound on the achievable ergodic secrecy rate of the users. The authors of [14] proposed a robust resource allocation algorithm for guaranteeing secure multiuser communication with energy harvesting receivers. However, the aforementioned works focused on guaranteeing DL communication security between a HD BS and associated DL users. The obtained results may not be applicable for securing UL transmission. In fact, guaranteeing UL transmission is not possible with an HD BS. Particularly, HD BSs cannot jam the eavesdroppers during UL transmission because they can either transmit or receive in a given time instant but not both.

To overcome these issues, an intuitive concept for improving the spectral efficiency is to employ full-duplex (FD) transceivers which can transmit and receive signals at the same time and in the frequency band. More importantly, an FD BS enables simultaneous secure DL and UL communication by transmitting AN in the DL to interfere potential eavesdroppers [25]. However, deploying wireless FD nodes has been generally considered impractical for a long time since the signal reception is severely impaired by the self-interference (SI) caused by the signal leakage due to the simultaneous transmission at the same node [20]. Recently, the authors of [2] developed a single-antenna FD transceiver prototype which achieves 110 dB SI cancellation offering a substantial system throughput improvement compared to HD transceivers. This has attracted significant attention from both academia and industry. Several FD prototypes using different SI cancellation techniques have been built and they demonstrate that FD operation can achieve higher throughput than HD for various system settings [1, 6, 11]. For instance, in [6], the authors designed an FD prototype equipped with multiple antennas. In [1], it could be shown that FD transmission can double the ergodic capacity of MIMO systems. Although [2, 11] reported that SI can be partially cancelled through analog circuits and digital signal processing, the residual SI still severely degrades the performance of FD systems if it is not properly controlled. Besides, in multiuser communication systems, co-channel interference (CCI) caused by the UL transmission impairs the DL transmission [18]. Moreover, the CCI becomes more severe if there are multiple DL and UL users in the communication system. Therefore, careful resource allocation is necessary and critical to fully exploit the potential performance gains enabled by FD communications. In fact, the unique challenges introduced by FD networks do not exist in HD networks, and thus, the conventional designs for HD networks cannot be directly applied to FD networks. Hence, in this chapter, we focus on resource allocation algorithm design for guaranteeing concurrent secure DL and UL transmission in multiuser FD systems.

The remainder of this chapter is organized as follows. In Sect. 10.2, we introduce the adopted FD system model. In Sect. 10.3, the resource allocation algorithm design for guaranteeing secure communication in FD systems is formulated as a non-convex optimization problem. The formulated problem is solved optimally in Sect. 10.4, and simulation results are provided in Sect. 10.5. In Sect. 10.6, we conclude with a brief summary of this chapter.

Notation We use boldface capital and lower case letters to denote matrices and vectors, respectively. $\mathbf{A}^H$, $\mathrm{Tr}(\mathbf{A})$, $\mathrm{Rank}(\mathbf{A})$, and $\det(\mathbf{A})$ denote the Hermitian transpose, trace, rank, and determinant of matrix $\mathbf{A}$, respectively; $\mathbf{A}^{-1}$ and $\mathbf{A}^{\dagger}$ represent the inverse and Moore–Penrose pseudoinverse of matrix $\mathbf{A}$, respectively; $\mathbf{A} \succeq \mathbf{0}$, $\mathbf{A} \succ \mathbf{0}$, and $\mathbf{A} \preceq \mathbf{0}$ indicate that $\mathbf{A}$ is a positive semidefinite, a positive definite, and a negative semidefinite matrix, respectively; $\mathbf{I}_N$ is the $N \times N$ identity matrix; $\mathbb{C}^{N \times M}$ denotes the set of all $N \times M$ matrices with complex entries; $\mathbb{H}^N$ denotes the set of all $N \times N$ Hermitian matrices; $|\cdot|$ and $\|\cdot\|$ denote the absolute value of a complex scalar and the Euclidean vector norm, respectively; $\mathcal{E}\{\cdot\}$ denotes statistical expectation; $\mathrm{diag}(\mathbf{X})$ returns a diagonal matrix having the main diagonal elements of $\mathbf{X}$ on its main diagonal. $[x]^+$ stands for $\max\{0, x\}$; the circularly symmetric complex Gaussian distribution with mean μ and variance σ^2 is denoted by $\mathcal{CN}(\mu, \sigma^2)$; and $\sim$ stands for "distributed as."

10.2   System Model

In this section, we present the considered MU-MIMO FD wireless communication system model.

10.2.1   Multiuser System Model

We consider a multiuser communication system [19]. The system consists of an FD BS, K legitimate DL users, J legitimate UL users, and a roaming user, cf. Fig. 10.1. The FD BS is equipped with $N_T > 1$ antennas to facilitate simultaneous DL transmission and UL reception in the same frequency band.[1] The $K + J$ legitimate users are single-antenna HD mobile communication devices to ensure low hardware complexity. The number of antennas at the FD BS is assumed to be larger than the number of UL users to facilitate reliable UL signal detection, i.e., $N_T \geq J$. Besides, the DL and the UL users are scheduled for simultaneous UL and DL transmission. Unlike the local legitimate signal-antenna users, the roaming user is a traveling wireless device belonging to another communication system and is equipped with $N_R > 1$ antennas. The multiple-antenna roaming user is searching for access to local wireless services. However, it is possible that the roaming user deliberately intercepts the information signal intended for the legitimate users if they are in the same service area. As a result, the roaming user is a potential eavesdropper which has to be taken into account for resource allocation algorithm design to guarantee communication security. In this chapter, we refer to the roaming user as a potential

---

[1]We note that transmitting and receiving signals simultaneously via the same antenna is feasible by exploiting a circulator [2].

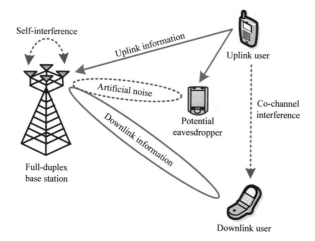

Fig. 10.1 A multiuser communication system with an FD BS, $K = 1$ HD DL user, $J = 1$ HD UL user, and one HD potential eavesdropper

eavesdropper and we assume $N_T > N_R$ for studying resource allocation algorithm design. Besides, in order to study the upper bound performance of the considered system, we assume that the FD BS has perfect channel state information (CSI) for resource allocation.

10.2.2   Channel Model

We focus on a frequency flat fading channel. In each scheduling time slot, the FD BS transmits K independent signal streams simultaneously at the same frequency to the K DL users. In particular, the information signal to DL user $k \in \{1, \ldots, K\}$ can be expressed as $\mathbf{x}_k = \mathbf{w}_k d_k^{DL}$, where $d_k^{DL} \in \mathbb{C}$ and $\mathbf{w}_k \in \mathbb{C}^{N_T \times 1}$ are the information bearing signal for DL user k and the corresponding beamforming vector, respectively. Without loss of generality, we assume $\mathcal{E}\{|d_k^{DL}|^2\} = 1, \forall k \in \{1, \ldots, K\}$.

However, the signal intended for the desired user may be eavesdropped by the roaming user. Hence, in order to ensure secure communication, the FD BS also transmits AN to interfere the reception of the roaming user (potential eavesdropper). Therefore, the transmit signal vector, $\mathbf{x} \in \mathbb{C}^{N_T \times 1}$, comprising K information streams and AN, is given by $\mathbf{x} = \sum_{k=1}^{K} \mathbf{x}_k + \mathbf{z}_{AN}$, where $\mathbf{z}_{AN} \in \mathbb{C}^{N_T \times 1}$ represents the AN vector generated by the FD BS to degrade the channel quality of the potential eavesdropper. In particular, $\mathbf{z}_{AN}$ is modeled as a complex Gaussian random vector with $\mathbf{z}_{AN} \sim \mathcal{CN}(\mathbf{0}, \mathbf{Z}_{AN})$, where $\mathbf{Z}_{AN} \in \mathbb{H}^{N_T}$, $\mathbf{Z}_{AN} \succeq \mathbf{0}$, denotes the covariance matrix of the AN. Therefore, the received signals at DL user $k \in \{1, \ldots, K\}$, the

FD BS, and the potential eavesdropper are given by

$$y_{\text{DL}_k} = \mathbf{h}_k^H \mathbf{x}_k + \underbrace{\sum_{i \neq k}^{K} \mathbf{h}_k^H \mathbf{x}_i}_{\substack{\text{multiuser} \\ \text{interference}}} + \underbrace{\mathbf{h}_k^H \mathbf{z}_{\text{AN}}}_{\substack{\text{artificial} \\ \text{noise}}} + \underbrace{\sum_{j=1}^{J} \sqrt{P_j} f_{j,k} d_j^{\text{UL}}}_{\substack{\text{co-channel} \\ \text{interference}}} + n_{\text{DL}_k}, \tag{10.1}$$

$$\mathbf{y}_{\text{BS}} = \sum_{j=1}^{J} \sqrt{P_j} \mathbf{g}_j d_j^{\text{UL}} + \underbrace{\mathbf{H}_{\text{SI}} \sum_{k=1}^{K} \mathbf{x}_k}_{\text{self-interference}} + \underbrace{\mathbf{H}_{\text{SI}} \mathbf{z}_{\text{AN}}}_{\substack{\text{artificial} \\ \text{noise}}} + \mathbf{n}_{\text{BS}}, \text{ and} \tag{10.2}$$

$$\mathbf{y}_{\text{E}} = \underbrace{\sum_{k=1}^{K} \mathbf{L}^H \mathbf{x}_k}_{\text{DL signals}} + \underbrace{\sum_{j=1}^{J} \sqrt{P_j} \mathbf{e}_j d_j^{\text{UL}}}_{\text{UL signals}} + \underbrace{\mathbf{L}^H \mathbf{z}_{\text{AN}}}_{\substack{\text{artificial} \\ \text{noise}}} + \mathbf{n}_{\text{E}}, \tag{10.3}$$

respectively. The DL channel between the FD BS and user k is denoted by $\mathbf{h}_k \in \mathbb{C}^{N_T \times 1}$ and $f_{j,k} \in \mathbb{C}$ represents the channel between UL user j and DL user k. Variables d_j^{UL}, $\mathcal{E}\{|d_j^{\text{UL}}|^2\} = 1$, and P_j are the data and transmit power sent from UL user j to the FD BS, respectively. Vector $\mathbf{g}_j \in \mathbb{C}^{N_T \times 1}$ denotes the channel between UL user j and the FD BS. Matrix $\mathbf{H}_{\text{SI}} \in \mathbb{C}^{N_T \times N_T}$ denotes the SI channel of the FD BS. Matrix $\mathbf{L} \in \mathbb{C}^{N_T \times N_R}$ denotes the channel between the FD BS and the potential eavesdropper. Vector $\mathbf{e}_j \in \mathbb{C}^{N_R \times 1}$ denotes the channel between UL user j and the potential eavesdropper. Variables $\mathbf{h}_k$, $f_{j,k}$, $\mathbf{g}_j$, $\mathbf{H}_{\text{SI}}$, $\mathbf{L}$, and $\mathbf{e}_j$ capture the joint effect of path loss and small scale fading. $\mathbf{n}_{\text{BS}} \sim \mathcal{CN}(\mathbf{0}, \sigma_{\text{BS}}^2 \mathbf{I}_{N_T})$, $n_{\text{DL}_k} \sim \mathcal{CN}(0, \sigma_k^2)$, and $\mathbf{n}_{\text{E}} \sim \mathcal{CN}(\mathbf{0}, \sigma_{\text{E}}^2 \mathbf{I}_{N_R})$ represent the additive white Gaussian noise (AWGN) at the FD BS, DL user k, and the potential eavesdropper, respectively. In (10.1), the term $\sum_{j=1}^{J} \sqrt{P_j} f_{j,k} d_j^{\text{UL}}$ denotes the aggregated CCI caused by the UL users to DL user k. In (10.2), the term $\mathbf{H}_{\text{SI}} \sum_{k=1}^{K} \mathbf{x}_k$ represents the SI.

10.3 Resource Allocation Problem Formulation

In this section, we first define the adopted performance metrics for the considered multiuser communication system. Then, we formulate the resource allocation problems for DL and UL transmit power minimization, respectively. For the sake of notational simplicity, we define the following variables: $\mathbf{H}_k = \mathbf{h}_k \mathbf{h}_k^H$, $k \in \{1, \ldots, K\}$, and $\mathbf{G}_j = \mathbf{g}_j \mathbf{g}_j^H$, $j \in \{1, \ldots, J\}$.

10.3.1   Achievable Rate and Secrecy Rate

The achievable rate (bit/s/Hz) of DL user k is given by

$$R_{DL_k} = \log_2(1 + \Gamma_k^{DL}) \text{ with} \tag{10.4}$$

$$\Gamma_k^{DL} = \frac{|\mathbf{h}_k^H \mathbf{w}_k|^2}{\sum_{r \neq k}^{K} |\mathbf{h}_k^H \mathbf{w}_r|^2 + \sum_{j=1}^{J} P_j |f_{j,k}|^2 + \mathrm{Tr}(\mathbf{H}_k \mathbf{Z}_{AN}) + \sigma_k^2}, \tag{10.5}$$

where Γ_k^{DL} is the receive signal-to-interference-plus-noise ratio (SINR) at DL user k. Besides, the achievable rate of UL user j is given by

$$R_{UL_j} = \log_2(1 + \Gamma_j^{UL}) \text{ with} \tag{10.6}$$

$$\Gamma_j^{UL} = \frac{P_j |\mathbf{g}_j^H \mathbf{v}_j|^2}{\sum_{n \neq j}^{J} P_n |\mathbf{g}_n^H \mathbf{v}_j|^2 + S_{SI_j} + \sigma_{BS}^2 \|\mathbf{v}_j\|^2}, \quad \text{and} \tag{10.7}$$

$$S_{SI_j} = \mathrm{Tr}\left(\rho \mathbf{V}_j \, \mathrm{diag}\left(\mathbf{H}_{SI} \mathbf{Z}_{AN} \mathbf{H}_{SI}^H + \sum_{k=1}^{K} \mathbf{H}_{SI} \mathbf{w}_k \mathbf{w}_k^H \mathbf{H}_{SI}^H\right)\right), \tag{10.8}$$

where Γ_j^{UL} is the receive SINR of UL user j at the FD BS. The variable $\mathbf{v}_j \in \mathbb{C}^{N_T \times 1}$ is the receive beamforming vector for decoding the information received from UL user j and we define $\mathbf{V}_j = \mathbf{v}_j \mathbf{v}_j^H, j \in \{1, \ldots, J\}$. In this chapter, zero-forcing receive beamforming (ZF-BF) is adopted. In this context, we note that ZF-BF closely approaches the performance of optimal minimum mean square error beamforming (MMSE-BF) when the noise term is not significant[2] [20] or the number of antennas is sufficiently large [16]. Besides, ZF-BF facilitates the design of a computationally efficient resource allocation algorithm. Hence, the receive beamformer for UL user j is chosen as $\mathbf{v}_j = (\mathbf{u}_j \mathbf{Q}^\dagger)^H$, where $\mathbf{u}_j = \underbrace{[0, \ldots, 0}_{(j-1)}, 1, \underbrace{0, \ldots, 0]}_{(J-j)}$, $\mathbf{Q}^\dagger = (\mathbf{Q}^H \mathbf{Q})^{-1} \mathbf{Q}^H$, and $\mathbf{Q} = [\mathbf{g}_1, \ldots, \mathbf{g}_J]$. Since SI cannot be cancelled perfectly in FD systems due to the limited dynamic range of the receiver even if the SI channel is perfectly known [5], we model the residual SI after cancellation at each receive antenna as an independent zero-mean Gaussian distortion noise whose variance is proportional to the received power of the antenna, i.e., the term $S_{SI_j} = \mathrm{Tr}\left(\rho \mathbf{V}_j \, \mathrm{diag}\left(\mathbf{H}_{SI} \mathbf{Z}_{AN} \mathbf{H}_{SI}^H + \sum_{k=1}^{K} \mathbf{H}_{SI} \mathbf{w}_k \mathbf{w}_k^H \mathbf{H}_{SI}^H\right)\right)$ in (10.7) and (10.8). We note that this SI model was first proposed in [5]. Besides, it was

---

[2] We note that the noise power at the BS is not expected to be the dominating factor for the system performance since BSs are usually equipped with a high quality low-noise amplifier (LNA).

shown in [13] that the adopted model accurately captures the combined effects of additive automatic gain control noise, non-linearities in the analog-to-digital converters and the gain control, and oscillator phase noise which are present in practical FD hardware. We refer to [5, Eq. (4)] for a more detailed discussion of the adopted SI model.

As discussed before, for guaranteeing communication security, the roaming user is treated as a potential eavesdropper who eavesdrops the information signals desired for all DL and UL users. Thereby, we design the resource allocation algorithm under a worst-case assumption for guaranteeing communication secrecy. In particular, we assume that the potential eavesdropper can cancel all multiuser interference before decoding the information of the desired user. Thus, under this assumption, the channel capacity between the FD BS and the potential eavesdropper for eavesdropping desired DL user k and the channel capacity between the UL user j and the potential eavesdropper for overhearing UL user j can be written as

$$C_{DL_k} = \log_2 \det(\mathbf{I}_{N_R} + \mathbf{X}^{-1}\mathbf{L}^H \mathbf{w}_k \mathbf{w}_k^H \mathbf{L}), \forall k, \quad \text{and} \tag{10.9}$$

$$C_{UL_j} = \log_2 \det(\mathbf{I}_{N_R} + P_j \mathbf{X}^{-1}\mathbf{e}_j \mathbf{e}_j^H), \forall j, \tag{10.10}$$

respectively, where $\mathbf{X} = \mathbf{L}^H \mathbf{Z}_{AN}\mathbf{L} + \sigma_E^2 \mathbf{I}_{N_R}$ denotes the interference-plus-noise covariance matrix for the potential eavesdropper. We emphasize that, unlike an HD BS, the FD BS can guarantee both DL security and UL security simultaneously via AN transmission. The achievable secrecy rates between the FD BS and DL user k and UL user j are given by

$$R_{DL_k}^{Sec} = \left[R_{DL_k} - C_{DL_k} \right]^+, \forall k, \quad \text{and} \quad R_{UL_j}^{Sec} = \left[R_{UL_j} - C_{UL_j} \right]^+, \forall j, \tag{10.11}$$

respectively.

10.3.2 Optimization Problem Formulation

In the following, we first introduce two individual system design optimization problems with conflicting design objectives. Then, we investigate the two system design objectives jointly under a multi-objective optimization framework [12, 15, 19]. In this chapter, we focus on the minimization of the total DL transmit power at the BS and the minimization of the total UL transmit power. In particular, since wireless BSs consume a considerable amount of energy, the associated energy cost has become a financial burden to the service providers. Hence, to reduce the energy consumption and its cost, it is necessary to design power efficient resource allocation schemes for reducing the transmit power of the BS. Thus, in this chapter, we aim to minimize the transmit power under secrecy and QoS constraints. In particular, for the secrecy constraints, the maximum information leakage to the potential eavesdroppers is constrained. In fact, the secrecy constraints imposing the maximum information leakage provide flexibility in controlling the security level of communication for different practical applications. For example, services involving confidential information, e.g., online banking, have more stringent requirements on low information leakage than ordinary services, e.g., video streaming. On the other hand, multimedia services, e.g., video streaming, have a higher information leakage tolerance than plain text services, e.g., E-mail and short message service (SMS), since it is hard to recover high quality multimedia content from limited eavesdropped information. Therefore, the considered problem formulation with secrecy constraints takes into account the different required security levels for facilitating more flexible resource allocation. In particular, the first considered objective is the minimization of the total DL transmit power at the FD BS and is given by

---

Problem 1 (Total DL Transmit Power Minimization)

$$\underset{\mathbf{Z}_{\mathrm{AN}}\in\mathbb{H}^{N_{\mathrm{T}}},\mathbf{w}_k,P_j}{\text{minimize}} \quad \sum_{k=1}^{K}\|\mathbf{w}_k\|^2 + \mathrm{Tr}(\mathbf{Z}_{\mathrm{AN}})$$

s.t. C1: $\dfrac{|\mathbf{h}_k^H \mathbf{w}_k|^2}{\sum\limits_{r\neq k}^{K}|\mathbf{h}_k^H \mathbf{w}_r|^2 + \sum\limits_{j=1}^{J} P_j|f_{j,k}|^2 + \mathrm{Tr}(\mathbf{H}_k \mathbf{Z}_{\mathrm{AN}}) + \sigma_k^2} \geq \Gamma_{\mathrm{req}_k}^{\mathrm{DL}}, \ \forall k, j,$

C2: $\dfrac{P_j|\mathbf{g}_j^H \mathbf{v}_j|^2}{\sum\limits_{n\neq j}^{J} P_n|\mathbf{g}_n^H \mathbf{v}_j|^2 + S_{\mathrm{SI}_j} + \sigma_{\mathrm{BS}}^2\|\mathbf{v}_j\|^2} \geq \Gamma_{\mathrm{req}_j}^{\mathrm{UL}}, \ \forall j,$

C3: $\log_2 \det(\mathbf{I}_{N_{\mathrm{R}}} + \mathbf{X}^{-1}\mathbf{L}^H \mathbf{w}_k \mathbf{w}_k^H \mathbf{L}) \leq R_{\mathrm{tol}_k}^{\mathrm{DL}}, \ \forall k,$

C4: $\log_2 \det(\mathbf{I}_{N_{\mathrm{R}}} + P_j \mathbf{X}^{-1}\mathbf{e}_j \mathbf{e}_j^H) \leq R_{\mathrm{tol}_j}^{\mathrm{UL}}, \ \forall j,$

C5: $P_j \geq 0, \ \forall j,$ C6: $\mathbf{Z}_{\mathrm{AN}} \succeq \mathbf{0}.$ (10.12)

The system design objective in (10.12) is to minimize the total DL transmit power which is comprised of the DL signal power and the AN power. Constants $\Gamma_{\mathrm{req}_k}^{\mathrm{DL}} > 0$ and $\Gamma_{\mathrm{req}_j}^{\mathrm{UL}} > 0$ in constraints C1 and C2 in (10.12) are the minimum required SINR for DL users $k \in \{1, \ldots, K\}$ and UL users $j \in \{1, \ldots, J\}$, respectively. $R_{\mathrm{tol}_k}^{\mathrm{DL}}$ and $R_{\mathrm{tol}_j}^{\mathrm{UL}}$ in C3 and C4, respectively, are pre-defined system parameters representing the maximum tolerable data rates at the potential eavesdropper for decoding the information of DL user k and UL user j, respectively. In fact, DL and UL security is guaranteed by constraints C3 and C4. In particular, if the above optimization problem is feasible, the proposed problem formulation guarantees that the secrecy rate for DL user k is bounded below by $R_{\mathrm{DL}_k}^{\mathrm{Sec}} \geq \log_2(1 + \Gamma_{\mathrm{req}_k}^{\mathrm{DL}}) - R_{\mathrm{tol}_k}^{\mathrm{DL}}$ and the secrecy rate for UL user j is bounded below by $R_{\mathrm{UL}_j}^{\mathrm{Sec}} \geq \log_2(1 + \Gamma_{\mathrm{req}_j}^{\mathrm{UL}}) - R_{\mathrm{tol}_j}^{\mathrm{UL}}$. We note that the maximization of the system secrecy throughput for the considered multiuser systems is generally NP-hard[3] [27]. Hence, we focus on the minimization of the total DL transmit power under secrecy constraints to obtain a tractable resource allocation design. Constraint C5 is the non-negative power constraint for UL user j. Constraint C6 and $\mathbf{Z}_{\mathrm{AN}} \in \mathbb{H}^{N_\mathrm{T}}$ are imposed since covariance matrix $\mathbf{Z}_{\mathrm{AN}}$ has to be a Hermitian positive semidefinite matrix. We note that the objective of Problem 1 is to minimize the total DL transmit power under constraints C1–C6 without regard for the consumed UL transmit powers.

Besides, as mobile devices are powered by batteries with limited energy storage capacity, minimizing the UL transmit power can prolong the lifetime of mobile devices. Therefore, the second system design objective is the minimization of total UL transmit power and can be mathematically formulated as

---

Problem 2 (Total UL Transmit Power Minimization)

$$\underset{\mathbf{Z}_{\mathrm{AN}} \in \mathbb{H}^{N_\mathrm{T}}, \mathbf{w}_k, P_j}{\text{minimize}} \quad \sum_{j=1}^{J} P_j$$

$$\text{s.t.} \quad \text{C1} - \text{C6}. \tag{10.13}$$

Problem 2 targets only the minimization of the total UL transmit power under constraints C1–C6 without taking into account the total consumed DL transmit power.

The objectives of Problems 1 and 2 are desirable for the system operator and the users, respectively. However, in secure FD wireless communication systems, these objectives conflict with each other. On the one hand, the DL information and AN transmission cause significant SI which impairs the UL signal reception. Hence, the UL users have to transmit with a higher power to compensate this interference to satisfy the minimum required receive SINR of the UL users at the FD BS. On the other hand, a high UL transmit power results in a strong CCI for DL signal reception and a higher risk of information leakage to the potential eavesdropper. Hence, the FD BS has to transmit both the DL information and the AN with higher power to ensure the QoS requirements of the DL users and the security requirements of the DL and UL users. However, this in turn causes high SI and gives rise to an escalating increase in transmit power for both UL and DL transmission. To overcome this problem, we resort to multi-objective optimization [12, 15]. In the literature, multi-

---

[3]Note that the maximization of the secrecy rate is also one possible system design objective. Yet, such a formulation may lead to exceedingly large energy consumption.

objective optimization is often adopted to study the trade-off between conflicting system design objectives via the concept of Pareto optimality [12, 15]. To facilitate our presentation, we denote the objective function of Problem i as $Q_i(\mathbf{w}_k, \mathbf{Z}_{AN}, P_j)$. The Pareto optimality of a resource allocation policy is defined in the following:

Definition 1 (Pareto Optimal [12]) A resource allocation policy, $\{\mathbf{w}_k, \mathbf{Z}_{AN}, P_j\}$, is Pareto optimal if and only if there does not exist any $\{\tilde{\mathbf{w}}_k, \tilde{\mathbf{Z}}_{AN}, \tilde{P}_j\}$ with

$$Q_i(\tilde{\mathbf{w}}_k, \tilde{\mathbf{Z}}_{AN}, \tilde{P}_j) < Q_i(\mathbf{w}_k, \mathbf{Z}_{AN}, P_j), \forall i \in \{1, 2\}. \tag{10.14}$$

In other words, a resource allocation policy is Pareto optimal if there is no other policy that improves at least one of the objectives without detriment to the other objective. In order to capture the complete Pareto optimal set, we formulate a third optimization problem to investigate the trade-off between Problems 1 and 2 by using the weighted Tchebycheff method [12]. The third problem formulation is given as

---

Problem 3 (Multi-Objective Optimization)

$$\underset{\mathbf{Z}_{AN} \in \mathbb{H}^{N_T}, \mathbf{w}_k, P_j}{\text{minimize}} \ \underset{i=1,2}{\max} \ \left\{ \varrho_i \Big(Q_i(\mathbf{w}_k, \mathbf{Z}_{AN}, P_j) - Q_i^* \Big) \right\}$$

$$\text{s.t.} \quad \text{C1 - C6}, \tag{10.15}$$

---

where $Q_1(\mathbf{w}_k, \mathbf{Z}_{AN}, P_j) = \sum_{k=1}^{K} \|\mathbf{w}_k\|^2 + \text{Tr}(\mathbf{Z}_{AN})$ and $Q_2(\mathbf{w}_k, \mathbf{Z}_{AN}, P_j) = \sum_{j=1}^{J} P_j$. Q_i^* is the optimal objective value of the i-th problem and is treated as a constant for Problem 3. Variable $\varrho_i \geq 0$, $\sum_i \varrho_i = 1$, specifies the priority of the i-th objective compared to the other objectives and reflects the preference of the system operator. By varying ϱ_i, a complete Pareto optimal set corresponding to a set of resource allocation policies can be obtained when (10.15) is solved. Thus, the operator can select a proper resource allocation policy from the set of available policies. Compared to other formulations for handing MOOPs in the literature (e.g., the weighted product method, the exponentially weighted criterion, and the ϵ-constraint method [12]), the weighted Tchebycheff method can achieve the complete Pareto optimal set with a lower computational complexity, despite the non-convexity (if any) of the considered problem. It is noted that Problem 3 is equivalent to Problem i when $\varrho_i = 1$ and $\varrho_j = 0$, $\forall j \neq i$. Here, equivalence means that both problems have the same optimal solution.

10.4 Solution of the Optimization Problem

Problems 1–3 are non-convex problems due to the non-convex constraints C3 and C4. To solve these problems efficiently, we first transform C3 and C4 into equivalent linear matrix inequality (LMI) constraints. Then, Problems 1–3 are solved optimally by semidefinite programming (SDP) relaxation.

To facilitate the SDP relaxation, we define $\mathbf{W}_k = \mathbf{w}_k \mathbf{w}_k^H$ and rewrite Problems 1–3 in the following equivalent forms:

Equivalent Problem 1

$$\underset{\mathbf{W}_k, \mathbf{Z}_{\text{AN}} \in \mathbb{H}^{N_{\text{T}}}, P_j}{\text{minimize}} \quad \sum_{k=1}^{K} \text{Tr}(\mathbf{W}_k) + \text{Tr}(\mathbf{Z}_{\text{AN}})$$

s.t. C1: $\dfrac{\text{Tr}(\mathbf{H}_k \mathbf{W}_k)}{\displaystyle\sum_{r \neq k}^{K} \text{Tr}(\mathbf{H}_k \mathbf{W}_r) + \sum_{j=1}^{J} P_j |f_{j,k}|^2 + \text{Tr}(\mathbf{H}_k \mathbf{Z}_{\text{AN}}) + \sigma_k^2} \geq \Gamma_{\text{req}_k}^{\text{DL}}, \; \forall k, j,$

C2: $\dfrac{P_j \text{Tr}(\mathbf{G}_j \mathbf{V}_j)}{\displaystyle\sum_{n \neq j}^{J} P_n \text{Tr}(\mathbf{G}_n \mathbf{V}_j) + I_{\text{SI}_j} + \sigma_{\text{BS}}^2 \text{Tr}(\mathbf{V}_j)} \geq \Gamma_{\text{req}_j}^{\text{UL}}, \; \forall j,$

C3: $\log_2 \det(\mathbf{I}_{N_{\text{R}}} + \mathbf{X}^{-1} \mathbf{L}^H \mathbf{W}_k \mathbf{L}) \leq R_{\text{tol}_k}^{\text{DL}}, \; \forall k,$

C4: $\log_2 \det(\mathbf{I}_{N_{\text{R}}} + P_j \mathbf{X}^{-1} \mathbf{e}_j \mathbf{e}_j^H) \leq R_{\text{tol}_j}^{\text{UL}}, \; \forall j,$

C5: $P_j \geq 0, \; \forall j,$ C6: $\mathbf{Z}_{\text{AN}} \succeq \mathbf{0},$

C7: $\mathbf{W}_k \succeq \mathbf{0}, \forall k,$ C8: $\text{Rank}(\mathbf{W}_k) \leq 1, \forall k,$ (10.16)

---

where

$$I_{\text{SI}_j} = \text{Tr}\left(\rho \mathbf{V}_j \, \text{diag}\left(\mathbf{H}_{\text{SI}} \mathbf{Z}_{\text{AN}} \mathbf{H}_{\text{SI}}^H + \sum_{k=1}^{K} \mathbf{H}_{\text{SI}} \mathbf{W}_k \mathbf{H}_{\text{SI}}^H \right) \right). \tag{10.17}$$

$\mathbf{W}_k \succeq \mathbf{0}$, $\mathbf{W}_k \in \mathbb{H}^{N_\mathrm{T}}$, and $\mathrm{Rank}(\mathbf{W}_k) \leq 1$ in (10.16) are imposed to guarantee that $\mathbf{W}_k = \mathbf{w}_k \mathbf{w}_k^H$ holds after optimization. Similarly, Problems 2 and 3 can be transformed, respectively, into:

Equivalent Problem 2

$$\underset{\mathbf{W}_k, \mathbf{Z}_{\mathrm{AN}} \in \mathbb{H}^{N_\mathrm{T}}, P_j}{\text{minimize}} \quad \sum_{j=1}^{J} P_j$$

$$\text{s.t. } \mathrm{C1} - \mathrm{C8}. \tag{10.18}$$

---

and

Equivalent Problem 3

$$\underset{\mathbf{W}_k, \mathbf{Z}_{\mathrm{AN}} \in \mathbb{H}^{N_\mathrm{T}}, P_j, \tau}{\text{minimize}} \quad \tau$$

$$\text{s.t.} \quad \mathrm{C1} - \mathrm{C8},$$

$$\mathrm{C9}: \varrho_i(Q_i - Q_i^*) \leq \tau, \forall i \in \{1, 2\}. \tag{10.19}$$

Note that in Equivalent Problem 3, τ is an auxiliary optimization variable and (10.19) is the equivalent epigraph representation of (10.15).

Since Problem 3 is a generalization of Problems 1 and 2, we focus on solving Problem 3. Now, we handle the non-convex constraints C3 and C4 by introducing the following proposition.

Proposition 1 *For $R_{\mathrm{tol}_k}^{\mathrm{DL}} > 0$ and $R_{\mathrm{tol}_j}^{\mathrm{UL}} > 0$, we have the following implications for constraints C3 and C4 of equivalent Problems 1–3, respectively:*

$$\mathrm{C3} \Rightarrow \widetilde{\mathrm{C3}}: \mathbf{L}^H \mathbf{W}_k \mathbf{L} \preceq \xi_k^{\mathrm{DL}} \mathbf{X}, \ \forall k, \tag{10.20}$$

$$\mathrm{C4} \Leftrightarrow \widetilde{\mathrm{C4}}: P_j \mathbf{e}_j \mathbf{e}_j^H \preceq \xi_j^{\mathrm{UL}} \mathbf{X}, \ \forall j, \tag{10.21}$$

where $\xi_k^{\mathrm{DL}} = 2^{R_{\mathrm{tol}_k}^{\mathrm{DL}}} - 1$ and $\xi_j^{\mathrm{UL}} = 2^{R_{\mathrm{tol}_j}^{\mathrm{UL}}} - 1$. We note that C3 and $\widetilde{\mathrm{C3}}$ are equivalent, respectively, if $\mathrm{Rank}(\mathbf{W}_k) \leq 1$. Besides, C4 and $\widetilde{\mathrm{C4}}$ are always equivalent.

Proof Please refer to Appendix 1. □

Note that $\widetilde{C3}$ and $\widetilde{C4}$ are convex LMI constraints which can be handled easily. The remaining non-convex constraint in (10.19) is the rank-one constraint C8. Solving such a rank-constrained problem is generally NP-hard [9]. Hence, to obtain a tractable solution, we relax constraint C8: $\text{Rank}(\mathbf{W}_k) \leq 1$ by removing it from the problem formulation, such that the considered problem becomes a convex SDP given by

$$
\underset{\mathbf{W}_k, \mathbf{Z}_{\text{AN}} \in \mathbb{H}^{N_{\text{T}}}, P_j, \tau}{\text{minimize}} \quad \tau
$$

$$
\text{s.t. } \text{C1, C2, C5, C6, C7,}
$$

$$
\widetilde{C3}: \mathbf{L}^H \mathbf{W}_k \mathbf{L} \preceq \xi_k^{\text{DL}} \mathbf{X}, \ \forall k,
$$

$$
\widetilde{C4}: P_j \mathbf{e}_j \mathbf{e}_j^H \preceq \xi_j^{\text{UL}} \mathbf{X}, \ \forall j,
$$

$$
\text{C9}: \varrho_i (Q_i - Q_i^*) \leq \tau, \ \forall i \in \{1, 2\}. \tag{10.22}
$$

The relaxed convex problem in (10.22) can be solved efficiently by standard convex program solvers such as CVX [7]. Besides, if the solution obtained for a relaxed SDP problem is a rank-one matrix, i.e., $\text{Rank}(\mathbf{W}_k) = 1$ for $\mathbf{W}_k \neq \mathbf{0}$, $\forall k$, then it is also the optimal solution of the original problem. Next, we verify the tightness of the adopted SDP relaxation in the following theorem.

Theorem 1 *Assuming the considered problem is feasible, for $\Gamma_{\text{req}_k}^{\text{DL}} > 0$, we can always obtain or construct a rank-one optimal matrix $\mathbf{W}_k^*$ which is an optimal solution for (10.22).*

Proof Please refer to Appendix 2. □

By Theorem 1, the optimal beamforming vector $\mathbf{w}_k^*$ can be recovered from $\mathbf{W}_k^*$ by performing eigenvalue decomposition of $\mathbf{W}_k^*$ and selection of the principle eigenvector as $\mathbf{w}_k^*$.

10.5 Simulation Results

In this section, we investigate the performance of the proposed multi-objective optimization based resource allocation scheme through simulations. The most important simulation parameters are specified in Table 10.1. There are $K = 3$ DL users, $J = 7$ UL users, and one potential eavesdropper in the considered cell. The users and the potential eavesdropper are randomly and uniformly distributed between the reference distance of 30 m and the maximum service distance of 600 m. The FD BS is located at the center of the cell and equipped with N_{T} antennas. The

Table 10.1 System parameters used in simulations

Carrier center frequency and system bandwidth	1.9 GHz and 1 MHz
Path loss exponent and SI cancellation constant, ρ	3.6 and -85 dB [2]
DL user noise power and UL BS noise power, σ_k^2 and σ_{BS}^2	-110 dBm
Potential eavesdropper noise power, σ_E^2, and BS antenna gain	-110 dBm and 10 dBi
Maximum tolerable eavesdropping data rate for DL users, $R_{tol_k}^{DL}$	0.1 bit/s/Hz
Maximum tolerable eavesdropping data rate for UL users, $R_{tol_j}^{UL}$	0.1 bit/s/Hz

potential eavesdropper is equipped with $N_R = 2$ antennas. The small scale fading of the DL channels, UL channels, CCI channels, and eavesdropping channels is modeled as independent and identically distributed Rayleigh fading. The multipath fading coefficients of the SI channel are generated as independent and identically distributed Rician random variables with Rician factor 5 dB. In addition, we assume that all DL users and all UL users require the same minimum SINRs, respectively, i.e., $\Gamma_{req_k}^{DL} = \Gamma_{req}^{DL}$ and $\Gamma_{req_j}^{UL} = \Gamma_{req}^{UL}$.

Besides, we also consider the performance of a baseline scheme for comparison. For the baseline scheme, we adopt ZF-BF as DL transmission scheme such that the multiuser interference is avoided at the legitimate DL users. In particular, the direction of beamformer $\mathbf{w}_k$ for legitimate user k is fixed and lies in the null space of the other legitimate DL users' channels. Then, we jointly optimize $\mathbf{Z}_{AN}$, P_j, and the power allocated to $\mathbf{w}_k$ under the MOOP formulation subject to the same constraints as in (10.22) via SDP relaxation.

10.5.1 Transmit Power Trade-off Region

In Fig. 10.2, we study the trade-off between the DL and the UL total transmit powers for different numbers of antennas at the FD BS. The trade-off region is obtained by solving (10.22) for different values of $0 \leq \varrho_i \leq 1, \forall i \in \{1, 2\}$, i.e., the ϱ_i are varied uniformly using a step size of 0.01 subject to $\sum_i \varrho_i = 1$. We assume a minimum required DL SINR of $\Gamma_{req}^{DL} = 10$ dB and a minimum required UL SINR of $\Gamma_{req}^{UL} = 5$ dB. It can be observed from Fig. 10.2 that the total UL transmit power is a monotonically decreasing function with respect to the total DL transmit power. In other words, minimizing the total UL power consumption leads to a higher power consumption in the DL and vice versa. This result confirms that the minimization of the total UL transmit power and the minimization of the total DL transmit power are conflicting design objectives. For the case of $N_T = 12$, 8 dB in UL transmit power can be saved by increasing the total DL transmit power by 5 dB. In addition, Fig. 10.2 also indicates that a significant amount of transmit power can be saved in the FD system by increasing the number of BS antennas. This is due to the fact that the extra degrees freedom offered by the additional antennas facilitate a more power efficient resource allocation. However, due to channel hardening, there is a

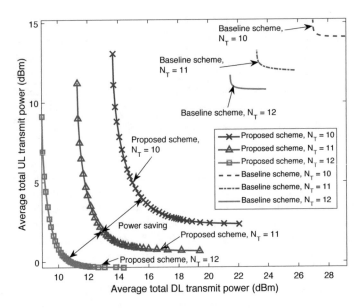

Fig. 10.2 Average system objective trade-off region achieved by the proposed resource allocation scheme. The double-sided arrows indicate the power saving due to additional antennas

diminishing return in the power saving as the number of antennas at the FD BS increases [20]. On the other hand, Fig. 10.2 also depicts the trade-off region for the baseline resource allocation scheme for comparison. As can be seen, the trade-off regions achieved by the baseline scheme are above the curves for the proposed optimal scheme. This indicates that the proposed resource allocation scheme is more power efficient than the baseline scheme for both DL and UL transmission. Indeed, the proposed resource allocation scheme can fully exploit the available degrees of freedom to perform globally optimal resource allocation. On the contrary, for the baseline scheme, the transmitter is incapable of fully utilizing the available degrees of freedom since the direction of the transmit beamformer $\mathbf{w}_k$ is fixed. Specifically, the fixed beamformer $\mathbf{w}_k$ can cause severe SI and increases the risk of eavesdropping which results in a high power consumption for UL transmission and the AN. Besides, the trade-off region for the baseline scheme is strictly smaller than that for the proposed optimal scheme. For instance, when $N_T = 12$, the baseline scheme can save only 1 dB of UL transmit power by increasing the total DL transmit power by 2.5 dB, due to the limited flexibility of the baseline scheme in handling the interference.

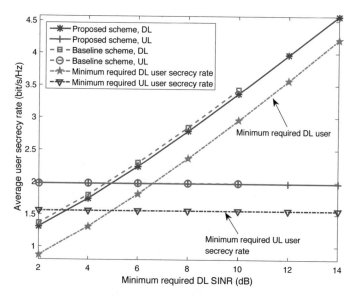

Fig. 10.3 Average user secrecy rate (bits/s/Hz) versus the minimum required DL SINR (dB), $\Gamma_{\text{req}}^{\text{DL}}$

10.5.2   Average User Secrecy Rate Versus Minimum Required SINR

Figure 10.3 depicts the average user secrecy rate of the DL and UL users versus the minimum required DL SINR, $\Gamma_{\text{req}}^{\text{DL}}$, for $R_{\text{tol}_k}^{\text{DL}} = R_{\text{tol}_j}^{\text{UL}} = 0.5$ bit/s/Hz and $\Gamma_{\text{req}}^{\text{UL}} = 5$ dB. The FD BS is equipped with $N_T = 10$ antennas. We select the resource allocation policy with $\varrho_1 = 0.1$ and $\varrho_2 = 0.9$.

The average user secrecy rates for the DL and UL users are calculated by averaging the total DL and the total UL secrecy rates, i.e., $\frac{\sum_{k=1}^{K} R_{\text{DL}_k}^{\text{Sec}}}{K}$ and $\frac{\sum_{j=1}^{J} R_{\text{UL}_j}^{\text{Sec}}}{J}$, respectively. As can be seen, that the average DL user secrecy rate increases with $\Gamma_{\text{req}}^{\text{DL}}$ since the channel capacity between DL user k and the potential eavesdropper is limited to $C_{\text{DL}_k} = 0.5$ bit/s/Hz. Besides, the average UL user secrecy rate depends only weakly on $\Gamma_{\text{req}}^{\text{DL}}$ for the proposed scheme. In addition, we compare the average DL and UL user secrecy rates of the proposed scheme with the minimum required DL and UL user secrecy rates, i.e., $\log_2(1 + \Gamma_{\text{req}}^{\text{DL}}) - R_{\text{tol}_k}^{\text{DL}}$ and $\log_2(1 + \Gamma_{\text{req}}^{\text{UL}}) - R_{\text{tol}_j}^{\text{UL}}$, respectively. As can be seen, the average user secrecy rate achieved by the proposed scheme fulfills the minimum required user secrecy rate in both DL and UL which confirms that the security of both links can be guaranteed simultaneously. This is due to the proposed optimization algorithm design. On the other hand, the baseline scheme achieves a slightly higher secrecy rate than the proposed scheme for $\Gamma_{\text{req}}^{\text{DL}}$ ranges from 2 to 10 dB. However, to accomplish this, the baseline scheme requires exceedingly large transmit powers at both the FD BS and the UL users compared to the proposed scheme, cf. Fig. 10.2. Besides, the superior performance of the baseline scheme in terms of secrecy rate also comes at the expense of an extremely high outage probability. In particular, the baseline scheme is always infeasible when $\Gamma_{\text{req}}^{\text{DL}}$ is larger than 10 dB.

10.6   Conclusions

In this chapter, we studied the power efficient resource allocation algorithm design for enabling secure MU-MIMO wireless communication with an FD BS. The algorithm design was formulated as a non-convex MOOP via the weighted Tchebycheff method. The proposed problem aimed at jointly minimizing the total DL and UL transmit powers for achieving simultaneous secure DL and UL transmission. The proposed MOOP was solved optimally by SDP relaxation. We proved that the globally optimal solution can always be obtained or constructed by solving at most two convex SDP optimization problems. Simulation results not only revealed the trade-off between the total DL and UL transmit power consumption, but also confirm that the proposed FD system provides substantial power savings over the baseline scheme. Furthermore, our results revealed that an FD BS can guarantee secure UL transmission which is not possible with an HD BS.

Appendix 1: Proof of Proposition 1

We start the proof by rewriting constraints C3 and C4 as follows:

$$\text{C3:} \qquad \det(\mathbf{I}_{N_R} + \mathbf{X}^{-1}\mathbf{L}^H\mathbf{W}_k\mathbf{L}) \leq 2^{R_{\text{tol}_k}^{\text{DL}}}$$

$$\overset{(a)}{\Longleftrightarrow} \det(\mathbf{I}_{N_R} + \mathbf{X}^{-1/2}\mathbf{L}^H\mathbf{W}_k\mathbf{L}\mathbf{X}^{-1/2}) \leq 2^{R_{\text{tol}_k}^{\text{DL}}}, \qquad (10.23)$$

$$\text{C4:} \qquad \det(\mathbf{I}_{N_R} + P_j\mathbf{X}^{-1}\mathbf{e}_j\mathbf{e}_j^H) \leq 2^{R_{\text{tol}_j}^{\text{UL}}}$$

$$\overset{(b)}{\Longleftrightarrow} \det(\mathbf{I}_{N_R} + P_j\mathbf{X}^{-1/2}\mathbf{e}_j\mathbf{e}_j^H\mathbf{X}^{-1/2}) \leq 2^{R_{\text{tol}_j}^{\text{UL}}}. \qquad (10.24)$$

(a) and (b) hold due to a basic matrix equality, namely $\det(\mathbf{I} + \mathbf{AB}) = \det(\mathbf{I} + \mathbf{BA})$. Then, we study a lower bound of (10.23) and (10.24) by applying the following Lemma.

Lemma 1 (Determinant Inequality [19]) *For any semidefinite matrix* $\mathbf{A} \succeq \mathbf{0}$, *the inequality* $\det(\mathbf{I} + \mathbf{A}) \geq 1 + \text{Tr}(\mathbf{A})$ *holds where equality holds if and only if* $\text{Rank}(\mathbf{A}) \leq 1$.

We note that $\mathbf{X}^{-1/2}\mathbf{L}^H\mathbf{W}_k\mathbf{L}\mathbf{X}^{-1/2} \succeq \mathbf{0}$ holds in (10.23). Thus, applying Lemma 1 to (10.23) yields

$$\det(\mathbf{I}_{N_R} + \mathbf{X}^{-1/2}\mathbf{L}^H\mathbf{W}_k\mathbf{L}\mathbf{X}^{-1/2}) \geq 1 + \mathrm{Tr}(\mathbf{X}^{-1/2}\mathbf{L}^H\mathbf{W}_k\mathbf{L}\mathbf{X}^{-1/2}). \quad (10.25)$$

As a result, by combining (10.23) and (10.25), we have the following implications:

$$\mathrm{Tr}(\mathbf{X}^{-1/2}\mathbf{L}^H\mathbf{W}_k\mathbf{L}\mathbf{X}^{-1/2}) \leq \xi_k^{\mathrm{DL}}$$

$$\overset{(c)}{\Longrightarrow} \lambda_{\max}(\mathbf{X}^{-1/2}\mathbf{L}^H\mathbf{W}_k\mathbf{L}\mathbf{X}^{-1/2}) \leq \xi_k^{\mathrm{DL}}$$

$$\Longleftrightarrow \mathbf{X}^{-1/2}\mathbf{L}^H\mathbf{W}_k\mathbf{L}\mathbf{X}^{-1/2} \preceq \xi_k^{\mathrm{DL}}\mathbf{I}_{N_R}$$

$$\Longleftrightarrow \mathbf{L}^H\mathbf{W}_k\mathbf{L} \preceq \xi_k^{\mathrm{DL}}\mathbf{X}, \quad (10.26)$$

where $\lambda_{\max}(\mathbf{A})$ denotes the maximum eigenvalue of matrix $\mathbf{A}$ and (c) is due to the fact that $\mathrm{Tr}(\mathbf{A}) \geq \lambda_{\max}(\mathbf{A})$ holds for any $\mathbf{A} \succeq \mathbf{0}$. Besides, if $\mathrm{Rank}(\mathbf{W}_k) \leq 1$, we have

$$\mathrm{Rank}(\mathbf{X}^{-1/2}\mathbf{L}^H\mathbf{W}_k\mathbf{L}\mathbf{X}^{-1/2})$$

$$\leq \min \left\{ \mathrm{Rank}(\mathbf{X}^{-1/2}\mathbf{L}^H), \mathrm{Rank}(\mathbf{W}_k\mathbf{L}\mathbf{X}^{-1/2}) \right\}$$

$$\leq \mathrm{Rank}(\mathbf{W}_k\mathbf{L}\mathbf{X}^{-1/2}) \leq 1. \quad (10.27)$$

Then, equality holds in (10.25). Besides, in (10.26), $\mathrm{Tr}(\mathbf{X}^{-1/2}\mathbf{L}^H\mathbf{W}_k\mathbf{L}\mathbf{X}^{-1/2}) \leq \xi_k^{\mathrm{DL}}$ is equivalent to $\lambda_{\max}(\mathbf{X}^{-1/2}\mathbf{L}^H\mathbf{W}_k\mathbf{L}\mathbf{X}^{-1/2}) \leq \xi_k^{\mathrm{DL}}$. Therefore, (10.23) and (10.26) are equivalent if $\mathrm{Rank}(\mathbf{W}_k) \leq 1$.

As for constraint C4, we note that $\mathrm{Rank}(P_j\mathbf{X}^{-1/2}\mathbf{e}_j\mathbf{e}_j^H\mathbf{X}^{-1/2}) \leq 1$ always holds. Therefore, by applying Lemma 1 to (10.24), we have

$$\det(\mathbf{I}_{N_R} + P_j\mathbf{X}^{-1/2}\mathbf{e}_j\mathbf{e}_j^H\mathbf{X}^{-1/2}) = 1 + \mathrm{Tr}(P_j\mathbf{X}^{-1/2}\mathbf{e}_j\mathbf{e}_j^H\mathbf{X}^{-1/2}). \quad (10.28)$$

Then, by combining (10.24) and (10.28), we have the following implications:

$$\mathrm{C4} \Longleftrightarrow \mathrm{Tr}(P_j\mathbf{X}^{-1/2}\mathbf{e}_j\mathbf{e}_j^H\mathbf{X}^{-1/2}) \leq \xi_j^{\mathrm{UL}}$$

$$\Longleftrightarrow \lambda_{\max}(P_j\mathbf{X}^{-1/2}\mathbf{e}_j\mathbf{e}_j^H\mathbf{X}^{-1/2}) \leq \xi_j^{\mathrm{UL}}$$

$$\Longleftrightarrow P_j\mathbf{X}^{-1/2}\mathbf{e}_j\mathbf{e}_j^H\mathbf{X}^{-1/2} \preceq \xi_j^{\mathrm{UL}}\mathbf{I}_{N_R}$$

$$\Longleftrightarrow P_j\mathbf{e}_j\mathbf{e}_j^H \preceq \xi_j^{\mathrm{UL}}\mathbf{X}. \quad (10.29)$$

Appendix 2: Proof of Theorem 1

The SDP relaxed version of equivalent Problem 3 in (10.22) is jointly convex with respect to the optimization variables and satisfies Slater's constraint qualification. Therefore, strong duality holds and solving the dual problem is equivalent to solving the primal problem [3]. For obtaining the dual problem, we first need the Lagrangian function of the primal problem in (10.22) which is given by

$$\mathcal{L} = \varrho_1 \pi_1 \sum_{k=1}^{K} \text{Tr}(\mathbf{W}_k) - \sum_{k=1}^{K} \text{Tr}(\mathbf{W}_k \mathbf{Y}_k) + \sum_{j=1}^{J} \mu_j \sum_{k=1}^{K} \text{Tr}(\rho \mathbf{W}_k \mathbf{H}_{\text{SI}}^H \mathbf{V}_j \mathbf{H}_{\text{SI}})$$

$$- \sum_{k=1}^{K} \text{Tr}\left(\frac{\lambda_k \mathbf{H}_k \mathbf{W}_k}{\Gamma_{\text{req}_k}^{\text{DL}}}\right) + \sum_{k=1}^{K} \text{Tr}(\mathbf{L}^H \mathbf{W}_k \mathbf{L} \mathbf{D}_k) + \Lambda. \tag{10.30}$$

Here, Λ denotes the collection of terms that only involve variables that are independent of $\mathbf{W}_k$. λ_k, μ_j, and π_i are the Lagrange multipliers associated with constraints C1, C2, and C9, respectively. Matrix $\mathbf{D}_k \in \mathbb{C}^{N_R \times N_R}$ is the Lagrange multiplier matrix for constraint $\widetilde{\text{C3}}$. Matrix $\mathbf{Y}_k \in \mathbb{C}^{N_T \times N_T}$ is the Lagrange multiplier matrix for the positive semidefinite constraint C7 on matrix $\mathbf{W}_k$. For notational simplicity, we define Ψ as the set of scalar Lagrange multipliers for constraints C1, C2, C5, and C9 and Φ as the set of matrix Lagrange multipliers for constraints $\widetilde{\text{C3}}$, $\widetilde{\text{C4}}$, C6, and C7. Thus, the dual problem for the SDP relaxed problem in (10.22) is given by

$$\underset{\Psi \geq 0, \Phi \geq 0}{\text{maximize}} \quad \underset{\mathbf{W}_k, \mathbf{Z}_{\text{AN}} \in \mathbb{H}^{N_T}, P_j, \tau}{\text{minimize}} \quad \mathcal{L}\left(\mathbf{W}_k, \mathbf{Z}_{\text{AN}}, P_j, \Psi, \Phi\right)$$

$$\text{s.t.} \sum_{i=1}^{2} \pi_i = 1. \tag{10.31}$$

Constraint $\sum_{i=1}^{2} \pi_i = 1$ is imposed to guarantee a bounded solution of the dual problem [3]. Then, we reveal the structure of the optimal $\mathbf{W}_k$ of (10.22) by studying the Karush–Kuhn–Tucker (KKT) conditions. The KKT conditions for the optimal $\mathbf{W}_k^*$ are given by

$$\mathbf{Y}_k^*, \mathbf{D}_k^* \geq 0, \quad \lambda_k^*, \mu_j^*, \pi_i^* \geq 0, \tag{10.32}$$

$$\mathbf{Y}_k^* \mathbf{W}_k^* = 0, \tag{10.33}$$

$$\nabla_{\mathbf{W}_k^*} \mathcal{L} = 0, \tag{10.34}$$

where $\mathbf{Y}_k^*$, $\mathbf{D}_k^*$, λ_k^*, μ_j^*, and π_i^* are the optimal Lagrange multipliers for dual problem (10.31). $\nabla_{\mathbf{W}_k^*}\mathcal{L}$ denotes the gradient of Lagrangian function $\mathcal{L}$ with respect to matrix $\mathbf{W}_k^*$. The KKT condition in (10.34) can be expressed as

$$\mathbf{Y}_k^* + \frac{\lambda_k^* \mathbf{H}_k}{\Gamma_{\mathrm{req}_k}^{\mathrm{DL}}}$$

$$= \varrho_1 \pi_1 \mathbf{I}_{N_{\mathrm{T}}} + \sum_{j=1}^{J} \mu_j^* \rho \mathbf{H}_{\mathrm{SI}}^H \mathrm{diag}(\mathbf{V}_j)\mathbf{H}_{\mathrm{SI}} + \mathbf{L}\mathbf{D}_k^*\mathbf{L}^H. \tag{10.35}$$

Now, we divide the proof into two cases according to the value of ϱ_1. First, for the case of $0 < \varrho_1 \leq 1$, we define

$$\mathbf{A}_k^* = \sum_{j=1}^{J} \mu_j^* \rho \mathbf{H}_{\mathrm{SI}}^H \mathrm{diag}(\mathbf{V}_j)\mathbf{H}_{\mathrm{SI}} + \mathbf{L}\mathbf{D}_k^*\mathbf{L}^H, \tag{10.36}$$

$$\mathbf{\Pi}_k^* = \varrho_1 \pi_1 \mathbf{I}_{N_{\mathrm{T}}} + \mathbf{A}_k^*, \tag{10.37}$$

for notational simplicity. Then, (10.35) implies

$$\mathbf{Y}^* = \mathbf{\Pi}_k^* - \frac{\lambda_k^* \mathbf{H}_k}{\Gamma_{\mathrm{req}_k}^{\mathrm{DL}}}. \tag{10.38}$$

Pre-multiplying both sides of (10.38) by $\mathbf{W}_k^*$, and utilizing (10.33), we have

$$\mathbf{W}_k^* \mathbf{\Pi}_k^* = \mathbf{W}_k^* \frac{\lambda_k^* \mathbf{H}_k}{\Gamma_{\mathrm{req}_k}^{\mathrm{DL}}}. \tag{10.39}$$

By applying basic inequalities for the rank of matrices, the following relation holds:

$$\mathrm{Rank}(\mathbf{W}_k^*) \overset{(a)}{=} \mathrm{Rank}(\mathbf{W}_k^* \mathbf{\Pi}_k^*) = \mathrm{Rank}\left(\mathbf{W}_k^* \frac{\lambda_k^* \mathbf{H}_k}{\Gamma_{\mathrm{req}_k}^{\mathrm{DL}}}\right)$$

$$\overset{(b)}{\leq} \min\left\{\mathrm{Rank}(\mathbf{W}_k^*), \mathrm{Rank}\left(\frac{\lambda_k^* \mathbf{H}_k}{\Gamma_{\mathrm{req}_k}^{\mathrm{DL}}}\right)\right\}$$

$$\overset{(c)}{\leq} \mathrm{Rank}\left(\frac{\lambda_k^* \mathbf{H}_k}{\Gamma_{\mathrm{req}_k}^{\mathrm{DL}}}\right) \leq 1, \tag{10.40}$$

where (a) is valid because $\mathbf{\Pi}_k^* \succ \mathbf{0}$, (b) is due to the basic result $\mathrm{Rank}(\mathbf{AB}) \leq \min\left\{\mathrm{Rank}(\mathbf{A}), \mathrm{Rank}(\mathbf{B})\right\}$, and (c) is due to the fact that $\min\{a, b\} \leq a$. We note that $\mathbf{W}_k^* \neq \mathbf{0}$ for $\Gamma_{\mathrm{req}_k}^{\mathrm{DL}} > 0$. Thus, $\mathrm{Rank}(\mathbf{W}_k^*) = 1$.

Then, for the case of $\varrho_1 = 0$, we show that we can always construct a rank-one optimal solution $\mathbf{W}_k^{**}$. We note that the problem in (10.22) with $\varrho_1 = 0$ is equivalent to a total UL transmit power minimization problem which is given by

$$
\begin{aligned}
\underset{\mathbf{W}_k, \mathbf{Z}_{\mathrm{AN}} \in \mathbb{H}^{N_\mathrm{T}}, P_j}{\text{minimize}} \quad & \sum_{j=1}^{J} P_j \\
\text{s.t.} \quad & \text{C1, C2, } \widetilde{\text{C3}}, \widetilde{\text{C4}}, \text{C5, C6, C7, C9.}
\end{aligned}
\tag{10.41}
$$

We first solve the above convex optimization problem and obtain the UL transmit power P_j^{**}, the DL beamforming matrix $\mathbf{W}_k^*$, and the AN covariance matrix $\mathbf{Z}_{\mathrm{AN}}^*$. If $\mathrm{Rank}(\mathbf{W}_k^*) = 1, \forall k$, then the globally optimal solution of problem (10.19) for $\varrho_1 = 0$ is achieved. Otherwise, we substitute P_j^{**} and $\mathbf{Z}_{\mathrm{AN}}^*$ into the following auxiliary problem:

$$
\begin{aligned}
\underset{\mathbf{W}_k \in \mathbb{H}^{N_\mathrm{T}}}{\text{minimize}} \quad & \sum_{k=1}^{K} \mathrm{Tr}(\mathbf{W}_k) + \mathrm{Tr}(\mathbf{Z}_{\mathrm{AN}}^*) \\
\text{s.t.} \quad & \text{C1, C2, } \widetilde{\text{C3}}, \widetilde{\text{C4}}, \text{C5, C6, C7, C9.}
\end{aligned}
\tag{10.42}
$$

Since the problem in (10.42) shares the feasible set of problem (10.41), problem (10.42) is also feasible. Now, we claim that for a given P_j^{**} and $\mathbf{Z}_{\mathrm{AN}}^*$ in (10.42), the solution $\mathbf{W}_k^{**}$ of (10.42) is a rank-one matrix. First, the gradient of the Lagrangian function for (10.42) with respect to $\mathbf{W}_k^{**}$ can be expressed as

$$
\mathbf{Y}^{**} = \mathbf{\Pi}_k^{**} - \frac{\lambda_k^{**} \mathbf{H}_k}{\Gamma_{\mathrm{req}_k}^{\mathrm{DL}}},
\tag{10.43}
$$

where

$$
\mathbf{\Pi}_k^{**} = \mathbf{I}_{N_\mathrm{T}} + \mathbf{A}_k^{**} \quad \text{and}
\tag{10.44}
$$

$$
\mathbf{A}_k^{**} = \sum_{j=1}^{J} \mu_j^{**} \rho \mathbf{H}_{\mathrm{SI}}^H \, \mathrm{diag}(\mathbf{V}_j) \mathbf{H}_{\mathrm{SI}} + \mathbf{L} \mathbf{D}_k^{**} \mathbf{L}^H.
\tag{10.45}
$$

$\mathbf{Y}_k^{**}$, $\mathbf{D}_k^{**}$, λ_k^{**}, and μ_j^{**} are the optimal Lagrange multipliers for the dual problem of (10.42). Pre-multiplying both sides of (10.43) by the optimal solution $\mathbf{W}_k^{**}$, we have

$$\mathbf{W}_k^{**}\mathbf{\Pi}_k^{**} = \mathbf{W}_k^{**}\frac{\lambda_k^{**}\mathbf{H}_k}{\Gamma_{\mathrm{req}_k}^{\mathrm{DL}}}. \tag{10.46}$$

We note that $\mathbf{\Pi}_k^{**}$ is a full-rank matrix, i.e., $\mathbf{\Pi}_k^{**} \succ \mathbf{0}$, and (10.46) has the same form as (10.39). Thus, we can follow the same approach as for the case of $0 < \varrho_i \le 1$ for showing that $\mathbf{W}_k^{**}$ is a rank-one matrix. Also, since $\mathbf{W}_k^{**}$ is a feasible solution of (10.41) for P_j^{**}, an optimal rank-one matrix $\mathbf{W}_k^{**}$ for the case of $\varrho_1 = 0$ is constructed.

References

1. D. Bharadia and S. Katti, "Full duplex mimo radios," in *Proc. USENIX Conf. on Network System Design.* USENIX, 2014, pp. 1–10.
2. D. Bharadia, E. McMilin, and S. F. d. r. Katti, "Acm sigcomm." pp. 375–386, 2013.
3. S. Boyd and L. C. o. Vandenberghe. Cambridge University Press, 2004.
4. X. Chen, D. W. K. Ng, W. H. Gerstacker, and H. H. Chen, "A survey on multiple-antenna techniques for physical layer security," *IEEE Commun Surveys & Tutorials*, vol. 19, pp. 1027–1053, 2017.
5. B. P. Day, A. R. Margetts, D. W. Bliss, and P. Schniter, "Full-duplex MIMO relaying achievable rates under limited dynamic range." *IEEE J Select Areas Commun.*, vol. 30, no. 8, pp. 1541–1553, 2012.
6. M. Duarte, A. Sabharwal, V. Aggarwal, R. Jana, K. K. Ramakrishnan, C. W. Rice, and N. K. Shankaranarayanan, "Design and characterization of a full-duplex multiantenna system for wifi networks." *IEEE Trans Veh. Technol.*, vol. 63, no. 3, pp. 1160–1177, 2014.
7. M. Grant, S. Boyd, and Y. C. Ye, *Matlab software for disciplined convex programming*, 2014. [Online]. Available: http://cvxr.com/cvx
8. D. Gesbert, M. Kountouris, R. W. Heath, C. B. Chae, and T. Salzer, "From single user to multiuser communications: Shifting the mimo paradigm," *IEEE Signal Process*, vol. 24, no. 5, pp. 36–46, 2007.
9. B. Gärtner and J. Matousek, *Approximation algorithms and semidefinite programming.* Science & Business Media, Springer, 2012.
10. Cisco, "Tech. Rep: Global mobile data traffic forecast update" *2016 to 2021 White Paper*. Cisco, 2017.
11. J. I. Choi, M. Jain, K. Srinivasan, P. Levis, and S. Katti, "Achieving single channel full duplex wireless communication," in *Proc. of the Sixteenth Annual Intern Conf. on Mobile Computing and Netw. ACM*, 2010, pp. 1–12.
12. R. T. Marler and J. S. Arora, "Survey of multi-objective optimization methods for engineering," *Structural and Multidisciplinary Optimization*, vol. 26, no. 6, pp. 369–395, 2004.
13. W. Namgoong, "Modeling and and analysis of nonlinearities and mismatches in ac-coupled direct-conversion receiver," *IEEE Trans Wireless Commun.*, vol. 4, pp. 163–173, 2005.
14. D. W. K. Ng, E. S. Lo, and R. Schober, "Robust beamforming for secure communication in systems with wireless information and power transfer," *IEEE Trans Wireless Commun.*, vol. 13, no. 8, pp. 4599–4615, 2014.

15. D. W. K. Ng, E. S. Lo, and R. Schober, "Multiobjective resource allocation for secure communication in cognitive radio networks with wireless information and power transfer," *IEEE Trans Veh Technol.*, vol. 65, no. 5, pp. 3166–3184, 2016.
16. H. Q. Ngo, H. A. Suraweera, M. Matthaiou, and E. G. Larsson, "Multipair full-duplex relaying with massive arrays and linear processing," *IEEE J Select Areas Commun*, vol. 32, no. 9, pp. 1721–1737, 2014.
17. D. Nguyen, L. N. Tran, P. Pirinen, and M. Latva-aho, "On the spectral efficiency of full-duplex small cell wireless systems," *IEEE Trans Wireless Commun.*, vol. 13, no. 9, pp. 4896–4910, 2014.
18. Y. Sun, D. W. K. Ng, and R. Schober, "Multi-objective optimization for power efficient full-duplex wireless communication systems," in *Proc. IEEE Global Commun.* Conf. IEEE, 2015, pp. 1–6.
19. Y. Sun, D. W. K. Ng, J. Zhu, and R. Schober, "Multi-objective optimization for robust power efficient and secure full-duplex wireless communication systems," *IEEE Trans Wireless Commun.*, vol. 15, no. 8, pp. 5511–5526, 2016.
20. D. Tse and P. Viswanath, *Fundamentals of wireless communication.* Cambridge University Press, 2005.
21. A. D. Wyner, "The wire-tap channel," *Bell system technical journal*, vol. 54, no. 8, pp. 1355–1387, 1975.
22. Q. Wu, G. Y. Li, W. Chen, D. W. K. Ng, and R. Schober, "An overview of sustainable green 5g networks," *IEEE Wireless Commun.*, vol. 24, no. 4, pp. 72–80, 2017.
23. Wong, V. W., Schober, R., Ng, D. W. K., Wang, L. C.: Key technologies for 5G wireless systems. Cambridge University Press (2017)
24. J. Zhang, L. Dai, S. Sun, and Z. Wang, "On the spectral efficiency of massive MIMO systems with low-resolution adcs," *IEEE Commun. Lett.*, vol. 20, no. 5, pp. 842–845, 2016.
25. F. Zhu, F. Gao, M. Yao, and H. Zou, "Joint information-and jamming-beamforming for physical layer security with full duplex base station," *IEEE Trans Signal Process*, vol. 62, no. 24, pp. 6391–6401, 2014.
26. J. Zhu, D. W. K. Ng, N. Wang, R. Schober, and V. K. A. a. Bhargava, "and design of secure massive mimo systems in the presence of hardware impairments. ieee trans," *Wireless Commun.*, vol. 16, pp. 2001–2016, 2017.
27. X. Zhou, Y. Zhang, and L. Song, *Physical layer security in wireless communications.* CRC Press, 2016.

Chapter 11
Integrated Full-Duplex Radios: System Concepts, Implementations, and Experimentation

Tingjun Chen, Jin Zhou, Gil Zussman, and Harish Krishnaswamy

Abstract This chapter reviews recent research on integrated full-duplex (FD) radio systems using complementary metal oxide semiconductor (CMOS) technology. After a brief review of challenges associated with integrated FD radios, several CMOS FD radio designs, particularly those developed at Columbia University, are discussed with self-interference (SI) suppression at antenna interface, and in RF, analog, and digital domains. This chapter also reviews the system design and implementation of two generations of FD radios developed within the Columbia FlexICoN project (Columbia full-duplex wireless: From integrated circuits to networks (FlexICoN) project, https://flexicon.ee.columbia.edu) using off-the-shelf components and a software-defined radio (SDR) platform. The performance evaluation of these FD radios at the node- and link-level is also reviewed.

11.1 Introduction

Recent demonstrations leveraging off-the-shelf components and/or commodity SDRs (such as [4–12]) have established the feasibility of FD wireless through SI suppression at the antenna interface and SI cancellation (SIC) in the RF, analog, and digital domains.

© Portions of this chapter are reprinted from [2, 3], with permission from IEEE.

Tingjun Chen and Jin Zhou contributed equally to this work. Gil Zussman and Harish Krishnaswamy contributed equally to this work.

T. Chen (✉) · G. Zussman · H. Krishnaswamy
Columbia University, New York, NY, USA
e-mail: tc2668@columbia.edu; gil.zussman@columbia.edu; hk2532@columbia.edu

J. Zhou
University of Illinois, Champaign, IL, USA
e-mail: jinzhou@illinois.edu

© Springer Nature Singapore Pte Ltd. 2020
H. Alves et al. (eds.), *Full-Duplex Communications for Future Wireless Networks*,
https://doi.org/10.1007/978-981-15-2969-6_11

From the recent advancement in CMOS IC technology, complex digital SIC algorithms that offer about 50 dB SIC can be readily implemented [7]. However, many existing FD demonstrations rely on RF/analog cancellers and antenna interfaces (such as ferrite circulators) that are bulky, use off-the-shelf components, and do not readily translate to compact and low-cost IC implementations. For example, a pair of separate TX and RX antennas was considered in [4, 6], a single antenna with a coaxial circulator was used in [7], and dually polarized antennas for achieving SI isolation was utilized in [8]. Moreover, these FD radios are often equipped with an RF canceller implemented using discrete components [7, 9], where the techniques applied are usually difficult to be realized in RFIC. Below, we first present challenges associated with the design and implementation of IC-based FD radios, followed by an overview of the research efforts in this area within the Columbia FlexICoN project.

11.1.1 Challenges Associated with Compact and Low-Cost Silicon-Based Implementation

Realizing RF/analog SIC and FD antenna interface in an IC is critical to bring FD wireless technology to mobile devices, such as handsets and tablets [3, 13–24]. Figure 11.1 depicts the block diagram of an FD radio that incorporates antenna, RF and digital SI suppression. The indicated transmitter and minimum received signal power levels are typical of Wi-Fi applications. Also depicted are various transceiver non-idealities that further complicate the SI suppression problem.

11.1.1.1 Achieving > 100 dB SI Suppression

The power levels indicated in Fig. 11.1 necessitate >110 dB SI suppression for Wi-Fi like applications. Such an extreme amount of cancellation must necessarily be achieved across multiple domains (here, antenna, RF, and digital), as >100 dB precision from a single stage or circuit is prohibitively complex and power inefficient. For example, [7] demonstrated 60 dB and 50 dB SIC in the RF/analog and digital domains, respectively, with +20 dBm average TX power and −90 dBm RX noise floor. The suppression must be judiciously distributed across the domains, as suppression in one domain relaxes the dynamic range requirements of the domains downstream. Furthermore, all cancellation circuits must be adaptively configured together—optimization of the performance of a single cancellation stage alone can result in residual SI that is sub-optimal for the cancellers downstream.

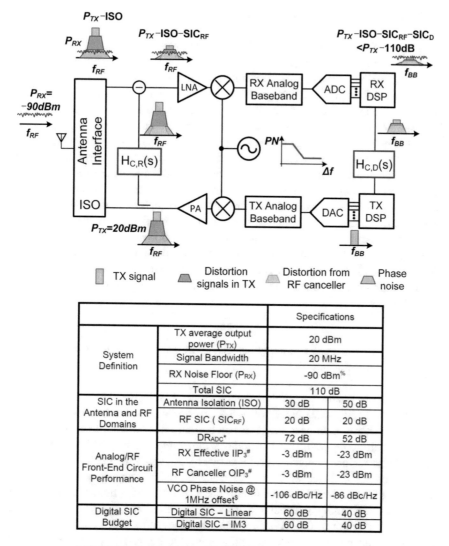

		Specifications	
System Definition	TX average output power (P_{TX})	20 dBm	
	Signal Bandwidth	20 MHz	
	RX Noise Floor (P_{RX})	-90 dBm%	
	Total SIC	110 dB	
SIC in the Antenna and RF Domains	Antenna Isolation (ISO)	30 dB	50 dB
	RF SIC (SIC_{RF})	20 dB	20 dB
Analog/RF Front-End Circuit Performance	DR_{ADC}*	72 dB	52 dB
	RX Effective IIP_3#	-3 dBm	-23 dBm
	RF Canceller OIP_3#	-3 dBm	-23 dBm
	VCO Phase Noise @ 1MHz offset$	-106 dBc/Hz	-86 dBc/Hz
Digital SIC Budget	Digital SIC – Linear	60 dB	40 dB
	Digital SIC – IM3	60 dB	40 dB

* 12 dB margin has been added to DR_{ADC}.

Assume the SI-induced IM3 at the ADC input has the same power

level as the residual SI after RF SIC.

% Assume 11dB implementation loss and noise figure from RX

$ Assume 10ns delay in the SI channel [41, 42].

Fig. 11.1 Block diagram of an FD radio featuring antenna, RF and digital SI suppression, along with a depiction of the various transceiver non-idealities that must also be managed for effective FD operation

11.1.1.2   Transceiver Non-idealities

The extremely powerful nature of the SI exacerbates the impact of non-idealities such as nonlinearity and phase noise, particularly for IC implementations. For instance, nonlinearity along the transmitter chain will introduce distortion products. Antenna and RF cancellation that tap from the output of the transmitter will suppress these distortion products, but linear digital cancellation will not as it operates on the undistorted digital signal. Depending on the amount of antenna and RF cancellation achieved, the analog receiver front-end may introduce distortion products as well, as may the RF cancellation circuitry. Nonlinear digital cancellation may be employed to recreate and cancel these distortion products, but the associated complexity and power consumption must be considered. Local oscillator (LO) phase noise can pose problems as well. If a common LO is used for the transmitter and the receiver, the phase noise in the transmitted and the received SI will be completely correlated, enabling its cancellation in the receiver downmixer. However, delay in the SI channel will decorrelate the phase noise, resulting in residual SI that cannot be canceled. Figure 11.1 depicts transceiver performance requirements calculated for two different SIC allocations across domains—one antenna-heavy and the other digital-heavy.

11.1.1.3   SI Channel Frequency Selectivity and Wideband RF/Analog SI Cancellation

The wireless SI channel can be extremely frequency selective. For example, a large amount of the SI signal power results from the antenna return loss in a shared antenna interface. Compact antennas can be quite narrowband, and the front-end filters that are commonly used in today's radios even more so. The wireless SI channel also includes reflections off nearby objects, which will feature a delay that depends on the distance of the object from the radio. This effect is particularly significant for hand-held devices due to the interaction with human body in close proximity. Performing wideband RF/analog cancellation requires recreating the wireless SI channel in the RF/analog domain. Conventional RF/analog cancellers feature a frequency-flat magnitude and phase response, and will therefore achieve cancellation only over a narrow bandwidth. Wideband SIC at RF based on time-domain equalization (essentially an RF finite impulse response [FIR] filter) has been reported in [7] using discrete components. However, the integration of nanosecond-scale RF delay lines on an IC is a formidable (perhaps impossible) challenge, and therefore alternate wideband RF/analog SIC techniques are required.

11.1.1.4   Compact FD Antenna Interfaces

FD radios employing a pair of antennas, one for transmit and one for receive, experience a direct trade-off between form factor and transmit–receive isolation arising from the antenna spacing and design. Therefore, techniques that can maintain or even enhance transmit–receive isolation, possibly through embedded cancellation, while maintaining a compact form factor are highly desirable. Compact FD antenna interfaces are also more readily compatible with MIMO and diversity applications, and promote channel reciprocity, which is useful at the higher layers. For highly form-factor-constrained mobile applications, single-antenna FD is required, necessitating the use of circulators. Traditionally, circulators have been implemented using ferrite materials, and are costly, bulky, and not compatible with IC technology. Novel techniques for high-performance non-magnetic integrated circulators are of high interest.

11.1.1.5 Adaptive Cancellation

The SIC in all domains must be reconfigurable and automatically adapt to changing operation conditions (e.g., supply voltage and temperature) and most importantly, a changing electromagnetic (EM) environment (i.e., wireless SI channel), given the high level of cancellation required. This requires the periodic (or perhaps even continuous) usage of pilot signals to characterize the SI channel, the implementation of reconfigurable cancellers (which is more challenging in the antenna and RF domains), and the development of canceller adaptation algorithms.

11.1.1.6 Resource Allocation and Rate Gains for Networks with Integrated FD Radios, and Rethinking MAC Protocols

The benefits of enabling FD are clear: the uplink (UL) and downlink (DL) rates can theoretically be doubled (in both random access networks, e.g., Wi-Fi, and small cell networks). That, of course, is true, provided that the SI is canceled such that it becomes negligible at the receiver. Hence, most of the research on FD at the higher layers has focused on designing protocols and assessing the capacity gains while using models of recent laboratory benchtop FD implementations (e.g., [7]) and assuming perfect SI cancellation. However, given the special characteristics of IC-based SI cancellers, there is a need to understand the capacity gains and develop resource allocation algorithms while taking into account these characteristics. These algorithms will then serve as building blocks for the redesign of MAC protocols for FD networks with integrated FD radios.

11.1.2 Overview of the Columbia FlexICoN Project

The Columbia full-duplex Wireless: from Integrated Circuits to Networks (FlexI-CoN) project [1, 2] was initiated in 2014 with a special focus on the development and implementation of IC-based FD radios. To utilize the benefits of these small-form-factor monolithically integrated FD system implementations, a careful redesign of both the physical and MAC layers and the joint optimization across these layers are needed. Moreover, the performance of the IC-based FD radios is evaluated using custom-designed prototypes and testbeds.

Figure 11.2 shows the overview of this interdisciplinary project with highlights on some of the advances on the IC design and testbed evaluations. In particular, we

1. developed novel antenna interfaces such as integrated CMOS non-reciprocal circulators that utilize time-variance to break Lorentz reciprocity;
2. developed several generations of FD radio ICs employing RF/analog SIC circuits to combat noise, distortion, and bandwidth limitations;
3. implemented FD radio prototypes and evaluated their performance using an SDR testbed in various network settings;
4. developed resource allocation and scheduling algorithms in the higher layers.

In this book chapter, we review some of the recent advances within the Columbia FlexICoN project on the design and implementation of CMOS circulators and IC-based FD radios (Sect. 11.2). We also review the design and implementation of two generations of FD radio prototype using an SDR testbed and their experimental evaluation (Sect. 11.3).

11.2   Integrated Full-Duplex Radios

As mentioned in Sect. 11.1.1, despite many challenges associated with compact and low-cost silicon-based FD radio implementation, many recent works have demonstrated integrated FD radios with SI suppression in different domains and at RF and millimeter-wave frequencies. Integrated wideband RF SIC receiver front-ends have been demonstrated using time-domain and frequency-domain equalization techniques [20, 25]. Compared to a time-domain approach, a frequency-domain-equalization-based design allows generation of large group delay on chip [25]. Other recently reported wideband RF SIC includes an FD receiver using baseband Hilbert transform equalization [26] and an FD transceiver using mixed-signal-based RF SIC [21]. The integrated Hilbert transform-based wideband RF SIC is similar to that in [25] and has the additional advantage of requiring a fewer number of canceller taps. The mixed-signal-based RF SIC [21] has great flexibility thanks to the powerful DSP but cannot cancel the noise and nonlinear distortion generated in the transmitter chain. At the antenna interface, integrated electrical-balance duplexers [27–30] have been demonstrated. However, these reciprocal antenna interfaces feature a fundamental minimum of 3 dB loss (typically higher when parasitic losses are factored in). A CMOS passive non-magnetic circulator has been reported lately in [31] and is integrated with an SI-canceling FD receiver in [18]. Complete integrated FD transceivers have also been reported and showed promising results at RF and millimeter-wave frequencies [15–17, 23, 24]. In this remaining of

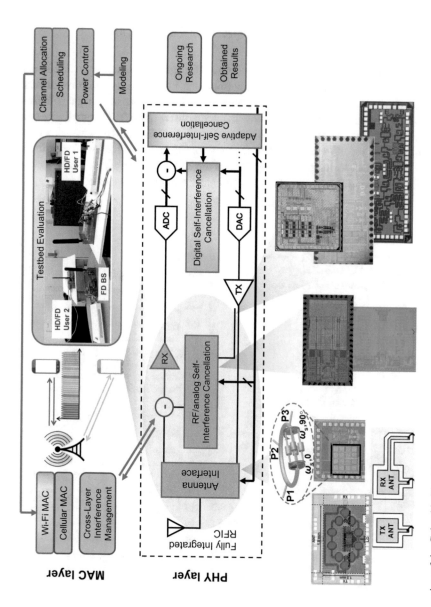

Fig. 11.2 Overview of the Columbia FlexICoN project

this section, we focus on integrated FD radio systems that developed at Columbia University.

11.2.1 Integrated RF Self-Interference Cancellation

To address the challenge related to integrated RF cancellation across a wide bandwidth, we proposed a frequency-domain approach in contrast to the conventional time-domain delay-based RF FIR approach [25]. To enhance the cancellation bandwidth, second-order reconfigurable bandpass filters (BPFs) with amplitude and phase control are introduced in the RF canceller (Fig. 11.3). An RF canceller with a reconfigurable second-order RF BPF features four degrees of freedom (amplitude, phase, quality factor, and center frequency of the BPF). *This enables the replication of not just the amplitude and phase of the antenna interface isolation [$H_{SI}(s)$] at*

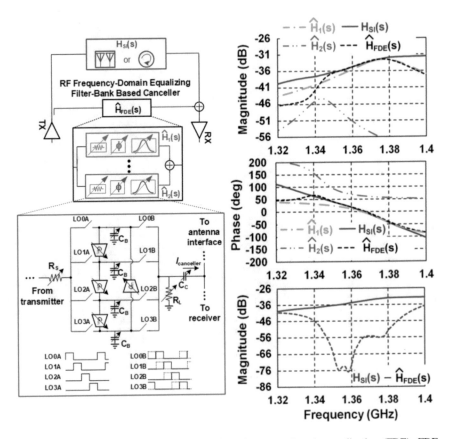

Fig. 11.3 Integrated wideband RF SIC based on frequency-domain equalization (FDE): FDE concept and two-port G_m-C N-path filter with embedded variable attenuation and phase shift

a frequency point, but also the slope of the amplitude and the slope of the phase (or group delay). The use of a bank of filters with independent BPF parameters enables such replication at multiple points in different sub-bands, further enhancing SIC bandwidth. Essentially, this approach is frequency-domain equalization (FDE) in the RF domain. In Fig. 11.3, which represents a theoretical computation on the measured isolation of a pair of 1.4 GHz antennas that are described in greater detail below, two BPFs with transfer functions $\hat{H}_1(s)$ and $\hat{H}_2(s)$ emulate the antenna interface isolation in two sub-bands resulting in an order of magnitude improvement in the SIC bandwidth over a conventional frequency-flat RF canceller.

For FDE, reconfigurable RF filters with very sharp frequency response (or high quality factor) are required. Assuming second-order BPFs with 10 MHz bandwidth are used for FDE-based SIC, a filter quality factor of 100 is required at 1 GHz carrier frequency. Furthermore, the required filter quality factor increases linearly with the carrier frequency assuming a fixed absolute filter bandwidth. The achievable quality factor of conventional LC-based integrated RF filters has been limited by the quality factor of the inductors and capacitors that are available on silicon. However, recent research advances have revived a switched-capacitor circuit-design technique known as the N-path filter that enables the implementation of reconfigurable, high-quality filters at RF in nanoscale CMOS IC technology [32]. Figure 11.3 depicts a two-port N-path filter, where R_S and R_L are the resistive loads at the transmit and receive sides, respectively. C_C weakly couples the cancellation signal to the receiver input for SIC. The quality factor of an N-path filter may be reconfigured via the baseband capacitor C_B, given fixed R_S and R_L. Through clockwise/counterclockwise (only counterclockwise connection is shown in Fig. 11.3 for simplicity) connected reconfigurable transconductors (G_m), an upward/downward frequency offset with respect to the switching frequency can be introduced without having to change the clock frequency [32]. Variable attenuation can be introduced by reconfiguring R_S and R_L relative to each other. *Interestingly, phase shifts can be embedded in a two-port N-path filter by phase shifting the clocks driving the switches on the output side relative to the input-side clocks as shown in Fig. 11.3 [25].* All in all, the ability to integrate reconfigurable high-quality RF filters on chip using switches and capacitors uniquely enables synthesis of nanosecond-scale delays through FDE over time-domain equalization.

A 0.8–1.4 GHz FD receiver IC prototype with the FDE-based RF SI canceller was designed and fabricated in a conventional 65 nm CMOS technology (Fig. 11.4) [25]. For the SIC measurement results shown in Fig. 11.4, we used a 1.4 GHz narrowband antenna-pair interface with peak isolation magnitude of 32 dB, peak isolation group delay of 9 ns, and 3 dB of isolation magnitude variation over 1.36–1.38 GHz. The SI canceller achieves a 20 dB cancellation bandwidth of 15/25 MHz (one/two filters) in Fig. 11.4. When a conventional frequency-flat amplitude- and phase-based canceller is used, the SIC bandwidth is about 3 MHz (>8×lower). The 20 MHz bandwidth over which the cancellation is achieved allows our FD receiver IC to support many advanced wireless standards including smallcell LTE and Wi-Fi.

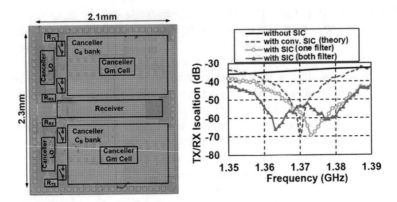

Fig. 11.4 Chip photo of the implemented 0.8–1.4 GHz 65 nm CMOS FD receiver with FDE-based SIC in the RF domain featuring a bank of two filters (left); transmit–receive isolation of an antenna pair without SIC, with conventional SIC (theoretical), and with the proposed FDE-based SIC (right)

11.2.2 Full-Duplex Receiver with Integrated Circulator and Analog Self-Interference Cancellation

Mobile applications constrained by form factors, particularly at RF frequencies where the wavelength is considerably higher, require single-antenna solutions. Single-antenna FD also ensures channel reciprocity and compatibility with antenna diversity and MIMO concepts. However, *conventional single-antenna FD interfaces, namely non-reciprocal circulators, rely on ferrite materials and biasing magnets, and are consequently bulky, expensive, and incompatible with silicon integration.* Reciprocal circuits, such as electrical-balance duplexers [27], have been considered, but are limited by the fundamental minimum 3 dB loss in both TX-antenna (ANT) and ANT-RX paths.

As mentioned earlier, non-reciprocity and circulation have conventionally been achieved using the magneto-optic Faraday effect in ferrite materials. However, it has recently been shown that violating time-invariance within a linear, passive material with symmetric permittivity and permeability tensors can introduce non-reciprocal wave propagation, enabling the construction of non-magnetic circulators [33, 34]. However, these initial efforts have resulted in designs that are either lossy, highly nonlinear, or comparable in size to the wavelength, and are fundamentally not amenable to silicon integration. *Recently, we introduced a new non-magnetic CMOS-compatible circulator concept based on the phase non-reciprocal behavior of linear, periodically time-varying (LPTV) two-port N-path filters that utilize staggered clock signals at the input and output [31].*

N-path filters, described earlier in the context of FDE, are a class of LPTV networks where the signal is periodically commutated through a bank of linear, time-invariant (LTI) networks. We found that when the non-overlapping clocks

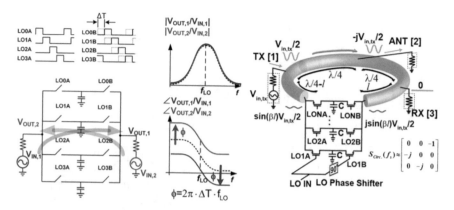

Fig. 11.5 Integrated non-magnetic circulator for single-antenna FD: non-reciprocity induced by phase-shifted N-path commutation (left); 3-port circulator structure obtained by placing the non-reciprocal two-port N-path filter with ±90° phase shift within a 3λ/4 transmission line loop (right)

driving the input and output switch sets of a two-port N-path filter are phase shifted with respect to each other, a non-reciprocal phase shift is produced for signals traveling in the forward and reverse directions as they see a different ordering of the commutating switches (Fig. 11.5). The magnitude response remains reciprocal and low-loss, similar to traditional N-path filters. To create non-reciprocal wave propagation, an N-path-filter with ±90° phase shift is placed inside a 3λ/4 transmission line loop (Fig. 11.5). This results in satisfaction of the boundary condition in one direction (−270° phase shift from the loop added with −90° from the N-path filter) and suppression of wave propagation in the other direction (−270°+90° = −180°), effectively producing unidirectional circulation. Additionally, a three-port circulator can be realized by placing ports anywhere along the loop as long as they maintain a λ/4 circumferential distance between them. Interestingly, maximum linearity with respect to the TX port is achieved if the RX port is placed adjacent to the N-path filter ($l = 0$), since the inherent TX–RX isolation suppresses the voltage swing on either side of the N-path filter, enhancing its linearity.

A prototype circulator based on these concepts operating over 610–850 MHz was implemented in a 65 nm CMOS process. Measurements reveal 1.7 dB loss in TX-ANT and ANT-RX transmission, and broadband isolation better than 15 dB between TX and RX (the narrowband isolation can be as high as 50 dB). The in-band ANT-RX IIP3 is +8.7 dBm while the in-band TX-ANT IIP3 is +27.5 dBm (OIP3 is +25.8 dBm), two orders of magnitude higher due to the suppression of swing across the N-path filter. The measured clock feedthrough to the ANT port is −57 dBm and IQ image rejection for TX-ANT transmission is 49 dB. Techniques such as device stacking in SOI CMOS can be explored to further enhance the TX-ANT linearity to meet the stringent requirements of commercial wireless standards. Clock feedthrough and IQ mismatch can be calibrated by sensing and injecting appropriate

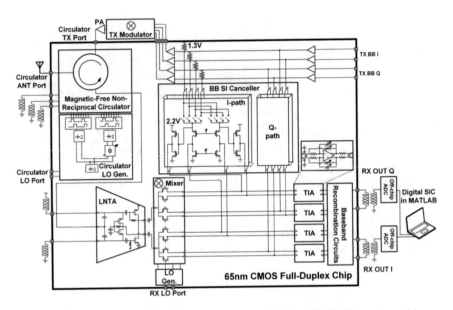

Fig. 11.6 Block diagram and schematic of the implemented 65 nm CMOS FD receiver with non-magnetic circulator and additional analog BB SI cancellation

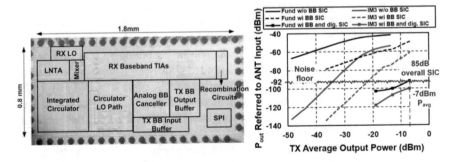

Fig. 11.7 Chip microphotograph of the implemented 65 nm CMOS FD receiver with non-magnetic circulator and additional analog BB SI cancellation (left), and measured two-tone SI test with SI suppression across circulator, analog BB and digital domains (right)

BB signals through the N-path filter capacitor nodes as shown previously in the literature [35].

A 610–850 MHz FD receiver IC prototype incorporating the non-magnetic N-path-filter-based passive circulator and additional analog baseband (BB) SI cancellation (shown in Figs. 11.6 and 11.7) was also designed and fabricated in the 65 nm CMOS process [36]. SI suppression of 42 dB is achieved across the circulator and analog BB SIC over a bandwidth of 12 MHz. Digital SIC has also been implemented in Matlab after capturing the BB signals using an oscilloscope (effectively an 8-bit 40 MSa/s ADC) (Fig. 11.7). The digital SIC cancels not only the main SI but also the IM3 distortion generated on the SI by the circulator, receiver, and canceller. A total 164 canceller coefficients are trained by 800 sample points. After digital SIC, the main SI tones are at the −92 dBm noise floor, while the SI IM3 tones are 8 dB below for −7 dBm TX average power. This corresponds to an overall SI suppression of 85 dB for the FD receiver.

11.3   Full-Duplex Testbed and Performance Evaluation

To experimentally evaluate the developed RF cancellers at the system-level, we prototyped two generations of FD radios using the National Instruments (NI) Universal Software Radio Peripheral (USRP) software-defined radios (SDRs). Figure 11.8 shows the block diagram of a prototyped FD radio with an RF SI canceller and a USRP SDR, where the RF SI canceller taps a reference signal at the output of the power amplifier (PA) and performs SIC at the input of the low-noise amplifier (LNA) at the RX side. Since interfacing the RFIC cancellers described in Sect. 11.2 to an SDR presents numerous technical challenges, we designed and implemented RF cancellers emulating their RFIC counterparts using discrete components on printed circuit boards (PCBs).

The Gen-1 FD radio [37] includes a frequency-flat amplitude- and phase-based RF canceller, which emulates its RFIC counterpart presented in [38]. An improved version of the Gen-1 RF SI canceller has been recently integrated in the open-access ORBIT testbed [39, 40] to support research. The Gen-2 FD radio [41, 42] includes a wideband frequency-domain equalization- (FDE-) based RF canceller, which emulates its RFIC counterpart presented in [25]. In this section, we review the design and implementation of the RF cancellers, and their integration with USRP SDRs and performance evaluation.

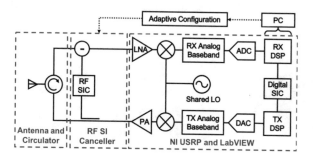

Fig. 11.8 Block diagram of a prototyped FD radio with an antenna, a circulator, an RF SI canceller, and a USRP SDR

11.3.1 Gen-1 Full-Duplex Radio with a Frequency-Flat Amplitude- and Phase-Based RF Canceller

Figure 11.9(a) shows the prototyped Gen-1 FD radio, which consists of an antenna, a circulator, a Gen-1 RF SI canceller, and an NI USRP-2932. The operating frequency of the NI USRP is configured to be 0.9 GHz, which is the same as the operating frequency of both the circulator and the RF SI canceller. The USRP is controlled from a PC that runs NI LabVIEW, which performs digital signal processing.

The implemented Gen-1 RF SI canceller depicted in Fig. 11.9(b) operates from 0.8 to 1.3 GHz and is configured by a SUB-20 controller through the USB interface. In particular, (1) the attenuator provides an attenuation range from 0 to 15.5 dB with a 0.5 dB resolution, and (2) the phase shifter is controlled by an 8-bit digital-to-analog converter (DAC) and covers the full 360° range with a resolution of about 1.5°. In the rest of the section, we describe the adaptive RF canceller configuration, the digital SIC algorithm, and the FD wireless link demonstration.

An Adaptive Gen-1 RF SI Canceller Configuration Scheme
We model the frequency-flat amplitude- and phase-based RF SI canceller depicted in Fig. 11.9(b) by a transfer function $H = Ae^{(-j\phi)}$, in which A and ϕ are the amplitude and phase controls of the RF canceller that need to be configured to match with that of the antenna interface at a desired frequency. Note that the Gen-1 RF SI canceller can only emulate the antenna interface at a single frequency point, resulting in narrowband RF SIC. As shown in Fig. 11.9(c), 40 dB RF SIC is achieved across 5 MHz bandwidth.

Ideally, only two measurements with different configurations of (A, ϕ) are needed to compute the optimal configuration, $(A^\star, \phi^\star)$. However, in practice, the USRP has an unknown RF front-end gain, denoted by G_{USRP}, which complicates the estimation of the amplitude and phase that the RF SI canceller should mimic. Therefore, we implemented an adaptive RF canceller configuration scheme using four measurements for the three unknowns A, ϕ, and G_{USRP}. After an initial

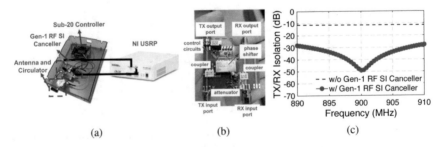

(a)   (b)   (c)

Fig. 11.9 (a) The Gen-1 FD radio composed of an antenna, a circulator, a conventional frequency-flat amplitude- and phased-based RF SI canceller, and an NI USRP, (b) the Gen-1 0.8–1.3 GHz frequency-flat amplitude and phase-based RF SI canceller, and (c) the measured TX/RX isolation where 40 dB RF SIC is achieved across 5 MHz bandwidth

configuration is obtained, a local tuning is followed to search for the optimal (A, ϕ). This scheme is described below and consists of two phases. While operating, if the FD radios encounter noticeable changes in the measured residual SI power level, the adaptive configuration scheme is triggered to recompute the canceller configuration.

1. *Initial Configuration*: To obtain the initial RF SI canceller configuration, denoted by (A_0, ϕ_0), the FD radio takes four measurements with different configurations of (A, ϕ): $(0, 0°)$, $(A', 0)$, $(A', 90°)$, and $(A', 270°)$, where A' is a known attenuation set manually. Denote by r_i $(i = 1, 2, 3, 4)$ the residual SI power with the ith configuration, (A_0, ϕ_0) can be obtained (in closed-form) by solving the following equations

$$G_{USRP} \cdot (r_1)^2 = (A_0)^2,$$

$$G_{USRP} \cdot (r_2)^2 = (A_0 \cos \phi_0 - A')^2 + (A_0 \sin \phi_0)^2,$$

$$G_{USRP} \cdot (r_3)^2 = (A_0 \cos \phi_0)^2 + (A' - A_0 \sin \phi_0)^2,$$

$$G_{USRP} \cdot (r_4)^2 = (A_0 \cos \phi_0)^2 + (A' + A_0 \sin \phi_0)^2.$$

From our experiments, the RF SI canceller with the initial configuration, (A_0, ϕ_0), can provide around 15 dB RF SIC across 5 MHz bandwidth.
2. *Local Adjustments*: After obtaining initial configuration, the FD radio performs a finer grained local tuning around (A_0, ϕ_0) to further improve the performance of RF SIC.

Digital SIC

The residual SI after isolation and cancellation in the antenna and RF domains is further suppressed in the digital domain. The digital SI canceller is modeled as a truncated Volterra series and is implemented based on a nonlinear tapped delay line to cancel both the main SI and the inter-modulation distortion generated on the SI. Specifically, the output of the discrete-time SI canceller, y_n, can be written as a function of the current and past TX digital baseband signals, x_n and x_{n-k} (k represents the delay index), i.e.,

$$y_n = \sum_{k=0}^{N} h_{1,k} x_{n-k} + \sum_{k=0}^{N} h_{2,k} x_{n-k}^2 + \sum_{k=0}^{N} h_{3,k} x_{n-k}^3, \tag{11.1}$$

in which N corresponds to the maximum delay in the SI channel and $h_{i,k}$ ($i = 1, 2, 3$) is the ith order digital canceller coefficient. Depending upon the SI channel, higher order nonlinear terms can be included (the model in (11.1) only includes up to the 3rd-order nonlinearity). Using a pilot data sequence, the digital SI canceller coefficients can be found by solving the least-square problem.

An FD Wireless Link Demonstration

In [37], we demonstrated an FD wireless link consisting of two Gen-1 FD radios. In particular, each FD radio transmits a 5 MHz multi-tone signal with 0 dBm average

TX power and +10 dBm peak TX power, and the SI signal is canceled to the −90 dBm USRP RX noise floor after SIC in both the RF and digital domains.

In particular, 40 dB SI suppression is provided by the circulator and the Gen-1 RF SI canceller before the USRP RX. The digital SIC achieves an additional 50 dB suppression, after which the desired signal is detected.

11.3.2 Gen-2 Full-Duplex Radio with a Frequency-Domain Equalization-Based RF Canceller

Figure 11.10a shows the prototyped Gen-2 FD radio, which consists of an antenna, a circulator, a Gen-2 wideband FDE-based RF SI canceller, and an NI USRP-2942.

Figure 11.10b shows the Gen-2 FDE-based RF SI canceller with 2 FDE taps implemented using discrete components. In particular, a reference signal is tapped from the TX input using a coupler and is split into two FDE taps through a power divider. Then, the signals after each FDE tap are combined and RF SIC is performed at the RX input. Each FDE tap consists of a reconfigurable 2nd-order BPF, as well as an attenuator and phase shifter for amplitude and phase controls. The BPF is implemented as an RLC filter with impedance transformation networks on both sides, and is optimized around 900 MHz operating frequency. The center frequency of the BPF in the ith FDE tap can be adjusted through a 4-bit digitally tunable capacitor in the RLC resonance tank. In order to achieve a high and adjustable BPF quality factor, impedance transformation networks including transmission-lines (T-Lines) and 5-bit digitally tunable capacitors are introduced. In addition, the programmable attenuator has a tuning range of 0–15.5 dB with a 0.5 dB resolution, and the passive phase shifter is controlled by a 8-bit digital-to-analog converter (DAC) and covers full 360° range.

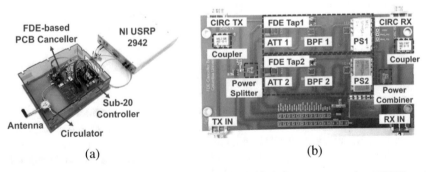

(a)   (b)

Fig. 11.10 (a) The Gen-2 FD radio composed of an antenna, a circulator, a wideband FDE-based RF SI canceller, and an NI USRP, (b) the Gen-2 wideband FDE-based RF SI canceller implemented using discrete components

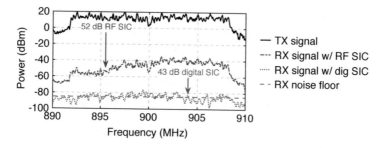

Fig. 11.11 Power spectrum of the received signal after SIC in the RF and digital domains with +10 dBm average TX power, 20 MHz bandwidth, and −85 dBm receiver noise floor

Optimized Gen-2 RF SI Canceller Configuration

Consider a wideband OFDM-based PHY layer and let $f_k (k = 1, 2, \ldots, K)$ be the frequency of the kth subcarrier. Denote by $H_{\text{SI}}(f_k)$ and $H^{\text{FDE}}(f_k)$ the frequency responses of the SI channel[1] and the FDE-based RF SI canceller, respectively. The optimized RF canceller configuration can be obtained by solving the following optimization problem:

$$\min : \sum_{k=1}^{K} \left| H_{\text{SI}}(f_k) - H^{\text{FDE}}(f_k) \right|^2$$

$$\text{s.t.} : \text{constraints on configuration parameters of } H^{\text{FDE}}(f_k), \ \forall k.$$

In particular, the goal is to select the RF canceller configuration (i.e., the values of the attenuator, phased shifter, and digitally tunable capacitors in each FDE tap) so that it best emulates the SI channel. In [42], we developed and experimentally validated a mathematical model of the frequency response of the implemented FDE-based RF SI canceller, which can be pre-computed and stored for obtaining the optimized RF canceller configuration. In practice, a finer grained local tuning is also included to further improve the RF SIC.

Overall SIC

Figure 11.11 shows the power spectrum of the received signal at the Gen-2 FD radio after RF SIC and digital SIC, respectively, where the digital SIC is implemented using the same method as described in Sect. 11.3.1. The results show that an average 95 dB SIC can be achieved by the Gen-2 FD radio across 20 MHz bandwidth, supporting a maximum average TX power of 10 dBm (a maximum peak TX power of 20 dBm) with the −85 dBm USRP receiver noise floor. In particular, 52 dB and 43 dB are obtained in the RF and digital domains, respectively.

---

[1]The SI channel here refers to the circulator TX–RX leakage after calibration of the USRP RF front-end.

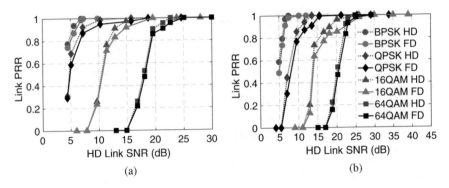

Fig. 11.12 HD and FD link packet reception ratio (PRR) with varying HD link SNR and modulation and coding schemes (MCSs). (**a**) Code rate 1/2. (**b**) Code rate 3/4

SNR-PRR Relationship at the Link-Level

We now evaluate the relationship between link SNR and link packet reception ratio (PRR). We set up a link with two Gen-2 FD radios at a fixed distance of 5 m with equal TX power. In order to evaluate the performance of our FD radios with the existence of the FDE-based RF canceller, we set an FD radio to operate in HD mode by turning on only its transmitter or receiver. We conduct the following experiment for each of the 12 modulation and coding schemes (MCSs) in both FD and HD modes,[2] with varying TX power levels and 20 MHz bandwidth. In particular, the packets are sent over the link simultaneously in FD mode or in alternating directions in HD mode (i.e., the two radios take turns and transmit to each other). In each experiment, both radios send a sequence of 50 OFDM streams, each OFDM stream contains 20 OFDM packets, and each OFDM packet is 800-Byte long.

We consider two metrics. The *HD (resp. FD) link SNR* is measured as the ratio between the average RX signal power in both directions and the RX noise floor when both radios operate in HD (resp. FD) mode. The *HD (resp. FD) link PRR* is computed as the fraction of packets successfully sent over the HD (resp. FD) link in each experiment. We observe from the experiments that the HD and FD link SNR and PRR values in both link directions are similar.

Figure 11.12 shows the relationship between link PRR values and HD link SNR values with varying MCSs. It can be seen that with sufficient link SNR values (e.g., 8 dB for BPSK-1/2 and 28 dB for 64QAM-3/4), the FDE-based FD radio achieves a link PRR of 100%. With insufficient link SNR values, the average FD link PRR is 6.5% lower than the HD link PRR across varying MCSs. This degradation is caused by the minimal link SNR difference when the radios operate in HD or FD mode [42]. Since packets are sent simultaneously in both directions on an FD link, this average PRR degradation is equivalent to an average FD link throughput gain

---

[2]We consider BPSK, QPSK, 16QAM, and 64QAM with coding rates of 1/2, 2/3, and 3/4.

of $1.87\times$ under the same MCS. In [42], we also experimentally evaluated the FD gains under various network settings. These results can serve as building blocks for developing higher layer (e.g., MAC) protocols.

11.4 Conclusion

Numerous recent research efforts within the Columbia FlexICoN project and elsewhere have been focusing on IC-based FD transceivers spanning RF to millimeter-wave, reconfigurable antenna cancellation, non-magnetic CMOS circulators, and MAC layer algorithms based on realistic hardware models. While exciting progress has been made in the last few years by the research community as a whole, several problems remain to be solved before FD wireless can be widely deployed and applied. Continued improvements are necessary in IC-based FD transceivers toward increased total cancellation over wide signal bandwidth and support for higher TX power levels through improved circulator and RX linearity. Incorporation of FD in large-scale phased-array transceivers and the extension of IC-based SIC concepts to MIMO transceivers is an important and formidable challenge, due to the large number of SI channels between each pair of TX and RX. Our preliminary results on incorporating FD in phased-array and MIMO transceiver ICs within the Columbia FlexICoN project were reported in [43–45]. At the higher layers, extensive research has focused on understanding the benefits and rate gains provided by FD under different network settings. Particularly, in [46, 47], we developed and evaluated distributed scheduling algorithms with provable performance guarantees in heterogeneous networks with both half- and FD users. However, there are still several important open problems related to efficient and fair resource allocation, as well as scheduling algorithms for both cellular and Wi-Fi type of random access networks, while taking into the account the special PHY layer characteristics of FD radios. It is also important to consider interference management in these networks jointly across different layers in the network stack.

Acknowledgements This work was supported in part by NSF grants ECCS-1547406, CNS-1650685, and CNS-1827923, the DARPA RF-FPGA, ACT, and SPAR programs, and two Qualcomm Innovation Fellowships. We thank Mahmood Baraani Dastjerdi, Jelena Diakonikolas, Negar Reiskarimian for their contributions.

References

1. "Columbia full-duplex wireless: From integrated circuits to networks (FlexICoN) project," https://flexicon.ee.columbia.edu, 2020.
2. H. Krishnaswamy, G. Zussman, J. Zhou, J. Marašević, T. Dinc, N. Reiskarimian, and T. Chen, "Full-duplex in a hand-held device – from fundamental physics to complex integrated circuits, systems and networks: An overview of the Columbia FlexICoN project," in *Proc. Asilomar Conference on Signals, Systems and Computers*, 2016.

3. J. Zhou, N. Reiskarimian, J. Diakonikolas, T. Dinc, T. Chen, G. Zussman, and H. Krishnaswamy, "Integrated full duplex radios," *IEEE Commun. Mag.*, vol. 55, no. 4, pp. 142–151, 2017.
4. J. I. Choi, M. Jain, K. Srinivasan, P. Levis, and S. Katti, "Achieving single channel, full duplex wireless communication," in *Proc. ACM MobiCom'10*, 2010.
5. M. Jain, J. I. Choi, T. Kim, D. Bharadia, S. Seth, K. Srinivasan, P. Levis, S. Katti, and P. Sinha, "Practical, real-time, full duplex wireless," in *Proc. ACM MobiCom'11*, 2011.
6. M. Duarte, C. Dick, and A. Sabharwal, "Experiment-driven characterization of full-duplex wireless systems," *IEEE Trans. Wireless Commun.*, vol. 11, no. 12, 2012.
7. D. Bharadia, E. McMilin, and S. Katti, "Full duplex radios," in *Proc. ACM SIGCOMM'13*, 2013.
8. M. Chung, M. S. Sim, J. Kim, D. K. Kim, and C.-B. Chae, "Prototyping real-time full duplex radios," *IEEE Commun. Mag.*, vol. 53, no. 9, pp. 56–63, 2015.
9. D. Korpi, Y.-S. Choi, T. Huusari, L. Anttila, S. Talwar, and M. Valkama, "Adaptive nonlinear digital self-interference cancellation for mobile inband full-duplex radio: Algorithms and RF measurements," in *Proc. IEEE GLOBECOM'15*, 2015.
10. D. Korpi, J. Tamminen, M. Turunen, T. Huusari, Y.-S. Choi, L. Anttila, S. Talwar, and M. Valkama, "Full-duplex mobile device: Pushing the limits," *IEEE Commun. Mag.*, vol. 54, no. 9, pp. 80–87, 2016.
11. M. Duarte, A. Sabharwal, V. Aggarwal, R. Jana, K. Ramakrishnan, C. W. Rice, and N. Shankaranarayanan, "Design and characterization of a full-duplex multiantenna system for WiFi networks," *IEEE Trans. Veh. Technol.*, vol. 63, no. 3, pp. 1160–1177, 2014.
12. M. S. Sim, M. Chung, D. Kim, J. Chung, D. K. Kim, and C.-B. Chae, "Nonlinear self-interference cancellation for full-duplex radios: From link-level and system-level performance perspectives," *IEEE Commun. Mag.*, vol. 55, no. 9, pp. 158–167, 2017.
13. B. Debaillie, D. van den Broek, C. Lavin, B. van Liempd, E. Klumperink, C. Palacios, J. Craninckx, and A. Parssinen, "Analog/RF solutions enabling compact full-duplex radios," *IEEE J. Sel. Areas Commun.*, vol. 32, no. 9, pp. 1662–1673, 2014.
14. N. Reiskarimian, T. Dinc, J. Zhou, T. Chen, M. B. Dastjerdi, J. Diakonikolas, G. Zussman, and H. Krishnaswamy, "One-way ramp to a two-way highway: Integrated magnetic-free nonreciprocal antenna interfaces for full-duplex wireless," *IEEE Microw. Mag.*, vol. 20, no. 2, pp. 56–75, 2019.
15. D. Yang, H. Yüksel, and A. Molnar, "A wideband highly integrated and widely tunable transceiver for in-band full-duplex communication," *IEEE J. Solid-State Circuits*, vol. 50, no. 5, pp. 1189–1202, 2015.
16. D.-J. van den Broek, E. A. Klumperink, and B. Nauta, "An in-band full-duplex radio receiver with a passive vector modulator downmixer for self-interference cancellation," *IEEE J. Solid-State Circuits*, vol. 50, no. 12, pp. 3003–3014, 2015.
17. T. Dinc, A. Chakrabarti, and H. Krishnaswamy, "A 60GHz CMOS full-duplex transceiver and link with polarization-based antenna and RF cancellation," *IEEE J. Solid-State Circuits*, vol. 51, no. 5, pp. 1125–1140, 2016.
18. N. Reiskarimian, J. Zhou, and H. Krishnaswamy, "A CMOS passive LPTV nonmagnetic circulator and its application in a full-duplex receiver," *IEEE J. Solid-State Circuits*, vol. 52, no. 5, pp. 1358–1372, 2017.
19. N. Reiskarimian, M. B. Dastjerdi, J. Zhou, and H. Krishnaswamy, "Analysis and design of commutation-based circulator-receivers for integrated full-duplex wireless," *IEEE J. Solid-State Circuits*, vol. 53, no. 8, pp. 2190–2201, 2018.
20. T. Zhang, A. Najafi, C. Su, and J. C. Rudell, "A 1.7-to-2.2GHz full-duplex transceiver system with > 50dB self-interference cancellation over 42MHz bandwidth," in *Proc. IEEE ISSCC'17*, 2017.
21. S. Ramakrishnan, L. Calderin, A. Niknejad, and B. Nikolić, "An FD/FDD transceiver with RX band thermal, quantization, and phase noise rejection and >64dB TX signal cancellation," in *Proc. IEEE RFIC'17*, 2017.

22. E. Kargaran, S. Tijani, G. Pini, D. Manstretta, and R. Castello, "Low power wideband receiver with RF self-interference cancellation for full-duplex and FDD wireless diversity," in *Proc. IEEE RFIC'17*, 2017.

23. T. Chi, J. S. Park, S. Li, and H. Wang, "A 64GHz full-duplex transceiver front-end with an on-chip multifeed self-interference-canceling antenna and an all-passive canceler supporting 4Gb/s modulation in one antenna footprint," in *Proc. IEEE ISSCC'18*, 2018.

24. K.-D. Chu, M. Katanbaf, T. Zhang, C. Su, and J. C. Rudell, "A broadband and deep-TX self-interference cancellation technique for full-duplex and frequency-domain-duplex transceiver applications," in *Proc. IEEE ISSCC'18*, 2018.

25. J. Zhou, T.-H. Chuang, T. Dinc, and H. Krishnaswamy, "Integrated wideband self-interference cancellation in the RF domain for FDD and full-duplex wireless," *IEEE J. Solid-State Circuits*, vol. 50, no. 12, pp. 3015–3031, 2015.

26. A. El Sayed, A. Ahmed, A. Mishra, A. Shirazi, S. Woo, Y.-S. Choi, S. Mirabbasi, and S. Shekhar, "A full-duplex receiver with 80MHz bandwidth self-interference cancellation circuit using baseband Hilbert transform equalization," in *Proc. IEEE RFIC'17*, 2017.

27. M. Mikhemar, H. Darabi, and A. A. Abidi, "A multiband RF antenna duplexer on CMOS: Design and performance," *IEEE J. Solid-State Circuits*, vol. 48, no. 9, pp. 2067–2077, Sept 2013.

28. B. van Liempd, B. Hershberg, B. Debaillie, P. Wambacq, and J. Craninckx, "An electrical-balance duplexer for in-band full-duplex with < −85dBm in-band distortion at +10dBm TX-power," in *Proc. IEEE ESSCIRC'15*, Sep. 2015, pp. 176–179.

29. S. H. Abdelhalem, P. S. Gudem, and L. E. Larson, "Tunable CMOS integrated duplexer with antenna impedance tracking and high isolation in the transmit and receive bands," *IEEE Trans. Microw. Theory Tech.*, vol. 62, no. 9, pp. 2092–2104, Sep. 2014.

30. M. Elkholy, M. Mikhemar, H. Darabi, and K. Entesari, "Low-loss integrated passive CMOS electrical balance duplexers with single-ended LNA," *IEEE Trans. Microw. Theory Tech.*, vol. 64, no. 5, pp. 1544–1559, May 2016.

31. N. Reiskarimian and H. Krishnaswamy, "Magnetic-free non-reciprocity based on staggered commutation," in *Nature Commun.*, vol. 7, no. 4, 2016.

32. M. Darvishi, R. van der Zee, E. A. Klumperink, and B. Nauta, "Widely tunable 4th order switched G_m-C band-pass filter based on N-path filters," *IEEE J. Solid-State Circuits*, vol. 47, no. 12, pp. 3105–3119, Dec 2012.

33. S. Qin, Q. Xu, and Y. E. Wang, "Nonreciprocal components with distributedly modulated capacitors," *IEEE Trans. Microw. Theory Tech.*, vol. 62, no. 10, pp. 2260–2272, Oct. 2014.

34. N. A. Estep, D. L. Sounas, and A. Alù, "Magnetless microwave circulators based on spatiotemporally modulated rings of coupled resonators," *IEEE Trans. Microw. Theory Tech.*, vol. 64, no. 2, pp. 502–518, Feb. 2016.

35. S. Jayasuriya, D. Yang, and A. Molnar, "A baseband technique for automated LO leakage suppression achieving < 80dBm in wideband passive mixer-first receivers," in *Proc. IEEE CICC'14*, 2014.

36. J. Zhou, N. Reiskarimian, and H. Krishnaswamy, "Receiver with integrated magnetic-free N-path-filter-based non-reciprocal circulator and baseband self-interference cancellation for full-duplex wireless," in *Proc. IEEE ISSCC'16*, 2016.

37. T. Chen, J. Zhou, N. Grimwood, R. Fogel, J. Marašević, H. Krishnaswamy, and G. Zussman, "Demo: Full-duplex wireless based on a small-form-factor analog self-interference canceller," in *Proc. ACM MobiHoc'16*, 2016.

38. J. Zhou, A. Chakrabarti, P. Kinget, and H. Krishnaswamy, "Low-noise active cancellation of transmitter leakage and transmitter noise in broadband wireless receivers for FDD/co-existence," *IEEE J. Solid-State Circuits*, vol. 49, no. 12, pp. 1–17, Dec. 2014.

39. T. Chen, M. Baraani Dastjerdi, J. Zhou, H. Krishnaswamy, and G. Zussman, "Open-access full-duplex wireless in the ORBIT testbed," *arXiv preprint arXiv:1801.03069*, 2018.

40. "Tutorial: Full-duplex wireless in the ORBIT testbed," http://www.orbit-lab.org/wiki/Tutorials/k0SDR/Tutorial25, 2017.

41. T. Chen, J. Zhou, M. Baraani Dastjerdi, J. Diakonikolas, H. Krishnaswamy, and G. Zussman, "Demo abstract: Full-duplex with a compact frequency domain equalization-based RF canceller," in *Proc. IEEE INFOCOM'17*, 2017.

42. T. Chen, M. B. Dastjerdi, J. Zhou, H. Krishnaswamy, and G. Zussman, "Wideband full-duplex wireless via frequency-domain equalization: Design and experimentation," in *Proc. ACM MobiCom'19*, 2019.

43. M. B. Dastjerdi, N. Reiskarimian, T. Chen, G. Zussman, and H. Krishnaswamy, "Full duplex circulator-receiver phased array employing self-interference cancellation via beamforming," in *Proc. IEEE RFIC'18*, 2018.

44. M. B. Dastjerdi, S. Jain, N. Reiskarimian, A. Natarajan, and H. Krishnaswamy, "Full-duplex 2x2 MIMO circulator-receiver with high TX power handling exploiting MIMO RF and shared-delay baseband self-interference cancellation," in *Proc. IEEE ISSCC'19*, 2019.

45. T. Chen, M. B. Dastjerdi, H. Krishnaswamy, and G. Zussman, "Wideband full-duplex phased array with joint transmit and receive beamforming: Optimization and rate gains," in *Proc. ACM MobiHoc'19*, 2019.

46. T. Chen, J. Diakonikolas, J. Ghaderi, and G. Zussman, "Hybrid scheduling in heterogeneous half- and full-duplex wireless networks," in *Proc. IEEE INFOCOM'18*, 2018.

47. ——, "Fairness and delay in heterogeneous half-and full-duplex wireless networks," in *Proc. Asilomar Conference on Signals, Systems and Computers*, 2018.

Printed in the United States
by Baker & Taylor Publisher Services